Making Jet Engines in World War II

Making Jet Engines in World War II

Britain, Germany, and the United States

HERMIONE GIFFARD

THE UNIVERSITY OF CHICAGO PRESS   CHICAGO AND LONDON

HERMIONE GIFFARD is a postdoctoral researcher in the Department of History and Art at Utrecht University, in the Netherlands.

The University of Chicago Press, Chicago 60637
The University of Chicago Press, Ltd., London
© 2016 by The University of Chicago
All rights reserved. Published 2016.
Printed in the United States of America

25 24 23 22 21 20 19 18 17 16   1 2 3 4 5

ISBN-13: 978-0-226-38859-5 (cloth)
ISBN-13: 978-0-226-38862-5 (e-book)
DOI: 10.7208/chicago/9780226388625.001.0001

Library of Congress Cataloging-in-Publication Data
Names: Giffard, Hermione, author.
Title: Making jet engines in World War II : Britain, Germany, and the United States / Hermione Giffard.
Description: Chicago ; London : The University of Chicago Press, 2016. | Includes bibliographical references and index.
Identifiers: LCCN 2016001733 | ISBN 9780226388595 (cloth : alk. paper) | ISBN 9780226388625 (e-book)
Subjects: LCSH: Jet engines—Great Britain. | Jet engines—Germany. | Jet engines—United States. | World War, 1939–1945—Equipment and supplies.
Classification: LCC D810.S2 G54 2016 | DDC 623.74/6044—dc23 LC record available at http://lccn.loc.gov/2016001733

DEDICATED TO MY FATHER, WHO GAVE ME THE TOOLS

Contents

Abbreviations

BMW	Bayrische Motorenwerke
Bramo	Brandenburgische Motorenwerke
GTCC	Gas Turbine Collaboration Committee
ICT	Internal Combustion Turbine
Jumo	Junkers Motorenwerke
MAP	British Ministry of Aircraft Production
Metrovick	Metropolitan Vickers
RAE	Royal Aircraft Establishment
RLM	Reichsluftfahrtministerium, German Air Ministry

Introduction

This book is about making new machines and about how we understand this. Its aim is to present a fresh history of the many jet engines that were in use and designed in Britain, Germany, and the United States by 1945. In the light of this, it seeks to stimulate historians to rethink the history of the emergence of technical novelty, especially but not only in the mid-twentieth century. Rather than tell the story of making the jet engine as the story of a few individual inventors, I focus on the previously neglected work of industrial firms in designing, developing, and producing them. Appealing to aeronautical engineers for their potentially high power-to-weight ratio (the increase in which was an all-consuming goal for extremely weight-conscious aero-engine makers) and their lack of propeller (which worked poorly at high altitudes), the attraction of the jet engine had to be balanced against the practical difficulties of making one work and producing it in large numbers. As this book shows in detail, the jet engine was not only the product of new ideas but also crucially of existing technical resources and skills that were redeployed and refined. In this picture, the famous jet engine inventors are understood as parts of larger inventive institutions that deeply influenced their work. These institutions are in turn understood as fitting into a broader context: the systems responsible for producing usable innovations in aeronautical technology (primarily for national air forces), systems that included national governments, commercial firms, and research institutes. There was little obvious correlation between the moment that any man became dedicated to the jet engine and the start of production of jet engines in any of the three countries studied here. Furthermore, the production of deployable jet engines by industrial firms was in all cases begun before development and thus innovation had been completed. This had a profound effect on what was developed, when, and how. The story told here is the story of the complex interaction between expectation, production, development, and innovation.

Paying attention to production and to industry has important interpretive consequences for the story of the jet engine. It transforms the standard story in which Germany surged ahead during the war while Britain was left behind and the United States failed entirely to advance. This book shows that Britain quickly decided to focus on the development of large numbers of new engine types for the postwar world and, despite a rushed and largely symbolic deployment of jets during the war, emerged from the war with by far the strongest portfolio of new engines of any country. National Socialist Germany, in contrast, produced a very large number of dangerous jet engines as a replacement for more expensive piston engines; their jet engines can be understood as an ersatz technology. The United States was in even less of a hurry to deploy jet engines during the war than Britain, and indeed, it did so only in very small numbers. It used the war to build up its capability to manufacture jet engines. American aero-engine firms, which had been explicitly excluded from jet engine development during the war, did not turn in the postwar period to confiscated German jet engines as a basis for future development but instead licensed British jet engines with wartime heritages. Looking in detail at the creation of the earliest jet engines explains why the engines produced by each nation's aero-engine firms took the shapes they did and underlines the many important continuities in design, tacit knowledge, management, and equipment that linked piston engine and jet engine production.

The stories that are commonly told about the creation of the first jet engines in Britain and Germany have done great violence to that early history—it has become a story of firsts and of single inventors, with the production of engines understood as little more than the principal measure of inventive success. The standard story focuses on two inventors from outside the aero-engine industry: Frank Whittle, a young officer in the Royal Air Force, and Hans von Ohain, a young German physicist. Their work supposedly culminated in the first jet flights by the Royal Air Force and Luftwaffe during World War II. According to this story, Whittle developed the idea of a jet engine in 1930 but was delayed in pursuing his idea by a lack of government interest and, as a result, was ultimately forced to form a private company to develop his jet engine. The first Whittle engine ran on April 12, 1937, and the first jet flight in Britain took place on May 15, 1941. The standard story argues that the delay in developing Whittle's idea let Germany overtake Britain in jet engines. Unlike Whittle, who struggled for many years, the story tells us that the inventor Hans von Ohain quickly found support for his work from the German aircraft company owner Ernst Heinkel in 1936. Von Ohain's first engine ran on the bench in 1937, and on August 27, 1939, a von Ohain engine powered a Heinkel airframe in Germany's (and the world's) first jet powered flight. Both Britain and Germany deployed their jet fighters in July 1944. Yet Germany produced far more jet engines than any other country during the war, proving, in the standard story, German superiority. In all measures, the United States lagged behind Germany and Britain in jet engines. According to the standard account, the United States Army Air Forces had to copy Whittle's engine in order to jumpstart American jet engine development because they lacked a jet inventor of their own. Despite this, no American jet aircraft were deployed to fight the war.

As well as telling a richer and more inclusive story of the making of the jet engine, this book will demonstrate that we need a new way to think about invention. This is perhaps surprising given that so much attention has been given to the subject, long a central concern of the history of technology. Yet treatment of the topic tends to be narrow, dominated by implicit and explicit models that are incompatible with the empirical evidence of the creation of some crucial midcentury technical developments, including the jet engine.

The first scholarly study of invention was carried out in the 1920s and 1930s by sociologists in the United States, who asked whether invention had individual or social causes. These sociologists studied invention as a way to explain broader questions of social change and insisted on the continuity of the nature of inventive activity over time.[1] Their work on causes survives today primarily in dismissive references to a fruitless, earlier "classical division" of invention.[2] In place of this early work, a popular narrative of invention arose in the second half of the twentieth century that combined the existing cult of inventor heroes with optimism about the industrial research lab—what I will call the transition narrative of invention. This narrative asserted that invention had changed from being the province of individual, independent inventors in the nineteenth century to being dominated in the twentieth by industrial research laboratories, which produced inventions at an increasing rate.[3]

Faith in industrial research and development labs as a source of invention crested in the interwar period.[4] After World War II, this enthusiasm gave way to the reassertion of the importance of the individual inventor, as the publicly celebrated nineteenth-century inventor also became the stereotypical independent inventor for late twentieth-century writers.[5] Key among them was the flamboyant American Thomas Edison, whose self-promotion has become legendary. The famous study of the sources of invention by John Jewkes, David Sawers, and Richard Stillerman of 1954, although taking aim primarily at the argument that laboratories had entirely replaced inventors in the twentieth century, ended up contributing to the resurgence of interest in the independent, individual inventor because it argued that individuals were an important source of inventions in the twentieth century.[6] Thomas Hughes, arguably the most influential theorist of invention in the late twentieth century, connected the rise of the serfdom of industrial research with the downfall of the free, independent inventor.[7] Invention in industry was increasingly equated with a narrow view of "research and development" in corporate laboratories following the model of academic science.[8] While the relationship between science and invention continues to be discussed and fought over, the relationship of the corporate research lab to invention has only rarely been challenged.[9] As the competition between individual and collective was sublimated during the Cold War into a story of competition between individual and corporate inventors, the unfortunate disappearance of independent inventors in the twentieth century began to be rendered in moral terms.[10]

Despite warnings from scholars like Lynn White that it was based on the uncritical acceptance of existing popular stories rather than on empirical detail, the transition narrative became part of the history of technology when the academic field was created.[11] It appeared in textbooks in the 1970s and can still be found there today.[12] The transition narrative has limited our understanding of how new machines emerge in two important ways. Firstly, its dominance has restricted our view of the creation of new things primarily to invention, or the very earliest work on a new idea that is generally proved important at a later time. The field certainly shares this with public discourse. Secondly, also mirroring the lack of public interest, it has generally directed attention away from the creativity of industry. The important shift historians of technology made to seriously study consumers and their influence on the shape of new machines has provided important research results but has allowed a gap to remain in our understanding of what happens in the space between inventors and users—a gap that has become ever wider as the creation of increasingly complex technologies involves an ever greater number of actors.[13]

The maintenance of the transition narrative (which has only recently begun to be challenged directly)[14] in the face of empirical evidence that contradicts it[15] has been accomplished through rigorous "boundary maintenance" work and shared, narrow categories.[16] For example, John Jewkes, David Sawers, and Richard Stillerman fit the story of the jet engine into the transition narrative as a late case of individual invention, not least because it doesn't look like a product of industrial research.[17] The invention of the jet engine may be closer to the transition narrative's understanding of individual invention than its picture of the corporate research laboratory, but it is poorly described by both. Indeed, both poles of the transition narrative have served to hem in our current understanding of the emergence of new technologies.

The space between invention and use hides a great many creative contributions without which the jet engine would not and could not have been created and deployed. In what follows, I expressly consider sources of creativity that lie outside of the transition narrative in both time and space. My focus is the manufacture of the world's first service jet engines. This extended process required the work of industry outsiders, as the standard story of the jet engine emphasizes, but many more industry insiders, who, contra the transition narrative, did not work in industrial research laboratories. Research laboratories are vital to this story, but they were generally the laboratories of state research institutions. These were not research labs associated with physics or chemistry, but rather with engineering and mathematics, with aviation and aerodynamics.

Despite the shortcomings of the transition narrative, the transition-narrative-friendly, heroic-inventor-focused story of British failure and German success with the jet engine has proven influential in professional histories. This also despite the fact that an outstanding and very different history of the jet engine appeared just after World War II. Already in 1950, the classical (by training) historian Robert Schlaifer published an account of the jet engine as part of a larger business study of government-industry relations in the development of aero-engines. Although Schlaifer's work is problematically based on the testimony of unidentified actors, the broader frame of his work offers a provocative starting point for a new history that examines the jet engine as a reflection of the strength of each nation's aero-engine industry.

The first professional account of the episode to appear after Schlaifer's was Edward Constant's *Origins of the Turbojet Revolution* (1980). Schlaifer's work features only as a reference. Constant sought to use the story of the jet engine to change the way that historians of technology think about invention by introducing the idea that new machines resulted from what he called (following Kuhn's earlier work) "technological revolutions" driven by "presumptive anomalies" identified by science. His framework has since been mentioned if not used by many scholars,[18] even though historians in the 1970s were already struggling against the assertion that it is primarily science that leads to new technologies (a topic that will be broached here only tangentially as science is not referred to in most of the narrative).[19] Constant's book's central thesis was that the timing of the turbojet revolution in Germany, Britain, and America—in that order (not the order in which national work began on jet engines)—can be explained by the character of the study of aerodynamics in each country, where national communities were variably disposed toward theoretical work on supersonic aerodynamics and practical work on subsonic flight. He argued that more theoretical work (as in Germany) led to earlier recognition of the presumptive anomaly that, in his model, led to the turbojet revolution. In stressing the key role of individuals, of outsiders, and of science to invention, however, Constant's account of the jet engine virtually reproduced the existing popular story, adding to it only a new interpretive framework and two additional jet engine "inventors"—who are, however, both present in Schlaifer's earlier account.

In addition to these two important comparative studies on the history of the jet engine are works that focus on or emphasize a single part of the story: the new account by Andrew Nahum of the British government's role in Whittle's work, Ralf Schabel's study of the politics of the jet engine in National Socialist Germany, and Michael Pavelec's tactical history of the weapons used during World War II. There are also many contributions to the literature by amateur historians, many of them ex-engineers and ex-pilots. This book makes extensive use of the detailed stories recovered by these sources but treats with circumspection any statements claiming to be historical interpretations, for many of these accounts hide assumptions about how new technologies come into being and who the most important actors in that process are.[20]

*   *   *

Several conceptual shifts are crucial to the reinterpretation of the history of the jet engine presented in this book. The first is to reverse the order in which the history of the jet engine has been told. Rather than move from invention, to development, to production, I discuss production then development, before giving an account of the inventive institutions where Whittle and von Ohain worked during the war. Throughout, the terms *invention*, *development*, and *production* will be used loosely, invoking their common definitions and being deployed as descriptive rather than analytical terms, in order to avoid existing assumptions about their (linear) relationship that historians have already effectively challenged.[21] Indeed, the activities are sometimes difficult to isolate empirically, for during much of the period covered by this book, all three occurred concurrently within a given engine project or company and across the entire aero-engine industry. This can be seen in that the same years appear in each chapter. Philip Scranton, writing on the development of American jet engines in the Cold War period, has described the creation of jet engines as "an a open, iterative, situated process of problem-stating and solving, often with multiple feedback loops that shift the grounds of inquiry, rather than as a series of discrete events."[22] I have purposefully chosen to tell the story of the creation of the jet engine in an unfamiliar way in order to prompt fresh questions about the episode, including about the relation of these activities.

Production forms the foundation of this account of the jet engine. By subverting the tendency to view production as the playing out of earlier development decisions, as a measure of success of other processes, the narrative order I adopt aims to provide a more compelling account of the episode, one in which production is shown as integral to the creation of the jet engine. I am not arguing for replacing one hierarchy with another, but rather for the necessity to view production, development, and invention as separate but related activities in which changes in one activity impacted the conception and execution of the others.[23] This mutual influence occurred in areas as fundamental as goal setting and competition for limited resources.

In all three countries, hopes of producing jet engines as soon as possible led production plans to be put into place often well before development had been completed. Rather than being an afterthought, planning for engine production fundamentally shaped other creative processes necessary to the creation of the world's first jet engines. Looking at the details of production demonstrates the changeability and responsiveness of production planning, the "concurrency" of development, production, and use, which Philip Scranton has already highlighted in the case of jet engine development and production during the Cold War.[24] By 1945, Britain, Germany, and the United States were all producing jet engines, and as the production decisions discussed in the first chapter will show, the paths that each nation followed had a profound effect on what each country produced, developed, and invented.

In Britain and the United States, early interest in production was relaxed when the new jet engines proved unready for production. Both nations' relative lack of output has strongly influenced ex post facto evaluations of both programs as failures. This account, in contrast, will emphasize both the intensity and the consequences of the early British dedication to jet engine production and show that the American program ought to be understood not as an inventive failure but as the result of a production-driven decision to produce a (supposedly) mature British design (as it already had in other cases, like that of radar and penicillin) in preference to a less developed American engine.

As this book shows, the well-known German lead in jet production, meanwhile, was not the result of that country's technical lead in aero-engines, nor of an early decision by the German government to produce jet engines. It was rather an outcome of the country's desperate position at the end of the war. The jet engines Germany produced required only a fraction of the man-hours and material that were needed to build a piston engine. They were designed for production in Germany's stretched, late-war economy using merciless, National Socialist methods. The nation's many jet engines were not wonder weapons, but ersatz ones. The first chapter shows that the new engines were not ultimately produced or deployed primarily because of their superior performance over piston engines but more for reasons of postwar competition, expediency, or scarcity. German jet engines sacrificed safety to enable large production numbers. Jet engine production in Germany was not, as some authors have portrayed it, the successful climax to a program of advanced technological development, but rather the large-scale manufacture of engines of desperation.

The book's second chapter describes the contributions of industry to the development of the world's first jet engines. It examines how existing aero-engine companies applied their skills, knowledge, and resources to developing a radically new type of aero-engine, acquiring new skills when they had to and not before. The range of industrial firms involved in the jet story—most of which do not feature in the standard account—included companies with an assortment of relevant expertise: engineering consultants, electrical companies and steam turbine builders, automobile companies, fuel companies, metal suppliers, and component manufacturers. In Britain and Germany, aero-engine companies were the central and coordinating actors among these many firms, which together carried out the majority of work developing jet engines during the war. By uncovering a previously hidden world of large-scale design and development activity, the industry-centered story of this chapter emphasizes the range of industrial skills and expertise that were essential to the creation of a service jet engine—skills that made the aero-engine industry every government's essential partner in creating new aero-engines.[25]

One group of firms with expertise that many people expected to be relevant to the jet engine was the steam turbine industry, which included the steam turbine divisions of large electrical companies. The work of the steam turbine firms that pursued jet engines, primarily in Britain and the United States, suggests that some existing skills and methods hindered rather than helped development of the new engines. The very different fates of the American General Electric Company and Westinghouse Corporation and the British Metropolitan Vickers company in the field of jet engines, for example, suggests that although steam turbine firms did successfully design gas turbines, a minority were capable of entering the specialized aero-engine industry, in no small part because they lacked experience in developing and producing complex and accurate aero-engines.

A key argument of this chapter is that it was continuities—not just in design, but also in personnel, method, and resources—that suited the aero-engine industry to tackle the new engine type and make it its own. One of the reasons that the aero-engine industry was able to take up the new type of engine so effectively was because the new engines could be designed to a significant extent by combining independent elements— many of which the industry was already familiar with. For example, many aero-engine firms had already developed superchargers that would allow their piston engines to operate at higher altitudes, where it was difficult to get enough oxygen for combustion directly from the atmosphere. These devices contained compressors and turbines—the same elements that are important in gas turbines.[26] The main components of each jet engine were often developed separately, and each had its own unique history.

Contra Constant, who emphasized the revolutionary nature of the new power plant, I emphasize the deep structural and institutional continuities between piston engine and jet engine development and follow their consequences in early jet engine design. Continuities were more central to facilitating the manufacture of the world's first jet engines than any major, revolutionary breaks. Yet, as I will show, continuities could also be toxic, leading firms down development paths that proved disastrous or never ending. They are no less important for that. In both cases, tracing continuities explains a great deal about engine design and the level of each firm's success.

The ambiguity of what an early gas turbine aero-engine consisted of was reflected in the variety of terms used to describe the early engines—in strong contrast to the dominant term *jet engine*, which today is in common use. The new aero-engines were frequently known as "gas turbine aero-engines" emphasizing the engine's central novel element. In other places, they were referred to as "internal combustion turbines," instead emphasizing continuity with internal combustion engines (of which piston engines were a leading example). Whether or not the gas turbine produced a propulsive jet or drove a propeller was an important distinction. The term *turbojet*, like the ungainly *gas turbine, jet propulsion aero-engine*, referred to a gas turbine that propelled an aircraft primarily using jet propulsion. This is what today is known as a jet engine. A *turboprop*, in contrast, used a gas turbine to power a propeller via a shaft (as in a piston engine).

A focus on continuities underlines the fact that the story told here is not that of a new type of engine creating a new industry. It is rather of the aero-engine industry creating a new type of engine. The youthfulness of many of the teams assembled to work on jet engines contrasted with established piston engine development teams, but this was a product of the availability of young graduates during a war in which engineers (like piston engines) were in high demand, rather than a reflection of the conservatism of the aero-engine industry. In fact, jet engine development teams were generally led by experienced piston engine and supercharger men. New jet engines took about the same amount of time to reach the point of first production as a new piston engine (about five years), contrary to optimistic, early hopes in industry and government.

Looking at the full story of the development of turbojet engines in the British and German aero-engine industries requires us to examine both the successes and the failed or abandoned engine designs. Doing so has important interpretive consequences, because it reveals not only how each company approached new problems, but also how each company and each country's goals changed over time. This can be seen, for example, in the crucial shifting of resources between piston engine and jet engine projects. The story of how Rolls-Royce became a successful maker of jet engines in Britain—a story that earns its own section in the book's second chapter—shows how that firm exercised all the resources at its command to succeed with the new engine type. Examining engines that were designed but not produced during the war is also valuable, because it counters the tendency to interpret low production numbers as indicating small development programs. The wartime numbers game gives no credit to firms for designing engines that became important only after the war, for example.

Because this book is concerned with the framing and understanding of the jet engine story as a whole, the details of the activities of some individuals and companies have been sacrificed in order to paint a broader picture. My focus on the aero-engine industry leads me to spend less time discussing the role of the government, for example, although government actors frequently enter the narrative. They have been studied in more detail elsewhere.[27] In addition, much detail on the technical aspects of the work can be found in existing sources, and a few examples must suffice here as illustrations. In the same way, the approaches of the different jet engine teams that are mentioned in the narrative, differences in the physical, theoretical, and mathematical machinery that they used, as well as the evolution of and connection of industry to aerodynamics and other sciences to theorize parts of the jet engine deserve a much more thorough examination and description than I can give, although they became more important later in time, as the technology developed.[28] Although the detailed design of many engines will be described below, the actual activities involved in the aero-engine development process fall outside the scope of this book. This book focuses instead on describing the institutional efforts needed to produce a new machine for use. A glossary gives additional details about the parts of a jet engine as well as the theories and language used to describe them.

After development, the book turns to innovation. This section of the book centers not on individuals but on "inventive institutions" in order to unpack how individual engineers worked as members of larger organizations—institutions on which the fate of their inventive ideas often depended. The third chapter thus studies the well-known Frank Whittle and Hans von Ohain as members of the institutions where they worked between 1936 and 1945: Power Jets and the Heinkel Aircraft Company, respectively. Both men willingly became part of private companies in pursuit of support for their inventive work on the world's first jet engines, and both men built careers in the aircraft industry during the war.

It is therefore striking how short the impact of the actual design work of these inventors and the institutions that promoted their design work (as opposed to their energy and inspiration) was, and this book makes clear that this was connected to the institutional instability of the firms where these men worked, which was in no small part a product of the war.

Both Power Jets and Heinkel hold central places in the history of the jet engine, despite the fact that neither company produced a jet engine for service use during the war (although both wanted to). They are grouped together here as "inventive institutions" because although they are historiographically overshadowed by the individual inventors Whittle and von Ohain, both companies were important sources and shapers of early jet engine design work. The jet engine work of both institutions thrived under particular circumstances. These had largely disappeared by the end of the war, by which time both firms found that their claims to exist as inventive institutions were increasingly ineffectual in the commercial marketplace. Taking these companies seriously as organizations shifts the discussion from treating institutions as a mere background for unconstrained individual inventors to looking at how the goals of institutions have shaped the emergence of technical novelty.

The point of telling the jet engine story with these inventive institutions at the end is to illustrate that invention should be understood not as the precursor to development and production, but much more broadly as an extended activity in the context of development and production. The work of the inventive firms that are the subject of chapter 3 was constantly impacted by the development and production decisions discussed in the two previous chapters. These decisions encompassed entire national efforts rather than just the work or outlook of a single firm.

After the discussion of inventive institutions in chapter 3, chapter 4 turns to the question of how the standard story of the jet engine that we know today was created and why it gained such wide public credibility. This is not meant as a commentary on the truth of the standard story, which is more than anything a particular selection of events from a bigger, more complex picture. Nor does it imply that misinformation was spread deliberately. Instead, it explores how the machinery of memory worked in the mid-twentieth century, and why the popular story emphasizes the particular subjects that it does.

Chapter 4 begins with an examination of how Frank Whittle's public persona as a British inventor-hero was constructed from the moment in 1944 when the existence of secret Anglo-American work on jets was first publicly revealed. In Britain, the government's strict wartime policy on the control of jet publicity created a gap in the public understanding of an increasingly visible and economically significant machine, which was filled by the already familiar story of heroic invention. In the United States, in contrast, public emphasis on the production and postwar success of British-derived engines at American companies distracted from the presumed inventive failure of American companies during the war and silenced stories of indigenous American work. The conscious British effort to promote British credit for the invention of the jet engine influenced the public story in both countries.

The narrative of British leadership that was circulated in technonationalist Britain early on was linked, in the United States and later in Germany, to a narrative of German invention that was (re)constructed decades after the end of the war. Emerging later than those of the victorious Allies, the story of Germany's jet achievements was shaped by earlier public accounts of Allied work. Hans von Ohain was fashioned as the German Whittle. (He still retains this position, even after Edward Constant's account challenged the symmetry, if not the content, of the two men's stories.) Many German jet engineers were active as engineers in the United States after the war, where they also supported efforts to recreate the history of Germany's first jet engine. It was in the United States that the story of the dual jet engine inventors (one British and one German), with which we are familiar today, emerged and reached a public audience for the first time.

The tales that arose in the decades after the war were well suited to the postwar needs of a range of groups, and set off a process of forgetting that resulted in the particular understanding of the jet story that has intentionally or unintentionally obscured the extensive activity of many different actors. The contrast between these stories inspired this book. Studying the emergence of the well-known story of the jet engine is important because it has also come to influence professional histories. Part of the reason that professional historians have adopted the salient features and emphasis of the dual inventor story is that they resonate with the standard treatment of invention in the history of technology.

From start to finish, this book challenges us to rethink our basic assumptions about how new machines were made in the twentieth century, how they were brought from idea into use, and to recalibrate our understanding of the importance of different actors in the process of technical change. By reversing the order in which the story of the jet is often related, I hope to shift our attention away from familiar individuals to the much less known role of industry, and to open up the story of technical novelty to a much wider array of actors, each of whom exerted influence in particular ways. This book uses the case of the jet engine to ask what a satisfactory account of the emergence of technical novelty should include.

Turbojet Production during World War II

National Socialist Germany built 6,569 jet engines during World War II—almost nine times as many as Britain, and more than twenty-two times as many as the United States produced in the same period (see tables 1.1 and 1.2).[1] Traditionally, this is taken to indicate an astonishing German technical success, all the more so given the dire state of the National Socialist economy at the end of the war. Not only is the country's production record hailed, but it is also taken as evidence for the usefulness and indeed superiority of Germany's engines.[2] Yet although Germany produced the highest number of engines, it also produced the lowest quality engines.[3] Material and labor shortages made cheap turbojets that would have been unacceptable in Britain or the United States attractive weapons for the Third Reich, where the jet engine was a substitute for expensive piston engines. This startling argument radically changes our understanding of the nature and power of early German jets: they were produced not because they were better than alternatives, but because they were easier to build. Their manufacture was well suited to and dependent on slave labor, as this chapter will show for the first time. Germany produced jet engines not despite the country's limitations, as most accounts make out, but as a response to them. Production concerns were at the heart of the German decision to deploy jet engines during 1944, and the engines moved to the very center of the National Socialist production system.

Germany did not owe its high production output to an early decision to produce jet engines. It was not the first country to decide to produce jet engines, but the last. Britain was first, and its government did so in 1940,

TABLE I.1 **Aero-engine production in Britain, Germany, and the United States, 1939–45**

	1939	1940	1941	1942	1943	1944	1945*
Britain	12,499	24,074	36,551	53,916	57,985	56,931	22,821
Germany	3,865	15,510	22,400	37,000	50,700	54,600	
USA		15,513	58,181	138,089	227,116	256,912	106,350

*Britain January–August; USA January–December
These figures may or may not include turbojet production; either way, piston engine production far outweighed turbojet production.
Source: Richard Overy, *The Air War: 1939–1945* (Dulles, VA: Brassey's, 2005), 150.

TABLE I.2 **Jet engine production in Britain, Germany, and the United States, 1939–45**

Company+	1943	1944*	1945**	Total
Rolls-Royce	9	98	553	660
De Havilland	—	at most 12		85
Total Britain				**745**
Jumo		2,660	3,350	6,010
BMW		189	370	559
Total Germany		**2,849**	**3,720**	**6,569**
General Electric				266
Allison	—			22
Allis-Chalmers	—			8
Total USA				**296**

Engine totals of recorded (delivered) production engines only; this does not include numerous experimental and de-velopment engines, nor reflect the extent of parts production.
+ Engine Types: Rolls-Royce: Welland and Derwent; de Havilland: Goblin; Jumo 004; BMW 003; GE I-16 marks 1 through 7 (produced 1944–45); Allison I-16-6 (1944–45); Allis-Chalmers H.1 (completed by April 1946)
* Germany: up to the end of 1944 including some in late 1943
** Germany: January–March 1945; Britain: January–December 1945
Sources: RRHT; William Hornby, *Factories and Plant* (London: HMSO, 1958), 245; Antony Kay, *German Jet Engine and Gas Turbine Development* (Shrewsbury: Airlife, 2002); IWM FD 5400/45, Jumo Report, TL-Geräte 004 B-1/4 Neubau und Reperatur; AFHR A2073; AFHR A2072.

with the intention of producing jet aircraft for their country's defense. The government's choice fell on a new engine design, the W.2, that, how-ever promising its possible future performance according to aerodynamic theory, was not yet ready for production. The prototype on which the Brit-ish production engine's design was based, the W.1 engine, took to the air only in 1941. In late 1942, the production design was almost abandoned because of problems in development, but it was retained and Britain's first production (scale production, not hand crafted) jet engines were de-ployed in mid-1944. During the war, in Britain as in Germany, alternative jet engines with superior performance were turned down for production, while less powerful engines better suited to production were accelerated.

In the event, Britain found that its air force did not need jet fighters to win the air war, and it instead devoted massive resources to developing jet engines for postwar use. Its interests came ultimately to rest on developing highly reliable and powerful jet engines, and the country ended the war with a range of advanced jet engine designs that dominated world jet production after 1945. Despite this change in goal, the early decision for production nevertheless shaped the British development program by leading it to support the development of centrifugal jet engines, although its focus shifted more toward axial jet engines over the course of the war. All of Germany's first service jet engines and most jet engines today are axial.

Centrifugal jet engines are distinguished by their radial or centrifugal compressors. They compress air by spinning and accelerating it from the engine's central axis to its outer edge. This added velocity is then converted to increased pressure. In the interwar period, the centrifugal compressors used in centrifugal jet engines were close to existing designs of turbo air compressors and could therefore be developed more quickly than unfamiliar axial compressors, although they were limited in performance by their radius. Nevertheless, compression ratios in early jet engines did not exceeded what a large radial compressor could achieve. Axial compressors compress air by forcing it along a straight path into a smaller and smaller space with each row or stage of compressor blades. While axial compressors promised theoretically to compress air to higher pressures than centrifugal compressors were able, the more complex axial compressor was almost entirely unknown to engineers in the interwar period (schematics of both types can be found in figure 1.1).

The decision to produce a British designed centrifugal jet engine in the United States was made in the hopes that the British engine offered a faster route to producing turbojet engines than an indigenous American one. It also led that country to invest resources in developing centrifugal jet engines. Yet the desire for production remained secondary to the establishment of a basis for building commercially competitive American jet engines, and the United States subsequently moved away from its early gamble on manufacturing to keep its eye on long-term development. For both the British and American programs, limited output actually reflected the strength of their war efforts: neither was forced to deploy jet aircraft out of military need. It was the weakness of the German position that demanded that country's existential focus on production.[4]

Although the extent of jet engine production in each country was thus not a measure of technical success, decisions to produce (and not to pro-

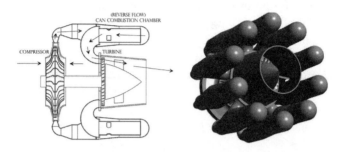

This is the basic Power Jets gas turbine, jet propulsion architecture with its double-sided, centrifugal impeller-compressor and reverse-flow combustion system, made up of individual combustors in which the air's direction is reversed twice, as indicated. This architecture was used in the Power Jets W.2B, the WR.1, and the Rolls-Royce Welland (approximately 158 cm long, 109 cm diameter).

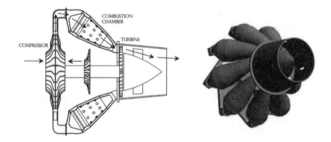

This architecture was used in the Rolls-Royce Derwent (approximately 213 cm long, 109 cm diameter); it has a double-sided, centrifugal impeller-compressor but, in contrast to the Power Jets architecture, has a straight-through combustion system.

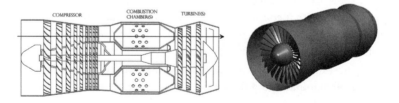

This axial turbojet architecture has an axial compressor (with multiple stages) powered by three turbines mounted on a single shaft. Early axial engines included the Metropolitan Vickers F.2 (approximately 404 cm long, 89 cm diameter), the Jumo 004 (approximately 386 cm long, 82 cm diameter) and the BMW 003 (approximately 363 cm long, 69 cm diameter). Each had different numbers of compressor stages and turbines; early axial compressors often used a single blade profile for all stages. *Source*: Models by Denie Technology duce) engines during the war nevertheless had a huge influence on the quality and nature of the world's jet engines in the short and medium term. By considering production first, I hope to clear away some important misconceptions about the jet engine and set the scene for a new, richer account of its development and invention in the subsequent chapters.

Britain's Turbojet Engines

Britain was the first country to decide to produce turbojet aero-engines. By the end of the war, three types of centrifugal turbojet engine had been produced in Britain: the Welland and the Derwent, both made by Rolls-Royce and the two engines deployed during the war, and the Goblin, made by de Havilland. Britain hoped to have jet aircraft in service in 1942, but during the war, the strength of Britain's war effort allowed it to deemphasize production when development problems arose. The Royal Air Force's first production engines arrived in 1943 and its first jet fighter flew in mid-1944.

From the start, the British government expected its war effort to be a technological one in which machines would compensate for the nation's small army. This tactical orientation and the creative spirit of Winston Churchill, Britain's prime minister from 1940, set the tone. Wartime Britain was, in the words of David Edgerton, a "cult of invention and inventor presided over by the Prime Minister."[5] In a reversal of much literature on Britain, Edgerton argues that while quirky British devices are frequently cited as "evidence of a peculiar technical resourcefulness in the face of material austerity, they were in fact compelling evidence of extravagant commitment to technical solutions, and of massive material and technical capacity."[6] The British pursuit of jet engines was one based on strength: both military and material.

Britain could afford to pursue jet engines during the war in part because the British Air Ministry's efforts at expanding aero-engine production during the years before the war were so successful. Between 1936 and 1945, the British government spent more than £100 million on expanding the nation's aircraft production capability.[7] Investment went to creating so-called shadow factories for the production of powerful Bristol Hercules piston engines and to building new factories for the production of Rolls-Royce piston aero-engines, including the important Merlin. The shadow factories were put under the management of Britain's motorcar industry, and they successfully produced engines according to

FIGURE 1.2 The sleeve-valved Bristol Hercules piston engine. The pictured engine is located in the Bristol Industrial Museum. The engine's diameter is approximately 140 cm, its length 135 cm. *Photograph*: Adrian Pingstone

Bristol's designs.[8] Rolls-Royce, in contrast, preferred to manage its factories itself, including the new factories the government built for it in Crewe and Glasgow, which began production in 1940.[9] The only shadow factory constructed to build Rolls-Royce engines was a factory run by Ford located in Trafford Park, near Manchester, which came online in late 1941.[10] During the war, 101,000 Bristol piston engines of six types and 112,000 Rolls-Royce piston engines of five types were delivered.[11]

It was in February 1940, with the war against Germany just a few months old and the Battle of Britain just months away, that the British Air Ministry decided to pursue the production of turbojet engines. At a meeting on February 26, 1940, the Air Ministry's director general of research and development, Air Vice Marshal Arthur Tedder, announced that he wanted to proceed with "development manufacture" of a jet engine designed by the small British firm Power Jets Ltd., which had designed Britain's first experimental jet engine that ran in 1937.[12] Tedder reasoned that if the engine were developed from the start with the requirements of production in mind, it would reach production earlier. Thus production was decided on even before Britain's first jet engine had flown. In fact, quite the reverse of what one might expect with production being decided on, the Air Ministry had placed a contract with Power Jets for its and Britain's first flight turbojet engine, the W.1 (which was never built in series), only months before, on July 12, 1939.[13] An experimental airframe to carry the engine, the Gloster E.28/39 (the twenty-eighth experimental airframe of 1939), was ordered on August 30, 1939, from the Gloster Aircraft Company.[14] Neither Power Jets nor Gloster was otherwise engaged on crucial war work, so the decision did not divert essential resources from the war effort. The jet project was immediately put under strict secrecy. The E.28/39 flew for the first time under power of the W.1 engine almost two years later, on May 15, 1941. Despite the desire of some British officials in 1939 to skip building an experimental aircraft altogether, the E.28 became an important and reliable test aircraft.[15]

Power Jets was contracted to design but not to build the country's first service jet engine, the firm's next design, the W.2. Although the British government has been blamed for taking the engine away from Power Jets, the company had neither the facilities nor skill to manufacture complete engines (see chapter 3). Instead, in February 1940, the government decided to enlist a respected engineering company, the Rover Car Company, to manage the large-scale production of the engine in the nation's first jet engine factory. Rover was not, as it has been pictured, entirely lacking in relevant experience. In fact, it had participated in the Air Ministry's shadow production scheme and had repaired other piston aero-engines.[16] On good terms with Rover's management, Power Jets' leaders had recommended the firm as a potential collaborator, albeit with Power Jets as the senior partner.[17] This wish was not however respected (Power Jets being the younger of the two firms), and by November 1940, Rover was working under direct government contract to build an experimental engine to Power Jets' plans.[18]

Committed to the project, Rover deployed some of its best resources to work on the fledgling jet engine program. It chose to establish turbojet production in one of the factories that had been assigned to it under the shadow production scheme: Bankfield Shed. The disused weaving mill in Barnoldswick was the most recent addition to the six shadow factories where the car firm was already carrying out piston-aero-engine production and repair work in 1940, and the firm's staff had already begun repairing Armstrong Siddeley Cheetah piston engines there.[19] The mill was about twice as large as a normal weaving shed, having approximately 165,000

FIGURE 1.3 Aerial view of Bankfield Shed in Barnoldswick. *Source*: Rolls-Royce Heritage Trust

square feet of space. Mr. Poppe, Rover's expert on quantity manufacture, was put in charge of the factory.[20] Rover decided to subcontract the engine's fuel and combustion systems to Joseph Lucas Ltd., a company that manufactured automotive components, including fuel pumps and injectors. Lucas went on to make fuel and combustion equipment for almost all of Britain's early jet engines.[21] Rover chose a second shadow site, Waterloo Mill in Clitheroe, as the center of its turbojet design and testing work.

That Rover's jet engine factories were also called "shadow factories" is misleading. Rover's work on jet engines included development and was thus crucially different from its previous participation in the government's shadow scheme, in which Rover had been charged with the production of already proven aero-engine designs. It was unusual, for example, that Waterloo Mill was staffed by a number of Rover's permanent staff. In the case of the jet engine, Rover was hired to make development units, and in recognition of its design work, the firm received a slightly higher than usual management fee for its turbojet factories of £16,000 per annum.[22] The confusing use of the term "shadow factory" to denote something quite different from earlier shadow manufacture mirrored a deeper confusion over Rover's role in the jet engine program (especially with respect to Power Jets), which ultimately led to Rover's withdrawal from the aero-engine field.

In February 1941, the government confirmed that Rover's Barnoldswick factory should be equipped to produce Power Jets engines at a rate of twenty per week. While this was not an overly high target as compared with piston engine production (thousands per week) or later German and American plans (hundreds per week), it nevertheless represented an enthusiastic and early investment in the new machine. At the same time, further demonstrating official eagerness, plans were also approved to provide for the production of an additional thirty engines per week using as much subcontracting as possible. The production total could be brought to about two hundred engines per month if needed.[23]

Which engine would be produced in the Barnoldswick factory was decided, however, only on March 5, 1941. It was decided that it would produce the Power Jets–designed W.2B (which replaced the original W.2 design). Fifty "development engines" were initially ordered from Rover. The W.2B engine was a scaled-up version of the W.1, which still had not flown. It was to fly for the first time in the E.28 on May 15, 1941. Whereas the W.1 was designed to produce 1,240 pounds of thrust (but it achieved only 850),[24] the W.2B was to reach the 1,600 pounds of thrust expected to be necessary for a useful fighter as outlined by Sir Henry Tizard, an early supporter of the jet engine, advisor to the British government, and chairman of Britain's Aeronautical Research Committee.[25] (Producing a given amount of thrust for a specified length of time became a key criterion for the acceptance of new jet engines.) In early 1941, the government demonstrated that it was serious about the new engines by placing contracts for an airframe to carry the W.2B engine.[26] On February 14, 1941, it ordered a mock-up and twelve prototypes of a new twin-engined turbojet fighter airframe from Gloster, the F.9/40 (the ninth fighter airframe of 1940), which was actually already under design.[27]

Production of a British jet engine had thus already been decided on when British aircraft production underwent a dramatic organizational change. In May 1940 (just before the start of the Battle of Britain), Churchill established the Ministry of Aircraft Production (MAP) to streamline aircraft production in Britain. Press baron Max Beaverbrook, who was appointed the first minister of aircraft production, immediately ordered that all development work be stopped. For fear that he would cancel the jet program and scatter its resources, jet engine development was initially concealed from the new minister. When it was revealed to him some weeks later, however, Beaverbrook did not decide to abandon the jet.[28] As he made clear, his interest did not represent a blind devotion to novelty. The jet was seen as an important part of the ministry's serious and constant pursuit of and support for improvements in military aviation, a commitment that dovetailed with the government's dedication to new machines as part of its military planning.[29] On April 18, 1941, a year later and still before the first flight of a British jet engine, Britain's Air Staff (the body of senior officers that ran the Royal Air Force) noted its desire to deploy turbojet aero-engines.[30]

Enthusiasm for the jet spread throughout the British government. In July 1941, Churchill and his scientific advisor, the recently ennobled Lord Cherwell, began to discuss the possible deployment of British turbojet aircraft in the summer of 1942. The two speculated about the possibility of putting the Gloster F.9/40 and the W.2B engine into full production even before they had been tested. By avoiding delays caused by what he called the "titivation" of "perfectionist" technicians, Churchill hoped to be able to deploy jet aircraft six months earlier than otherwise.[31] This interval could be crucial, since he feared that Britain might come under imminent threat from high-altitude German bombers—turbojets work efficiently at high altitudes where even supercharged piston engines begin to lose power because of the decrease in atmospheric oxygen, which is required for combustion. Yet a jet fighter would be deployed in addition to maintaining the deployment of existing, piston-engined aircraft. The prime minister was confident that the production of jet aircraft could be pursued without adversely affecting the country's air defenses; he had been assured that about one thousand of the new jet aircraft could be produced without sacrificing the crucial production of piston-engined aircraft.[32] The interest from Downing Street further increased the pressure on jet engine development. Whittle recalled feeling "like a hunted man."[33] By the end of October 1941, Rover had already been given a contract for 550 production W.2B engines, and investment in production capacity for eighty F.9/40 airframes a month had been authorized by the government at Gloster.[34]

Yet despite the government's interest, the Power Jets W.2B engine was still disappointingly far from being ready for production. The first W.2B engines produced much less than the 1,600 pounds they had been designed for. In order to allow flight trials to proceed, MAP decided in late 1941 to accept Rover W.2B engines when they achieved a rating of 1,000 pounds thrust (although this was only 62.5 per cent of design thrust).[35] Even then, it was not until May 31, 1942, that Rover's first underpowered W.2B engine was dispatched to Gloster for taxiing (rather than flight) trials in an F.9/40 prototype. With the engine still unreliable, Rover had to run five different W.2B engines for a total time of 45 hours and 45 minutes to get a single one to pass at 1,050 pounds thrust, the ministry's acceptance test for the trials, which consisted of about two and a half hours of continuous running.[36] The requirement for reliability remained crucial for every new British jet engine. In mid-1942, the W.2B engine was not yet powerful enough to propel the airframe that had been designed for it, and it remained below design expectations for the rest of the year. With the engine that the factory was supposed to be producing in series mired in development problems, the Barnoldswick factory was limited in 1942 to making a small number of development jet engines, each of which varied from each other in design.[37]

Not all of MAP's employees were carried away by enthusiasm for the innovative jet. Although Air Marshal Francis J. Linnell, the ministry's controller of research and development, was eager to expand Rover's production capacity in late 1941, the Ministry's deputy director general for engine development and production, Major George Bulman, cautioned that until the engine made more progress toward production standards, "it is impracticable ... to earmark further productive capacity" for it.[38] Bulman's skepticism toward the new engines was based largely on his experience in World War I as a member of the Aeronautical Inspection Department, when the Air Ministry had ordered 10,000 ABC Dragonfly engines before the engine design had been built or tested.[39] The Dragonfly proved so poor that only the end of the war had prevented the "ghastly tragedy" of deploying it, in Bulman's words. Bulman believed that the Dragonfly mess had occurred because the engineers who could have foreseen problems with the engine "were not strong enough to counter the arguments of the uninitiated enthusiasts"—a problem that he feared would be repeated with the jet.[40] Power Jets vilified Bulman as a determined enemy of the jet, whereas Bulman saw himself as the chief voice of restraint preventing an immature engine from being foisted on the Royal Air Force through the enthusiasm of the Ministry's research directorate. His judgment that it was an "appalling risk" to try to produce a new engine off the drawing board ultimately proved correct.[41]

Although large-scale production of the W.2B engine remained a far-off possibility, as its design was still in flux, Rover continued to plan for series production of the engine as stipulated in their contract.[42] This was no inconsequential matter during a war in which production capacity proved decisive. During 1942, MAP's jet supporters had to defend keeping resources locked up in the ministry's jet engine factory to the newly formed, skeptical Ministry of Production. Linnell's argument in April 1942 was that the ability to produce jet engines in the future justified the cost of reserving production capacity for what he confessed was "an engine we cannot yet produce."[43]

Yet circumstances didn't bode well for the W.2B engine in 1942: cooperation between Power Jets and Rover had lapsed into acrimony, and rumors began circulating in mid-1942 that the W.2B engine would be canceled entirely. Linnell moved to quash the speculation, insisting the engine would be made. But although Linnell defended the ministry's investment, he was chary of investing any more in turbojet aero-engines until some progress was demonstrated.[44] In August 1942, he rejected Rover's request to expand the factories at Barnoldswick and Clitheroe, which Rover argued were necessary to fill the production program that had been set for the company a year before.[45] The government had already invested a significant amount in its two jet engine factories and had little to show for it. Total government expenditure was estimated in mid-December 1942 (with no jet engines close to being ready for service use) to be £1.5 million.[46] At the end of 1942, Barnoldswick had about 1,250 employees and Clitheroe 250,[47] yet the future of the W.2B was hardly clear.

Competition in Britain

The W.2B engine wasn't the only design in Britain at the time. A strong competitor to the W.2B engine emerged in 1941–42 and looked likely to reach production readiness first: the H.1 turbojet engine designed by the independent aero-engine designer Frank Halford. The de Havilland Aircraft Company, for which Halford was a routine consultant, was to manufacture it. The engine was designed to reach 3,000 pounds of thrust, almost twice as much as the W.2B. This performance was calculated to allow a jet fighter with a single engine to outperform the contemporary Typhoon and Spitfire aircraft (powered by piston engines)—the surest way to justify the introduction of a new type during the war.[48] In mid-1942, the H.1 seemed to offer greater thrust, reliability, and development potential than the W.2B, despite having been designed less than a year earlier. Bulman, enthusiastic about the engine's prospects and confident in the skill of Halford, an earlier colleague, asked de Havilland in July 1942 to prepare a preliminary production plan for twenty engines per week—the same production rate as at the already established Barnoldswick factory.[49]

While de Havilland's engine division set about planning their jet engine production organization, the firm's aircraft division designed a fighter airframe to carry the engine. Unlike the conventional F.9/40 fighter airframe designed by Gloster, the E.6/41 or DH 100 had a twin-boom tail, which kept its tail plane out of the backward exhaust of its single, centrally located turbojet engine. It was made from both wood and metal, using the same methods as the firm's successful Mosquito and Hornet airframes.[50] On April 11, 1942, MAP placed an order for two prototypes at a cost of £40,000; a third airframe, to be built to operational standards including the provision of weapons, was added three days later. Permission was given to build the airframe at de Havilland's main works at Hatfield, provided that it did not impact production of the hugely important Mosquito.[51]

Delay in building de Havilland's E.6/41 airframe meant that the H.1 engine was first tested in a Gloster F.9/40 fighter airframe prototype. Due to development problems with the underpowered W.2B engine, two H.1 engines ended up powering the airframe meant for the W.2B engine on its first flight on March 5, 1943, and for a time, the future of the Gloster F.9/40 airframe appeared to rest with de Havilland's turbojet.[52] This didn't exactly mean that the H.1 engine was developmentally ahead of the W.2B engine, however, for while the F.9/40 could fly powered by two H.1 engines, each producing two thousand pounds of thrust, or 67 percent of the H.1 engine's design thrust,[53] it could not fly with two W.2B engines until that design had been developed to reach 1,450 pounds of thrust, or 90 percent of its design thrust. Further showing the Ministry's faith in the H.1 engine, but not the radical design of de Havilland's E.6/41 airframe, Gloster was asked in 1942 to design a more conventional single-engined fighter for the H.1 engine. The design, to MAP specification E.5/42, was known as the Gloster Ace. It was never ordered into production.[54] Nevertheless, by the end of 1942, Britain's MAP had ordered two jet engines and two jet airframes intended for service use.

Halford's H.1 engine was poised to take the place of the Power Jets W.2B in 1942, yet it continued under the shadow of the Power Jets engine. Linnell took a more circumspect approach to the new aero-engine for fear of repeating "our experiences with the W.2B."[55] The bar was set higher for MAP investment in jet engine production at de Havilland than it had been at Barnoldswick. Only in September 1942, when the second H.1 engine duplicated the performance of the first on test, did Linnell finally agree to Bulman's recommendation that de Havilland be granted a contract for the pilot production of one hundred H.1 engines at a rate of five engines per week.[56] Just as had happened with the W.2B engine at Power Jets and Rover, development of the H.1 proceeded in parallel with the construction of the production organization for the engine. Eager to help de Havilland rise to the challenge of building a powerful aero-engine (it had only produced relatively small piston engines before, its largest, the Gispy Queen, produced 200 horsepower compared with the more than 1,000 of the Rolls-Royce Merlin), Bulman arranged in August 1942 for "engine planning engineers" from Armstrong Siddeley Motors, another British aero-engine company, to help de Havilland plan for the production of the firm's first large aero-engine.[57]

At the end of 1942, competition between the H.1 and W.2B programs became even more explicit as the two programs began competing for the same production resources. (Rover's factory was already producing some parts for the H.1.)[58] MAP's new deputy director general of engine production, Air Commodore F. R. Banks, who took over some of Bulman's duties in October 1942,[59] ensured de Havilland that it would get machine tools from Barnoldswick. The existing factory at Barnoldswick was both a blessing and a curse because it linked de Havilland's production progress with the ministry's decision about producing the W.2B engine. As long as the future of the W.2B was in doubt, and thus a complete jet engine factory was potentially going begging, the Air Supply Board refused to give de Havilland permission to acquire additional new machine tools for aero-engine production.[60]

Abandoning the W.2B

There was a strong case for abandoning the W.2B engine design in 1942 because of its developmental problems. On November 2, 1942, Linnell chaired a meeting "to consider the future of the W.2B design and the possibility of abandoning [it]."[61] The meeting, attended by members of MAP and the staff of its research establishment who were closely involved in the jet engine program, concluded that the Ministry's gamble on the W.2B engine had been a failure and that preparations for its production should be ended. The W.2B powered Meteor, as the F.9/40 service type was called, was to be reduced to a training role, in which it would introduce the Royal Air Force to jet aircraft. The material already "under fabrication" at Barnoldswick (estimated as enough for between 150 and 200 engines) was to be finished as "development and training engines" under a modification of Rover's first development contract for fifty engines, while its later contract for 550 production engines was canceled.[62] The existing order for Meteor airframes was drastically reduced from eighty a month to forty-six total.[63]

This seemed to be the unfortunate end of the W.2B engine, and Linnell drew conclusions from the case to guide future ministry policy. In an early note to Air Marshal Wilfrid Freeman, who had returned to MAP to fill the newly created post of chief executive in mid-October 1942, he blamed the failure on "undue optimism" about the design, on trying to produce a new type from the drawing board, and (doubtless referring to the disagreements between Power Jets and Rover) on placing production at a firm other than the design firm. Linnell went on to suggest that Rolls-Royce take over both the country's work on jet engines and Power Jets Ltd., leaving the "salvage of the W.2B engine" at Barnoldswick to Rover. Work on the H.1 engine would meanwhile continue at de Havilland.[64] The turbojet engine program had begun before Freeman's departure from MAP on November 5, 1940, when he had left to serve as vice chief of the air staff.[65] Not everyone agreed with Linnell's plan for the redistribution of the country's turbojet development and production resources, however. Power Jets and its supporters within MAP, particularly the first deputy director of scientific research, Harold Roxbee Cox (see figure 1.4), defended Power Jets' independence. Ultimately, it was decided that Power Jets would be left alone while Rolls-Royce would take over both of the Rover turbojet factories instead.[66] Rolls-Royce filtered its management into the factories over the first three months of 1943.[67]

After taking over the Rover factories in Barnoldswick and Clitheroe, Rolls-Royce decided to finish the W.2B engines already under construction there. Rover had manufactured just thirty-two development W.2B engines before the end of 1942.[68] For Rolls-Royce, completing Rover's contract for W.2B engines was advantageous because it would introduce the aero-engine firm to jet engine production and would keep the Barnoldswick factory busy, thus maintaining a claim to the labor and machines that had been assembled there, which would otherwise be reallocated.

With the change in management at Barnoldswick, the W.2B engine moved quickly toward production readiness. By February 20, 1943, it had been put through a 100-hour type test, which included 26 starts, 106 accelerations, and 100 hours of running at 1,250 pounds thrust (78 percent design) without replacing any parts—a significant milestone for an engine that months before had been judged too unreliable for service use. Before the end of February, the strongly motivated Rolls-Royce firm, which sought to move into turbojet production, had also put the engine into a flying test bed, a piston-engined Wellington, for a twenty-five-hour flight—another important step for an engine that had never been airborne.[69]

FIGURE 1.4 Harold Roxbee Cox, the Ministry of Aircraft Production official in charge of jet propulsion, in 1946. *Source*: British Pathé

Production of the Meteor airframe at Gloster, meanwhile, remained on low priority. The total number of airframes on order had shrunk to twenty-seven in January 1943, which included seven prototype F.9/40 fighter airframes.[70] Linnell reported to Rolls-Royce that there was no hope of resurrecting the pared-down Meteor contract at the cost of the production of proven types until the power and reliability of the W.2B jet engine had been proven.[71] If the W.2B engine continued to produce too little thrust, the Meteor would end up being slower than existing piston-engined fighters—a performance hardly likely to recommend the jet fighter to the Royal Air Force. At 1,450 pounds thrust, the W.2B-Meteor pair would be good enough for a jet trainer. MAP judged that it would take a full 1,600 pounds to make the Meteor a competitive fighter above 35,000 feet, and 1,750 pounds thrust to make it an "outstanding" fighter.[72] By April 30, 1943, however, the W.2B engine's progress prompted the Air Ministry to increase the order for twenty-seven Meteor airframes to 120; soon after it was raised to three hundred.[73]

A few weeks later, on June 12, 1943, two W.2B engines, each producing 1,450 pounds thrust, powered a Gloster F.9/40 fighter airframe for the first time.[74] Rolls-Royce produced twenty-five development engines before beginning limited production of W.2B service engines, known as Welland engines, around November 1943. The Welland engines were cleared for flight at 1,600 pounds thrust with a respectable time between overhauls of 150 hours.[75] By the end of November 1944, one hundred Welland engines had been manufactured.[76] Stanley Hooker of Rolls-Royce later reflected that "although this was peanuts compared with our 1,000 Merlins [piston engines] a week, it did at least get the Meteor into the war."[77]

The Welland-powered Meteor would wait to be deployed on active service until after July 25, 1944, when the first German jet fighter, the Messerschmitt 262, was met in the air by a British Mosquito (a piston-engined aircraft).[78] The flight of the Me 262 called for a demonstration of Britain's jet capability in return, and the Welland-powered Meteor was the only British jet fighter that could go into service at the time (the H.1 engine, which was then struggling to develop higher power, lacked an airframe).[79] Despite the desire to deploy jet fighters, only twenty of the aircraft were predicted to be ready in 1944. In the event, these were to be sent not against German jet aircraft but against a new menace closer to home: the German flying bomb, or V1, which the Royal Air Force had been fighting since June 1944. Lord Cherwell observed that in view of the jet production record so far, "the whole of our effort in this field will yield scant results against the flying bomb," perhaps "the last major opportunity for these aircraft in this war."[80] The Royal Air Force's 616 Squadron flew the first Welland-powered Meteors against the V1 on July 27, 1944.[81]

The government kept up the pressure for jets: Churchill wrote to Sir Stafford Cripps, the minister of aircraft production, on July 30, 1944, saying, "I hope you will make a special drive to turn out as many [Meteors] as you can. . . . If we are caught behind-hand in jet-propulsion aircraft, it will be a serious reflection on MAP."[82] Success against the V1s has been celebrated as one of the major achievements of the early operational Meteors, which did not see action against German jet aircraft during the war. Yet the operational record of the Meteor against the V1 was no better than the aircraft's piston-engined counterpart. Whereas the few early Meteors eliminated some forty V1s, (faster) Sabre-engined Hawker Tempests successfully destroyed "many hundreds."[83] The V1, however, presented a tactical and propaganda opening for the British jet fighter that was aimed as much at America's emergent turbojet program as at the Germans.

Rolls-Royce's lack of further interest in the design of the Power Jets

W.2B engine, which was not only low powered but was also difficult to produce,[84] became clear when Rolls-Royce decided not to produce Power Jets' subsequent engine design, the W.2/500. Instead, Rolls-Royce rapidly designed and produced its own centrifugal turbojet engine, the B/37.[85] This provided crucial design expertise for the firm, which was set on producing jet engines after the war. By June 1943, the first prototype of the B/37 turbojet engine had been completed at Barnoldswick.[86] The engine prototype quickly reached its type tests, and a supportive MAP built a new factory for Rolls-Royce of about 200,000 square feet at Newcastle-under-Lyme (near Stoke, as the factory was often referred to) for production of the engine.[87]

The first B/37 series engine, later known as the Derwent, emerged from the Newcastle factory in November 1944, in time for deployment in Europe before the end of the war.[88] By January 30, 1945, thirteen Derwent engines with ratings of approximately 2,000 pounds had been delivered to Gloster for installation in Meteor airframes.[89] The Newcastle factory produced five hundred Derwent engines before December 1945, after which the factory was closed down. The factory at Barnoldswick produced fifty Derwent engines, starting in December 1944.[90] By the end of the war in Europe, the Royal Air Force had fifty-eight active Meteors, thirty-five of which were Mk Is, powered by Rolls-Royce's first production jet engine, the Welland. The remainder, Meteor Mk IIIs, were powered by the Rolls-Royce Derwent.[91]

The brief series production of the W.2B engine in 1943–44 (like the creation and the production of the Derwent after it) was a direct result of the decision to give control over the Barnoldswick factory to Rolls-Royce. Production of the H.1 engine at de Havilland, in contrast, had a serious delay imposed on it when it lost access to the plant at Barnoldswick. In late 1942, when the decision about the factory's future was made in Rolls-Royce's favor, de Havilland's general manager responsible for engines, J. J. Parkes, appealed directly to Freeman to reverse the decision. He argued that "it is generally recognised that the H.1 is appreciably ahead of all contemporary development both in progress, reliability and power output. . . . It would seem therefore to be most unfortunate to lose any advantage and opportunity for its earliest production." Parkes estimated that the necessity of finding replacements for the machinery located in Barnoldswick would delay de Havilland's production plans by six to twelve months.[92] Despite Parkes's claims, however, the H.1 engine was no closer to production than the W.2B engine at the time, and may even have been further from it.

FIGURE 1.5 Derwent production at the Rolls-Royce factory at Newcastle-under-Lyme.
Source: Rolls-Royce Heritage Trust

The first de Havilland E.6/41 airframe prototype first flew with a de
Havilland H.1 engine (rated at 2,225 pounds thrust, or 74 percent of design) on September 20, 1943, three months after Rolls-Royce's W.2B engines powered a Gloster F.9/40 airframe.[93] Nevertheless, the de Havilland
E.6/41-H.1 combination seemed to perform better. In early flight testing,
a Royal Air Force pilot reported to MAP his preference for the de Havilland pair over the Meteor-Welland combination. MAP, interested in providing the best possible aircraft and engines to the Royal Air Force, re-
acted swiftly. Pending confirmation from other pilots, Freeman requested
in March 1944 that his ministry investigate "immediate production possibilities" for the de Havilland combination as "an urgent matter."[94] Because de Havilland could not take on the extra airframe work, however,
Freeman placed a production order for 120 E.6/41 airframes at English
Electric (in Preston) on May 13, 1944.[95]

Although MAP had placed an order for 230 H.1 engines with de
Havilland on September 25, 1943,[96] compressor failure and combustion
problems kept the H.1 engine from reliably achieving its design goal of
3,000 pounds thrust. By July 1944, de Havilland had still failed to get
the H.1 turbojet to reach even 2,400 pounds thrust (80 percent of its design thrust), and as a result, MAP called for an "urgent revision" to the

hands-off policy that it had been following with the H.1 engine.[97] The Meteor airframes that had previously been ordered to carry H.1 engines (the Meteor Mk II) were officially canceled on July 25, 1944. The Meteors (Mk I and III) would exclusively take Rolls-Royce engines; all of de Havilland's H.1 engines were thenceforth reserved for the E.6/41.[98] On January 20, 1945, the H.1 engine finally completed a 100-hour service type test at 2,700 pounds thrust,[99] and by May 1945, de Havilland had produced twenty-seven engines, fifteen of which were delivered to MAP with a rating of 2,700 pounds thrust.[100] The firm produced a total of eighty-five engines before the end of 1945.[101]

Despite government and pilot enthusiasm, production of the H.1 engine, designated the Goblin, and the E.6/41 airframe, known as the Vampire, were delayed by commitments to other engines and aircraft. The Barnoldswick factory was dedicated to W.2B production. Labor for aeroengine production was preferentially sent to the new Rolls-Royce jet engine factory in Newcastle-under-Lyme rather than to de Havilland's in Cannons Park, London.[102] Production of the E.6/41 at English Electric was prevented by that firm's commitment to Halifax bomber production. The first Goblin-powered Vampire flew on April 20, 1945.[103] It went into service with the Royal Air Force in April 1946.[104]

Britain enthusiastically pursued the production of jet engines for service use from 1940, but during the war, it changed its production program repeatedly in an attempt to produce a reliable, high-powered engine as soon as possible. The W.2B engine, nearly canceled in 1942, ultimately went into service due to Rolls-Royce's efforts. That firm's opportunistic entry to the turbojet field marked the end of Power Jets' preeminence. In fact, the H.1 engine, a more powerful engine, beat the W.2B engine into the air, but its production was delayed by development problems and the struggle for resources, particularly against machines already proven in battle. The desire for production dictated that the British program took on a particular shape and valued particular types of expertise over others, but the imperative to manufacture engines was ultimately put into second place behind producing a safe engine with superior performance.

The late war push for jet production in Britain might be understood as an attempt by the British government to catch up with German deployment, yet the British government was less concerned by the threat posed by the German jet fighter than with assuring that Britain got credit for its extensive wartime work on the jet. Deploying a British jet aircraft before the end of the war would prove to the British public, in Freeman's words, that "the Hun is not always ahead of us in secret weapons."[105] At the same time, it would demonstrate the success of the nation's wartime work and make a strong statement about British priority to firms in the United States, which had also begun developing jets during the war.

American Turbojet Engines

The United States was the second nation to decide to produce jet fighters. During 1941, the United States Army Air Forces started serious turbojet development projects at the nation's leading steam turbine firms under the supervision of the National Advisory Committee for Aeronautics. To American officials, however, the jet engine being developed in Britain appeared to offer a quicker path to production and deployment than the newly started American projects. The British design was not meant as a replacement for them; it was explicitly specified that work on the British engine was not to affect other jet work in the United States.[106] Despite optimistic hopes, however, the United States produced few engines during the war and deployed no jet aircraft. Yet although its version of the British W.2B engine never went into service, America successfully built up its indigenous turbojet development and production capacity during the war, which it successfully put toward dominating the world market after the war. This brief section focuses on the nation's first production plans, and because they were based on expectations for British jet engine designs, it focuses on Anglo-American collaboration in the field of turbojet engines.

The United States had heard about the British turbojet engine program from the British Technical and Scientific Mission, known as the Tizard Mission, in autumn 1940.[107] In addition to the technical cooperation so close to the chairman, Sir Henry Tizard's, heart, the mission was keen on enlisting American production help in producing British-designed weapons, particularly in the manufacture of microwave radar.[108] Indeed, discussions in Britain had similarly concluded that "America's best contribution to the provision of jet propelled aircraft would be the provision of engines."[109] It has not been recognized that some in Britain and the United States saw the United States as a potential production center for jet engines.

In May 1941, when the first British turbojet flight took place in England, the United States had not yet entered World War II. The enthusiasm of the United States Army Air Force's General Henry "Hap" Arnold on hearing about Britain's progress with jet propulsion (likely during his trip to England between April 11 and 27, 1941), led the United States military attaché in London to request complete information on the W.2B engine, which American authorities referred to as the Whittle engine, in June 1941.[110] The next month, in July 1941, the British North American Supply Committee agreed that information on the Power Jets W.2B engine could be shared with the United States government.[111] This was followed by information about all of the jet engines under development in Britain, and during 1943, missions from the American military and aircraft industry were allowed to tour the British factories producing jet engines and airframes.[112]

Around September 4, 1941, Arnold informed the British Ministry of Aircraft Production that he wanted to produce the W.2B engine in the United States. He requested permission to produce the British engine as well as the necessary blueprints, an actual engine, an airframe for testing, and British turbojet engineers to facilitate American work.[113] Manufacturing permission was granted for production during the war "in order to assist the joint defence plans of our respective Governments."[114] Arnold later recalled in his autobiography that "I . . . knew the first jet plane might not be produced in this country for a couple of years, and, when it did come out, would probably not have long enough 'legs' to participate in our combat missions. However, I was of the opinion that if the British had jets in production, the chances were good that the Germans did also."[115] In hindsight, at least, Arnold was in no hurry to build an American jet. Even if the jet engine was of little use to the American war effort, producing it had its own logic. Arnold was convinced that the higher engine powers required by the American Air Force in the future would only be achievable by jet engines.[116] As it turned out, his embrace of the W.2B engine was based on a misconception about how close the British engine was to production.

In meetings on September 4 and 5, 1941, the United States Army Air Forces tasked the supercharger division of the General Electric Company, a trusted contractor that had previously produced exhaust-driven superchargers, with producing the British W.2B engine. Rover shipped its current design drawings (which were not yet final) to General Electric in September 1941.[117] The British government had similarly suggested that General Electric be given the production job but gave its blessing if Arnold wanted to put more than one firm on the job "in order to achieve rapid production."[118] General Electric began its work on the turbojet by building a copy of the W.2B engine, which was known as the Type I. Building the Type I was to prove more difficult than initially expected because the blueprints from Britain were found to be "incomplete in some very important aspects," and the electrical firm had never made an aircraft engine before.[119] Nevertheless, on December 23, 1941, the United States Army Air Forces approved an order for fifteen Type I engines from General Electric. In fulfillment of the contract, the firm included fourteen Type I-A engines, which, according to the firm, increasingly included changes based on General Electric's expertise.[120]

Although the engine design was coming from Britain, the United States Army Air Forces chose an American airframe maker to design and build an airframe for the engine to be built by General Electric. Its choice landed on the Bell Aircraft Company, which received a contract for three experimental XP-59A airframes on October 3, 1941. Not a significant American airframe builder at the time, Bell was chosen in part because it had spare capacity and in part because its location in Buffalo, New York, was near General Electric's supercharger division at Lynn, Massachusetts.[121]

A key reason for American enthusiasm about the British engine was that Britain was already planning for imminent large-scale production of the engines, which seemed to imply that Britain's jet engine design was much closer to a service engine than any of those in the United States. In a letter to Arnold in July 1941, Vannevar Bush, head of the United States Office of Scientific Research and Development and the National Advisory Committee for Aeronautics, made clear why he supported adopting the British scheme (he had discussed the jet engine with Tizard): "It becomes evident that the Whittle engine [the W.2B] is a satisfactory development and that it is approaching production, although we yet do not know just how satisfactory it is. Certainly if it is now in such state that the British plans call for large production in five months, it is extraordinarily advanced and no time should be lost on the matter."[122] Although Arnold pushed for production, Bush's qualifications turned out to be prescient.

General Electric and Bell set about the engine project with vigor. Strict secrecy was maintained throughout; General Electric's other divisions didn't even know about the secret project.[123] The first flight of an American-built jet aircraft, the Bell XP-59A powered by two General Electric Type I-A engines, took place on October 18, 1942. This feat took place in California some six months before the first flight of a British-produced W.2B engine in a Gloster F.9/40 jet fighter airframe. The fact that General Electric had started with Rover's production drawings for the W.2B engine in October 1941 and still beat Britain's producers to the first flight made the lack of progress with the W.2B at Rover even more poignant in Britain. Although optimistic plans circulated by the United States Army Air Forces in mid-1942 suggested producing up to 1,000 Type I engines per month,[124] in July 1942, General Electric's production capacity at West Lynn was only fifteen engines per week.[125]

Production and Improvement in the United States

Just as Power Jets and Rover struggled with the W.2B, so too did General Electric struggle to increase the power of its derivative jet engine. The firm's Type I-A produced 1,250 pounds of thrust rather than 1,600. The first jet engine to be put into large-scale production in the United States was General Electric's follow-up to the Type I, the General Electric I-16 (later renamed the J31), which produced 1,600 pounds of thrust and was the first engine design to contain General Electric's own design features.[126] Contrary to initial expectations, a representative of the United States Army Air Forces remarked, "the original [W.2B] drawings did not represent an operable engine and much engineering has been required to reach the present stage symbolized by the Type I-16 engine."[127] Nevertheless, there was still hope in July 1943 that the army would be able to deploy jet aircraft during the war. Accordingly, in October 1943, General Electric was given a further production contract for 550 I-16 engines plus spares at a cost of $26,812,500. At the same time, the firm was also developing a more powerful centrifugal engine, the I-40 (to produce between 2,500 and 5,000 pounds thrust), which the Army Air Forces selected for production by May 28, 1943, after preliminary design studies had been completed.[128] Despite the fact that in October 1943, General Electric possessed the capacity to allow it to produce 1,000 Type I-16 engines before June 1945, the government deemphasized production of the engine, ordering only enough of the underpowered engines to 'maintain the facilities', while allowing some skilled labor to be transferred to the pressing development work on the more powerful I-40 (the J33). The I-40 engine, a centrifugal scheme like the Type I but able to produce 4,000 pounds thrust, was on highest priority by February 16, 1944.[129] General Electric's I-16 contract was initially maintained for the sake of the Bell P-59 program, but after poor performance consigned the P-59-I-16 combination, known as the Airacomet, to the role of a "transitional trainer" in early

1944, production of the I-16 was gradually abandoned. Before production of the I-16 was finally stopped in 1945, General Electric had built a total of 266 I-16 engines of various marks.[130]

Although General Electric's Type I turbojet lagged behind expectations, the United States Army Air Forces continued to add to General Electric's jet engine production capacity. The Army Air Forces had already paid for the material and tools that the initial Type I engine project would require. New facilities were designed around production of the I-16 engine, although it was already planned to convert the facilities to the production of the I-40 engine as soon as the design was ready. The Army Air Forces chose to deal with technical uncertainty by converting facilities at a later date because officials judged that waiting to build production facilities until the I-40 had passed its acceptance tests would result in an unacceptable gap of one to two years before the I-40 could be in production.[131]

Although the army was enthusiastic about the I-40, General Electric repeatedly failed to meet its production promises for the engine. Officials of the Army Air Forces at Wright Field, Ohio, blamed the repeated delays on a lack of communication between General Electric's plants.[132] On April 6, 1944, Wright Field's production engineering section recommended that a second producer be found for the I-40 and suggested the Allison Division of General Motors for the job.[133] Allison estimated that with a slight decrease in the production of its highly successful V-1710 and experimental V-3420 piston engines, it could transfer over 400,000–500,000 square feet of space in one of its government-owned factories in Indianapolis, Indiana, to jet engine production. This would be enough to build about 500 jet engines per month.[134] On November 26, 1944, the United States Army Air Forces confirmed a contract to obtain 2,000 I-40 engines from Allison at a cost of $112,834,000. The first production engine was to be ready in March 1945.[135]

Allison's jet engine production plans were ambitious. On December 6, 1944, the firm submitted another bid for 5,000 I-40 engines for a total of $150 million, with the first twenty-five engines to be delivered in July 1945. To meet the Army Air Forces' previous order, Allison had constructed facilities with a capacity of 2,500 jet engines per month.[136] It was confident in January 1945 that it would reach a rate of 1,000 engines per month by January 1946, and 2,500 per month by July 1946.[137] When the I-40 production program was scaled down after the end of the war, the Army Air Forces maintained Allison's production contract for the I-40 and canceled General Electric's instead. This decision was motivated by Allison's lower price and the desire to "insure [*sic*] participation in the postwar aircraft program by old-line engine manufactures such as Allison."[138] The decision freed up General Electric to work on jet engine development.

Prospects for the H.1

The flight of the first H.1 engine in March 1943 drew American attention to Britain's second centrifugal jet engine.[139] The H.1 turbojet seemed to promise a faster route to a more powerful engine than General Electric's development of the W.2B engine, a goal that contemporary rumors of German jet and rocket developments made even more urgent.[140] On April 14, 1943, the Bureau of Aeronautics (the procurement agency of the United States Navy), the United States Army Air Forces, the British Air Commission in Washington, DC, and Bell Aircraft Company all agreed that the H.1 should be produced in the United States by Allis-Chalmers, a steam turbine manufacturer, which was already producing turbo superchargers for the navy's aircraft and had begun designing a turbojet engine of its own as part of the country's indigenous jet engine effort (described in the next chapter) in late 1941.[141]

Given the problems at de Havilland in Britain, the American decision in 1943 to produce H.1 engines was surprising. By July 2, 1943, Allis-Chalmers had volunteered to abandon its own axial-ducted fan turbojet in favor of the H.1 engine.[142] The British warned in July 1943 that the H.1 was not ready for production, yet on October 26, 1943, Allis-Chalmers submitted a proposal to the navy for production of the H.1 engine that was approved in April 1944 (the air force later ordered H.1 engines too), almost a year before de Havilland finished its first production engine. The decision to begin production at Allis-Chalmers of what was not yet a fully developed jet engine is further confusing because the firm had made clear that it was no longer interested in turbojet development, as would be required for an immature engine. Indeed, Allis-Chalmers agreed to produce the H.1 only on the condition that the engine was already "technically satisfactory."[143] In addition, it demanded additional facilities and the "right to purchase as many rough or partially machined parts from DeHavilland [*sic*] as the Ministry of Aircraft Production would permit."[144] This yoked the American production program to the British program. In contrast to the hopes of those who saw the United States as the only possibility for the engine's large-scale production during the war,[145] American production of the H.1 engine (unlike the W.2B-inspired program) added very little to British work, yet suffered from the same delays due to technical setbacks and reduced priority in Britain. Allis-Chalmers had built only eight H.1 engines when the cessation in 1945 of reverse Lend-Lease, through which Allis-Chalmers had been receiving H.1 parts, led to the halting of H.1 production and Allis-Chalmers' departure from the aero-engine industry.[146]

Commitment to the H.1 engine had consequences for the American airframe industry as well. Lockheed began work on a single-engined airframe for the H.1 engine, the P-80, in May 1943.[147] Designed and built in a rush during 1943, the airframe's prototype, the XP-80 *Lulu-Belle*, had to wait for the delivery of an H.1 engine from Britain before its first test flights.[148] Following repeated appeals by American officials in Britain and a major effort at de Havilland, the first H.1 engine arrived at Lockheed Aircraft Corporation in Burbank, California, on November 2, 1943 (it then produced 2,300 pounds thrust).[149] A second H.1 engine (necessitated by damage caused to the first engine by the collapse of the intake duct of the XP-80 in which it was being tested), arrived on January 8, 1944.[150] There were so few flight-worthy H.1 engines available at the time that the second H.1 engine shipped to the United States had actually been intended for installation in de Havilland's second E.6/41 prototype.[151] Because of the difficulty of getting H.1 engines, Lockheed redesigned the P-80 in 1944 to carry General Electric's more plentiful I-40 engines. The redesigned P-80 airframe flew with General Electric engines on June 11, 1944.[152] Built in small numbers during the war but not deployed, the P-80 became an important jet fighter after the war.

The American production program was begun with the promise of a quick start by adopting the British-designed W.2B. General Electric's first version of the engine proved unimpressive, yet the initial spur to production got American firms into the jet engine production business. Emphasis on simultaneous development and production contributed both to the construction of manufacturing facilities in America and the American industry's postwar success.

Turbojet Engine Production in Germany

Despite the fears of the British and American governments, Germany was actually the third country to decide to produce jet engines. Unlike Britain and the United States, however, the country's production program could

draw on foundations laid by a long-term development program begun in
1938. The country's earliest initiatives (discussed in chapter 2) were overseen by the relatively low-level Office for Special Propulsion Systems
(Sondertriebwerke) in the Technical Office of the German Air Ministry
(Reichsluftfahrtministerium, RLM). It is important to distinguish these
early efforts from the urgent jet engine production program initiated in
1943 by Erhard Milch, state secretary for air and from 1941 head of the
RLM's Technical Office. The two had very different goals and involved
different technical decisions. The work that followed the earliest impetus
of the Office for Special Propulsion Systems was a necessary but not sufficient prerequisite for the later production program. The Technical Office's early jet contracts meant that the Air Ministry was able to order jet
engines into production during the war, but they in no way dictated when
or indeed if it would do so—nor the costs of ordering them. In 1943, the
RLM's large-scale production order significantly shifted the program's
nature from the more leisurely development foreseen by the RLM's Office for Special Propulsion Systems to the pressing requirements of immediate production.

The RLM's Office for Special Propulsion Systems, which also supported work on rocket propulsion, granted the country's first jet engine
development contracts to its leading aero-engine firms in 1937–38. (The
companies given contracts at this date did not include the Heinkel Aircraft Company, which had no engine division. Heinkel is known for its
early work on jet engines, which was, however, privately funded. See chapter 3.) Germany's aero-engine firms continued to develop jet engines
from 1938, although the work was not a high priority. State Secretary
Milch did not meet the groups working on turbojets until 1942.[153]

Helmut Schelp, the civil servant who took charge of the RLM's Office
for Special Propulsion Systems in 1941, viewed jet engine development
as a long-term activity that would come to fruition only after the war.
He wrote a sixteen-year national plan for German jet engine development, in which Germany's jet engines gradually achieved higher powers
and longer ranges—both ultimately to exceed that possible with piston
engines. The desperate nature of Milch's program is revealed in that neither of the first generation engines that Milch ordered into production
(the BMW 003, government number 109-003, and the Junkers Motoren
004, or 109-004) was thought suitable to be a service engine by Schelp.[154]
All of Germany's aero-engine firms decided early on to use axial compressors in their jet engines in order to minimize engine frontal area—a fea-

ture considered essential for high-speed flight—and potentially achieve higher efficiency than the alternative, larger-radial compressors.[155] At the start of development there was, after all, no hurry to produce jet engines, and in that calculation, the theoretical superiority of axial compressors justified the extra time that would be needed for development.

In mid-1942, Schelp inaugurated work to prepare the two most advanced jet engines then available in Germany—those designed by BMW and Junkers Motoren (Jumo) for small-scale, experimental production. This process would both allow the RLM to meet the challenges arising from putting a "development in completely new territory" into series production and give the Luftwaffe time to try out the new engines.[156] It was not yet clear that there would be a need for both engines, but pursuing two alternatives was part of Schelp's strategy to reduce the risk of the novel development program and to acquire as much experience with the new engines as quickly as possible.[157]

Concerned foremost with tackling "the difficulties that occur with the start of any production run," Schelp instructed Jumo to prepare for series production of its jet engine, sacrificing performance and fuel consumption if necessary. Under the urgency of the RLM's later production order from Milch, the firm continued developing its engine primarily with respect to production imperatives rather than those of performance.[158]

Like other accelerated weapons programs that National Socialist Germany pursued later in the war, the leadership's interest in deploying jet aircraft during the war rested on a failure, in this case, the failure to develop new, faster piston-engined aircraft.[159] From late 1941, Milch had increased production of Germany's aircraft industry by rationalizing the increased production of the Luftwaffe's "tried and tested" fighters: the piston engine powered Bf 109 and Fw 190.[160] By continuing to produce the same aircraft as before, he could take advantage of learning effects to decrease the time and cost of production as well as use.[161] The Bf 109 was designed by Messerschmitt well before the war, yet made up "by far the largest share" of the fighter aircraft produced by National Socialist Germany after 1942. Hitler gave his approval to stop production of the airframe only in late March 1945.[162] Some 24,000 Bf 109 airframes were mass-produced between 1942 and the end of 1944. Through the more efficient production of older, familiar types, the monthly output of aircraft in Germany more than doubled between 1942 and 1943 with a minimal increase in labor and no additional aluminum.[163] A less welcome consequence of this production drive, however, was the Luftwaffe's increasing technological inferiority as compared to its enemies. Showing the trade-offs that the regime accepted—improvements aimed at increasing the Bf 109's speed in level flight to compete with Allied aircraft, for example—were achieved by sacrificing the aircraft's handling characteristics.[164]

It was agreed that big advances in performance could only come through new designs, yet the failure of Germany's development projects left Milch with few ways to improve the speed of the Luftwaffe's airplanes. During 1942, the Me 210 and He 177 bombers and the Me 309, a wholly new fighter development, proved to be inferior to the aircraft they were to replace. The following year, in 1943, the Me 209 (which unlike the failed Me 309 was based on the Bf 109 airframe) failed to better the already outclassed Bf 109.[165] Milch had little faith in the potential of any of the large piston engines then being developed by German firms, despite the sanguine views of some of his advisors in the RLM (in November 1942 one remarked that they shouldn't worry about their development position because no nation possessed an engine that produced more than 2,000 horsepower).[166] Milch considered all three of the nation's most advanced piston engines to be physically too large, lacking in power, difficult to build, and beset by challenging development problems. So desperate was Milch that, if it got to the point of readiness for manufacture, he was willing to put Jumo's problem-prone 222 (the best of the group) into production, "despite the fact that the inclusion of every new piston engine in the production program is painful because of the recognized requirements of mass production" (extra capacity was to come from ship motor production). Nevertheless, he doubted that the 222 would ever exceed 3,000 horsepower, which he judged would soon be too little.[167] Milch dismissed Daimler-Benz's 606 piston engine, which was made up of two coupled 601 V12 piston engines and was producing about 3,000 horsepower, as a "terribly primitive, makeshift engine, unlikely to be deployed for use on the front and to be gotten rid of at the earliest possible moment. Herman Göring (head of the RLM) angrily referred to it as a monstrosity ("Missgeburt"). BMW's high-powered piston engine, the 803, which was essentially two coupled BMW 801 double-star engines, Milch admitted might eventually develop 4,000 horsepower, but he found its construction to be far too complicated.[168]

In view of this development mess, the RLM saw jet propulsion as a way to leapfrog the enemy's more powerful piston engines, which German piston engines were unable to compete with.[169] Forced thus to switch to jet

FIGURE 1.6 The Daimler-Benz 610 piston engine. In order to produce more power, the 610 engine was two coupled Daimler-Benz 605 engines geared to provide power to a single shaft. This shows how gigantic piston engines were becoming in order to increase their power output; the engine is more than 2 meters long and more than a meter wide. The engine pictured is located at the Deutsches Museum's Flugwerft Schleissheim.

Daimler-Benz's 606 piston engine (mentioned in the text) was made from two coupled 601 engines. The 605 engine was developed from the 601; both were V12 piston engines. *Photographs*: Hermione Giffard

power earlier than anticipated, Milch was able to take advantage of the earlier development program of the RLM's Office for Special Propulsion Systems to begin a drive for jet engine production.[170] He ordered Germany's first jet engines into series production in August 1943.[171] However, several months before, in February 1943, he had already instructed Jumo to plan for production of its 109-004 axial jet engine.[172] In January 1944, Jumo's output was planned to reach a rate of 1,000 jet engines per month by the end of the year. Monthly production of Germany's principle piston aero-engines by Daimler-Benz, BMW, and Jumo together had reached 4,000 for the first time around March 1943; the contrast shows the relative ease of producing jet engines as well as the scale of the RLM's plans. By the end of 1945, Jumo was supposed to reach an output of 3,000 jet engines per month.[173] Yet no German factory ever produced more than 815 jet engines in a month before the end of the war.[174]

In addition to the Jumo engine, Milch also ordered BMW's 109-003 jet engine into production. The BMW order was briefly canceled in December 1943, due to fears of duplication, before being reinstated in January

1944, in response to the fear of disorganization at Jumo, where jet engine production had still not begun.[175] The second engine, as it had for Schelp, served as a way to reduce the risk of the program.

The RLM's Aircraft Development Group, like its Special Propulsion Office, had commissioned two jet airframes before 1941 (although at that time it was unclear what the performance of their engines might be): the Messerschmitt 262 and the Arado 234. Both airframes were quicker to build than their novel power plants, and their flight testing was held up by the lack of jet engines. When the RLM began to worry that the supply of Jumo 004 engines would not be enough for its jet fighter and bomber programs, the Ar 234 was redesigned as a jet bomber with four of BMW's jet engines rather than the two Jumo jet engines of the original reconnaissance version. The redesign was completed by July 1944,[176] but at that time the "development condition" ("Entwicklungszustand") of the 003 was judged to endanger the bomber program entirely.[177]

As Schelp had foreseen, there were many problems in bringing Jumo's and BMW's novel jet engines to the point of series production. The technical challenges of preparing new assembly lines were exacerbated by the RLM's inability to get resources for the new program, like the Ministry's 1943 aircraft production program more generally. Despite the increased pressure to move aircraft production into new, bombproof factories in 1943, for example, the Luftwaffe's share of national construction projects dropped to half the level of the previous four years.[178] In an economy ruled by scheming self-interest, the ministry's access to material suffered, caught as it was between Göring's increasing ineffectiveness and the efforts of Albert Speer. Speer, the minister of armaments (Reichsminister für Rüstung und Kriegsproduktion) from February 1942, sought to take over aircraft production.[179]

Further exacerbating the challenges of the RLM's crash jet engine production program was the fact that Jumo's main works at Dessau, where production of the 004 was first set up, was heavily bombed in late 1943, and the firm claimed that it was unable to get sufficient resources to rebuild its factory.[180] The physical problems the firm faced may have been compounded by poor management: in May 1944, Milch reflected that "there was complete confusion" at Jumo, "the entire planning for jet engines is a first-class mess" ("eine Schweinerei erster Klasse").[181]

Production of Jumo 004 engines finally began in mid-1944, not in Dessau as first planned, but at the large Jumo factory at Muldenstein, a few miles south.[182] Muldenstein became the center for the design, develop-

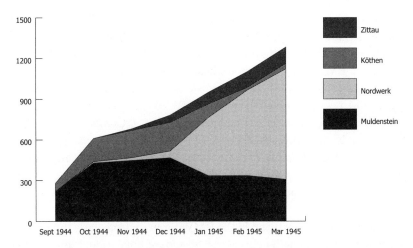

FIGURE 1.7 Jumo 004 jet engine production by month and factory, 1944–45. Data are shown for the main production factories. Production prior to September 1944 is not included because detailed monthly information is not available for that period. The Nordwerk factory, which expanded production so rapidly, is discussed starting on page 51. *Source*: IWM FD 5400/45

ment, production, and testing of Jumo's turbojet engines. The factory's planned capacity was allegedly 1,200 turbojet engines per month,[183] but according to production figures its peak output was 467 engines in December 1944.[184] The factory's early start, however, meant that it produced the largest number of 004 engines of any German factory during the war: 45 percent of the total. Jumo's factories in Köthen (near Muldenstein) and Zittau also began producing jet engines in late 1944, and Dessau finally began producing 004 engines in March 1945[185] (see figure 1.7). The firm made plans to begin production in some of its other factories, including at Magdeburg and Strasbourg, but neither of these produced turbojets before the end of the war. Magdeburg was put out of action by bombing, and the Strasbourg plant had not yet begun production when it was taken over by the Allies in January 1945.[186] Some other Jumo factories had begun producing parts for the engines, however.[187]

BMW's development work was some nine months behind the work at Jumo. This was due, at least in part, to the firm's decision to pursue a riskier line of development than that of the more conservative 004 (the 003 design include more unknown components) in order to achieve better performance. The firm's turbojet production was initially concentrated in what had been Bramo's main development factory in Spandau, which BMW acquired in 1939 (see chapter 2). It was in Spandau that the first 003 engines were built, but the factory was bombed with nearby Berlin in late 1943. BMW tried, like Jumo, to set up several factories for turbojet production, but only a single factory, BMW's license production factory at Zühlsdorf, ever assembled complete BMW 003 series engines during the war. The first production 003 engines appeared in October 1944.[188] Production at Zühlsdorf reached a peak of 150 engines in March 1945, in contrast to its planned output of some 1,500 engines per month.[189] Because of its inability to free up capacity to produce turbojets, BMW relied heavily on other firms for parts production.[190] All of the sheet metal work for the engines, for example, was subcontracted to other firms.[191]

The SS was also involved in the BMW program from early on. In mid-September 1943, Schelp approached SS-Obergruppenführer Jüttner, the head of the SS's Main Leadership Office (SS-Führungshauptamt) for help with production. Jüttner recommended to his boss, the head of the SS, Heinrich Himmler that "in view of the terror raids [Terrorangriffe] by the enemy's air forces, the RLM should be given all necessary assistance" to ensure that the new jet fighters would be able to be deployed in defense of the Reich—the SS was not yet looking to take over turbojet production.[192] Parts were subsequently produced for the BMW 003 engine at the SS-Kraftfahrttechnischen Versuchsanstalt in Oranienburg, not far from Zühlsdorf. Before beginning production, the SS carried out tests to make sure that concentration camp inmates could produce the required parts. Slave labor was typically used for heavy work, but the SS calculated that even for the precision parts required for the 003 turbojet, slave laborers could in some cases produce parts in half the time that BMW had calculated. With the goal of starting production of engine parts by the end of September 1943, preparatory work commenced even before Himmler had given his approval. In addition to the approximately two thousand workers who eventually worked on turbojets at BMW's main jet engine plant, one thousand concentration camp inmates produced parts for BMW 003 engines at the SS facility until the end of the war.[193]

Production Underground

Manufacturing in surface factories was affected by the increasing chaos of the war-torn country's production system, and during 1943, jet engine production too was moved underground. Earlier in the war, the RLM had responded to the danger of Allied bombing by dispersing aircraft production across Germany. More than one hundred dispersal factories sprouted up across the country's territory, particularly in the eastern part of France near Strasbourg and the western part of Czechoslovakia. Yet by making production thus dependent on the country's failing transportation system, dispersion contributed to undermining the output of aircraft. By recentralizing production, large underground factories could help avoid the high costs in efficiency and quality that had resulted from dispersion aboveground.[194] The Allied targeting of (dispersed) airframe and aeroengine production in Germany in late 1943 and early 1944 also added urgency to efforts to move aircraft production underground.

The destruction of German factories during "Big Week" in February 1944, a series of Allied air raids that was impressive if not overly effective in halting aircraft production, led Milch to begin giving up responsibility for aircraft production to Speer. The state secretary believed that the only way to increase the output of aircraft was to give up control of production; he blamed Speer, who had amassed production resources under his control since 1942, for starving aircraft production of resources. With the support of Karl-Otto Saur, Speer's deputy, Milch suggested to Speer that an executive committee be formed to rebuild and streamline aircraft production.[195] The committee, which became known as the Fighter Staff (Jägerstab), was established on March 1, 1944. It was headed by Saur and included Milch.[196] The Fighter Staff's main purpose was to increase the output of fighter aircraft and to protect it from Allied bombing through dispersal, underground where possible. Himmler's SS, which had power in armaments production because of its monopoly over the increasingly indispensable slave labor force of the country's concentration camps, was represented on the staff by the engineer Hans Kammler, who had earned a reputation for both ruthlessness and competence by accelerating underground V2 production in late 1943.[197]

After the Fighter Staff began its autocratic work, aircraft production climbed steadily upward. Already dramatically increased under Milch, output in terms of airframe weight doubled between February 1944 and July 1944.[198] Yet this was accomplished without any dramatic reorganization of aircraft production. Speer argued in 1944 that the surge in fighter production came not from the restoration of resources previously denied to the RLM—as Göring and Milch insisted—but rather through the use of the RLM's own reserves.[199] Yet Göring and Milch were correct: the increased availability of resources, including "raw materials, labor, food and transport" allowed measures that the RLM had earlier put in place to finally take effect. In fact, Speer's "production miracle" of early 1944 followed the lines already laid down in the RLM's 1943 production program, churning out thousands of out-of-date, single-engined Bf 109 and Fw 190 piston-engined fighters, which after five years of production had become increasingly inexpensive and quick to produce.[200] As Milch had done earlier, Speer accepted technical obsolescence as the price for higher production numbers.

It was well before any significant number of series turbojet engines had appeared in Germany, in April 1944, that Speer's ever-expanding Armaments Ministry took over responsibility for air armaments. Speer's Fighter Staff then turned its attention to hastening the production of turbojet engines for service use using the same draconian methods as for other weapons. First, the staff moved the production of turbojet engines underground. In April 1944, the Fighter Staff convinced Hitler that space could be found for the production of Jumo 213 piston engines (the company's latest piston engine) and Jumo 004 jet engines in an existing factory, alongside the V2.[201] This solution not only avoided the prohibitive (according to Speer) cost of building a massive, new, concrete factory for aero-engine production, but also avoided the further delay of setting up a new factory.

The previous year, on August 26, 1943, Speer's Armament's Ministry had decided to move the production of V2 missiles from the recently bombed factory in Peenemünde on the Baltic coast to an underground site near Nordhausen, in central Germany. Before it was chosen by Speer's Armaments Ministry as a dispersal site, the tunnel complex in the Harz Mountains had been in use as an oil storage depot. The depot's approximately 1.27 million square feet of tunnel space, part of a larger plan that was never realized, had been excavated between 1936 and 1942. The factory was laid out like a ladder, with two long, horizontal tunnels large enough for dual railroad tracks that gave access from the side of the mountain to forty-three internal rungs and three half rungs (see figure 1.3). When the oil depot was requisitioned for weapons production, the most significant extra excavation required was to extend the second access tunnel (one side of the ladder) to the south side of the mountain and to increase the height of some of the galleries. Many of the tunnels also needed cement floors; rungs numbered seventeen to forty-five had circular cross sections to accommodate petrol tanks and "required considerable filling to bring their floors level with those of the main service tunnels." The galleries numbered zero to sixteen, in contrast, had been designed to house oil tanks and already had flat floors.[202]

Under Kammler's command, the work required to convert the oil depot into a weapons factory was carried out in a ferociously short time by concentration camp prisoners. Nevertheless, if the tunnels hadn't been mostly excavated before the site was selected (undoubtedly an important consideration), the factory would not have been able to begin production in less than four months, as it did. (Not unlike how Schep's program laid the groundwork for the quick realization of turbojet production under Milch and then Speer.) The first V2 rockets emerged from the factory on January 1, 1944.[203] After construction work on the factory was completed, concentration camp prisoners were kept on site to manufacture weapons in the new factory. Sleeping at first inside the factory's tunnels, they were ultimately housed in prison camp Dora, which was later built on the surface.[204] Manufacture in the factory came to epitomize and amplify the coercive nature of weapons production under Speer. In the Third Reich, production became defined through the combination of Speer's growing control over material resources and the ruthless exploitation of slave laborers under Himmler.[205]

In June or July 1944, Kammler, acting on behalf of the Fighter Staff, consolidated V2 production into the complex's southern tunnels (numbers twenty-one to forty-six), which became known as the Mittelwerk, in order to make twenty northern tunnels, referred to as the Nordwerk, available for the production of Jumo aero-engines.[206] Jumo aero-engine production was allotted about 560,000 square feet of the original factory, while V2 production retained about 650,000 square feet. Production of the V1 was also moved into the factory in mid-1944, after Allied bombing hit the aboveground Volkswagen factory at Fallersleben, where the V1 flying bomb was being built. Production of the V1 occupied about 56,000 square feet in the Mittelwerk factory, but the V1 also continued to be produced aboveground until early 1945. The V1 and V2 were differentially successful at fitting into Speer's production-oriented regime: the output of V2 rockets was limited by the difficulty of their production to about 800, whereas the easier to build V1 (little precision work was required) was built in increasing numbers until the end of the war. About 300 V1s were produced in Mittelwerk in July 1944; seven months later, in March 1945, the factory produced about 2,200.[207]

The regime's underground factories are most famous from histories of the V2. While these accounts often mention in passing that jet engine production occurred beside V2 production, no history of the jet engine shows an understanding of the scale or significance of jet engine production that occurred underground or that jet engine production came to exemplify the values of the regime.[208] Nordwerk began producing Jumo jet engines in October 1944 using some parts made in the factory and some made outside by subcontractors. This single factory, easy to control closely, produced fully one-third of all German jet engines built during the war.[209] Production of the 004 engine began earlier in two of Jumo's factories aboveground, but the output from the Nordwerk rose much more quickly (see figure 1.7). When output from Jumo's surface factories was reduced by bombing in January 1945, production of the Jumo 004 engine at the protected Nordwerk factory continued to rise. In February 1945, Nordwerk alone produced more than half of the jet engines that were built in Germany that month. It was astounding that a factory that began production in October 1944 reached an output of 815 engines in March 1945. The accelerating production of jet engines suggests that they, like the V1, were well suited to production, which was pursued with barbaric effectiveness in the grim underground factory.

In addition to installing jet engine production alongside critical weapons production in the valuable underground factory near Nordhausen, in spring 1944, Speer's ministry also made plans for a chain of new underground weapons factories that would provide millions more square feet of factory space for weapons production, including of jet engines. A group of seven more weapons factories was planned in the same mountain as the existing factory. They were known collectively as Mittelbau. The new excavations were planned on a scale that dwarfed the Mittelwerk and Nordwerk factories. They were to provide factory space for jet engine production, aircraft component production, a liquid oxygen plant (for V2 rockets), and synthetic oil storage and production facilities. The Harz Mountains, in which the oil depot had been dug, were composed of anhydrite, which allowed for the creation of continuous, unlined, and extremely high galleries. Furthermore, the excavated rock could be used to manufacture sulfuric acid at I. G. Farben's nearby Leuna Factory. The new factories were designed using a grid layout rather than the ladder layout of the less hastily excavated Mittelwerk. This layout facilitated faster construction by providing multiple access passages from the side of the mountain through which rock could be removed. The factories were laid out, whenever possible, in collaboration with the firms that were to use them, supporting Michael Thad Allen's argument that industry cooperated in or at least acquiesced to the SS's brutal production regime.[210]

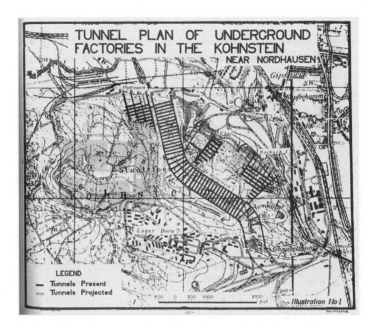

FIGURE 1.8 Underground factories in the Kohnstein Mountain near Nordhausen. Completed (black) and planned (gray) tunnel complexes at the end of the war are indicated. North is roughly vertical. The completed factory running diagonally through the middle of the image is the Nordwerk/Mittelwerk tunnel complex. The Dora concentration camp was above ground, to the southwest of the Mittelwerk factory. *Source*: Combined Intelligence Objectives Sub-Committee File No. XXXII-17

No fewer than three of the seven projected factories in Mittelbau were meant for Junkers turbojet engines. The biggest of these, known variously as Project Zinnstein, Bau 12, or Hydrawerk, was projected to reach 2.7 million square feet. When the war ended, 134,000 square feet of the factory had been excavated and a partially completed turbojet engine assembly line was in place in the factory. Further Jumo jet engine production was also planned for a new factory known as Malachit that was being excavated in a sandstone mountain near Halberstadt-Langenstein, located nearby. A least 120,000 square feet were already in use by Junkers there at the end of the war, and up to 400,000 square feet might ultimately have been available. Roy Fedden, an important British aero-engine designer, reported to the British aviation public in 1945 that "a big chain of bomb-proof underground factories was being built up when the war ended ... earmarked almost entirely for jet and rocket-unit production."[211]

If its plans are anything to go by, Speer's Armaments Ministry was intent on providing as many factories for jet engine production as possible.

Altogether something like 6.4 million square feet were planned for Jumo 004 production (of which about 1.31 million square feet had been excavated by the end of the war), almost the same amount of factory space as in BMW's main works at Allach.[212] For comparison, Rolls-Royce's total factory space in 1944 came to about 7.3 million square feet. Its jet engine production represented only about 730,000 square feet of this or 10 percent.[213] The limited success of Speer's ministry's later efforts to build underground factories demonstrates the importance of the fact that the Nordhausen oil depot's tunnels were largely excavated before Speer's ministry took an interest in the site. Even where the ministry was able to use existing mines, creating underground factories took time and significant resources—both of which were in short supply in the Germany economy.

BMW's 003 jet engine reached series production a few months after Jumo's more conservative engine. The Fighter Staff pressured BMW to move its aero-engine production, including producing jet engines, underground. It was even suggested at one point that BMW's jet engines be produced in the Mittelwerk factory. Instead, in June 1944, the Fighter Staff assigned several potash mines to BMW for jet engine production. Among these was a large underground mine at Heiligenrode, near BMW's factory at Eisenach. Through the work of the Fighter Staff, Organisation Todt, and the SS, the mine was prepared for turbojet engine production by a group of concentration camp prisoners and prisoners of war in March 1945 using methods no less brutal than the well-known case of Kammler's Mittelwerk factory. As in the tunnels of the earlier factory, large swaths of concrete floor were laid underground at a brutal pace.[214] The Heiligenrode plant never produced any jet engines, however. It was close to commencing engine production in 1945 using the first approximately 130,000 square feet of the almost 431,000 square feet available in the mine, when it was captured by American troops.[215] Other smaller mines nearby, including ones at Abteroda, Ploernintz, and Bad Salzungen, were allocated for component production and assembly—the last was never used.[216]

Despite plans that suggested that the Messerschmitt 262 jet airframe and later the Heinkel 162 jet airframe would be also produced in the Mittelwerk factory,[217] no airframes were produced there during the war. Kammler's stated intention to produce almost all of the country's jet engines and airframes in the Mittelwerk factory was symptomatic of the SS's bid to control crucial weapons production toward the end of the war, which began with Kammler's takeover of V2 production in Autumn 1944.[218] Instead of in Mittelwerk, production of the Me 262 ended up taking place in various dispersed forest factories and a few underground factories. Some of the dispersal sites chosen near Messerschmitt's Augsburg factory famously used the nearby autobahn as a runway.[219] Toward the end of the war, a giant underground factory dug from two sandstone mines near Kahla began to produce Fw 190 and Me 262 airframes. The Kahla factory was masterminded by Fritz Sauckel, general plenipotentiary for labor deployment since March 1942, who dictated that an airstrip be constructed on the top of the mountain in which the factory was hidden, the Walpersberg.[220] This was criticized for rendering the factory visible to enemies from the air, but it was never attacked. The eastern sandstone mine was ultimately to have provided 322,800 square feet of factory space to Messerschmitt.[221]

The output of German jet engines and airframes in the last desperate year of the war was extraordinary, suggesting the extent of the regime's support for the project and the centrality of the project to the regime's survival. Production of the Me 262 peaked at almost three hundred airframes in February 1945, several months after fighter production in general had begun to decline (see figure 1.11). While German piston engine production began to decline even earlier, in May 1944, in no small part due to the disarray of Germany's weapons production,[222] turbojet engine production rose from August 1944 until the end of the war, thanks to the significant contribution of the Nordwerk factory. In its six months of operation, the single Nordwerk factory produced 1,955 Jumo 004 jet engines underground.[223] By the end of 1944, just months after series jet engine production had begun in Germany, with piston engine output falling, the new engines already represented an astonishing 20 percent of German aero-engine production.[224]

There were crucial reasons that the beleaguered regime was able to produce jet engines in such quantities at a time when it was unable to produce other aero-engines. This gets at the heart of how production can influence development and invention. The initial production order from the RLM compelled Jumo and BMW engineers to design jet engines that could be produced with the minimal amount of material and labor that was made available to the industry. The jet engine was adopted by the RLM in the first place because of the developmental failure of Germany's piston engines, but Germany's ability to produce so many jet engines during the war was due to the success of Germany's aero-engine engineers in exploiting the engine's inherent advantages to match their designs to the unique and desperate conditions of production in the country's late-war weapons economy. Production requirements continued to be a crucial focus under Speer, even as more resources, like those of the Nordwerk factory, were diverted to jet engine production.

The Remarkable Production of Ersatz Aero-Engines

Increasing the ease and reducing the cost of aircraft production had been a goal of the RLM since the beginning of the war, but as the war continued, savings in the cost of production became as important to the country's ability to wage war and defend itself as improvements in weapons. As Speer's "armaments miracle" and Milch's earlier production drive demonstrated, small improvements constantly made to aircraft designs were sacrificed because a larger number of aircraft could be produced if a design remained unchanged in series production. Following Hitler's order of June 19, 1944, which forbade work on weapons that would not be ready for production within six months,[225] Speer explained to industry that no design changes would be allowed to weapons that were in series production with three exceptions: changes would be allowed if they offered an overwhelming advantage in battle, rectified a flaw that made the weapons useless, or reduced the cost of production by simplifying it, through conserving strategic materials or reducing weight.[226] So important was reducing the cost of production of weapons that Speer made an allowance for it even if it created additional costs in the short run. Rumors of amazing new weapons fortified the country's will to fight as the war dragged on, but where possible, Speer shifted production to less impressive weapons that were quicker, easier, and less costly to make.[227] All of the "Volks"-weapons produced in the last months of the war—the Volkspistole, Volksgewehr, and Volksjäger—followed these principles; they were primitive but quick to produce.[228] Large numbers of weapons offered Speer's struggling country an important advantage in battle even if they were of low quality.[229]

As Speer gained firm control of Germany's weapons production, his ideas about how best to increase weapons output combined with his wide influence to radicalize weapons production in the Third Reich. His belief that he could orchestrate production miracles rested on his suspicion of the willingness and competence of Germany's industrial leadership; he was particularly mistrustful of what he saw as their knee-jerk opposition to mass production. By installing handpicked managers across industry, who were familiar with mass-production methods, Speer hoped to break what he saw as the debilitating complicity of conservative industrialists and bureaucrats in order to introduce radical production systems and new, war-changing weapons. No matter the evidence of success or failure on the ground, his ministry continued to apply the same coercive methods to new weapons projects.[230] The "fiasco" of trying to produce the Mark XXVI U-boat according to mass-production methods that ignored naval construction experience, for example, did not cause Speer's ministry to reevaluate its methods, but only to intensify its authoritarian pressure on U-boat production.[231] Not all weapons could be produced by Speer's system equally well (so his record of success was variable), but in the jet engine, Speer found a weapon that fit exactly into his well-publicized script for increasing weapons production. Not only did it represent a radical new technology that the Luftwaffe already looked to to transform its prospects in the air war, but the jet engine also yielded to the ministry's authoritarian, mass-production methods.

The requirements of production in a very particular production system shaped the country's jet engines. Under pressure from above, Germany's aero-engine engineers exploited the jet engine's inherent properties to design turbojets suited to that production system. Unlike piston engines, early jet engines could be made from sheet metal and still function. Jet engine turbine blades could be made from relatively cheap and plentiful steels if scarce high-temperature alloys were not available, although this gave them a dangerously meager, expected service life of only thirty hours.[232] Such compromises were deemed acceptable if they eased production.

Many authors have suggested that the replacement of strategic metals in Germany's turbojet engines was a virtuoso technical accomplishment,[233] yet efforts to reduce the strategic metal content of piston aero-engines were already standard in the first years of the war. Aero-engine production had used about 130 tons of nickel a month at the start of 1940 as compared to only around 50 tons a month in 1942—a reduction of almost two-thirds despite an increase in overall engine production over the same period.[234] In 1943, the RLM had made an effort to reduce its requirements for scarce metals through the development of more "economic manufacturing methods" and the "rational" usage of materials.[235] Part of the reason for the literature's fixation on the reduction of metals in the Jumo engine is because the strategic metals content of the production version was significantly less than in the firm's prototypes and pre-production versions. The first production version, the 004B-0, was about

100 kg lighter than the preproduction version, 004A, and used less nickel and chrome; later production versions had even less.[236] This progression was, however, planned from the start rather than being a response to an acute change in the supply situation of strategic metals. The first Jumo engine was intentionally designed without thought for weight or metal content so that the company could begin testing sooner rather than spend extra time reducing the machine's weight.[237] The commonly noted reduction in strategic metals, while significant, was a consequence of the fact that Jumo chose to develop its engine quickly (a decision reflected in its conservative design that privileged the use of relatively familiar components over performance). Yet the crucial point is not even broached in the literature: the large-scale production of jet engines saved strategic metals.

As important as the fact that they could be built from metals that were available in the limited German war economy, however, was the fact that jet engines required overall less material to make than piston engines. Just moving to turbojets saved metal. The Daimler-Benz 603 piston engine produced 1,750 horsepower for a dry weight of 2,030 pounds, for example, whereas Jumo's first production jet engine weighed only 1,585 pounds dry and produced the rough equivalent of 2,580 horsepower when flying at 400 mph.

Labor was another key shortage in the German war economy that jet engines helped to address. The new engines were designed to be constructed from simplified parts, allowing them to be built efficiently by the unskilled and slave labor that was increasingly essential to the economy. Complex forged parts were replaced by ones folded out of sheet metal that were easier and faster to make, although these parts increased the frequency of fatal, catastrophic engine failures.[238] Both the Jumo and BMW engines were simplified, sacrificing performance, so that production could be carried out on standard machines by unreliable and unskilled workers.[239] The 003 engine was designed so as to be able to be produced on an assembly line but required specialized machinery only for shaping its main shafts (as opposed to having a machine for each separate operation—the waste of metal caused by the use of unspecialized machines elsewhere on the line was judged less of a problem for the war economy than getting these machines). To "make up for the very low craftsmanship of the workpeople" that were expected to build the engines, the firm's designers relied on high quality tools and gauges.[240]

In addition to requiring less skilled labor, the times ultimately achieved

FIGURE 1.9 Images showing how the turbine blades for the BMW 003 jet engine were folded from sheet metal. The composition of the blades was: Cr 17%, Ni 15%, Mo 2%, Ta-Nb 1.15%, C 0.1%. *Source*: Combined Intelligence Objectives Sub-Committee File No. XXX-59

for manufacturing a single turbojet engine were a factor of two to three less than for piston engines. The crucial Jumo 213, Daimler-Benz 603, and BMW 801 piston engines all took approximately 3,000 man-hours to build (although BMW's Augsburg plant cut production time of the BMW 801 to 1,250 hours through the introduction of special purpose machinery later in the war),[241] whereas by the end of the war, the production version of the Jumo 004 engine ultimately required a mere 700 man-hours to build (down from about 980 in May 1943).[242] In Britain, for comparison, Ford's carefully planned and highly specialized production lines in Manchester allowed the company to reduce over three years the production time required for a Merlin piston engine from 10,000 hours to 2,727 hours.[243] The RLM demanded that BMW reduce their turbojet production time to 500 hours.[244] The company reduced its engine construction time (and its strategic metal consumption) largely through the substitution of sheet metal parts for parts that required more time for machining or casting. The engine's air-cooled turbine blades, for example, were ultimately made from folded sheet metal.[245] Even at the 1,000 man-hours required for production in practice, the production time for turbojets represented an important savings over the production of the Reich's primary piston engines. This clearly wasn't a decisive priority in Britain, where the Welland took many times longer to produce than the production-optimized, if less reliable, Jumo 004. Of course, a comparatively small number of the British engines was produced.[246]

Further adding to the production advantages of Germany's jet engines, the turbojet could also be readily adapted to the increasingly dispersed nature of Germany's production system.[247] The "unique modular design" of the Jumo 004 allowed various engine components such as the "compressor/inlet housing assembly, the turbine/combustor assembly, and the exhaust nozzle" to be manufactured in different factories, in some cases far from the place where assembly and testing took place.[248]

Unlike the more gradually changing material and labor situation, turbojets were also a solution to Germany's dramatic loss of aviation-fuel-refining capacity during the last years of the war. Unlike piston engines, turbojets could run on less refined and more plentiful diesel oil. Using more widely available, lower grade fuel also helped to offset the fact that early jet engines consumed much more fuel per kilometer flown than contemporary piston engines.[249] The Jumo 004 was designed to burn diesel oil from the beginning (at first for safety reasons), and the BMW 003 was converted to diesel fuel in 1943 without significant problems.[250] Indeed, the fact that jet engines used diesel enabled the transfer of pilots from Luftwaffe squadrons downed for lack of fuel to Galland's jet fighter squadron.[251] The fact that jet engines did not need refined aviation fuel and the expectation that Germany's Luftwaffe would begin to rely on jets might have influenced Speer's decision in early 1944 not to spend resources on expanding the country's hydrogenation capacity, which had been severely reduced by foreign bombing.[252] Nevertheless, the jet engine's use of diesel fuel offered no advantage in the matter of fuel distribution, which could not avoid the increasing chaos across Germany.

That jet engines, unlike piston engines, could be made virtually out of sheet metal and run on lower-quality fuel should be understood not as leading to the downgrading or distortion of an ideal aero-engine type but rather as a reason for the increasing emphasis on their production in a war economy stretched to the breaking point. The turbojet's ability to be shaped to Germany's increasingly desperate production regime under Speer is illustrated by the fact that it took the Nordwerk factory only six months during the last year of a hugely destructive war to achieve a numerical monthly output of Jumo 004 engines that was equivalent to the unrivaled monthly output of Merlin piston engines achieved by Ford in Britain after two years of production.[253] Germany's rising jet engine production figures achieved at the very end of the war, when production numbers across Germany were falling, were evidence not so much of technical superiority (as some authors have argued) as of the turbojet engine's ability to be mass-produced under conditions of extreme scarcity and using brutal, authoritarian labor practices. Even early in their developmental lives, Germany's production-optimized jet engines produced more thrust per pound of steel and man-hour of labor than any other

FIGURE 1.10 Photograph of a Jumo 004 engine installed in an Me 262 (the engine's outer covering has been removed). The aircraft pictured is located in the Deutsches Museum. *Photograph*: Hermione Giffard

German power plant, while allowing forced and unskilled workers to be used as effectively as possible in manufacture. Rather than faltering as the war drew to a close, more than half of German turbojet engines were produced during the first quarter of 1945. A total of about 6,010 Jumo 004 engines were produced before the end of the war, while BMW assembled approximately 559 production 003 engines. Parts for many more BMW engines were manufactured but never used.[254]

That production concerns were supremely important to Speer's ministry is suggested by the fact that design changes made to ease the production of weapons were made at the cost of dangerous operational shortcomings. The decision to manufacture dangerous jet engines in large numbers was different from the familiar trade-off between quality and quantity inherent in the need to freeze designs to enable mass production. In the case of the jet engine, lower quality (made worse by shoddy construction work) was accepted—even recommended—as a way to ease production regardless of the known and dangerous consequences of doing so.[255] The operational cost of these engines of desperation was high. Routine flights were often fatal. The fraction of German jet engines that made it off the ground and into service had extremely short service lives, and many exploded in use.[256] Problems with unsafe and poorly constructed engines, combined with manufacturing problems of the Me 262 airframes in which they were installed, meant that of a total 1,433 Me 262 airframes complete by the end of April 1945, only about 358 aircraft became operational.[257]

Just how hard the leaders of Germany's war effort were trying to maximize the nation's ability to produce weapons toward the end of the war is indicated by the regime's effort to have a new, smaller jet airframe designed in late 1944, when desperation was running high. The cost of aeroengines only made up a fraction of the total cost of an aircraft to the Luftwaffe; the airframe to carry those engines generally cost much more. The Messerschmitt 262 airframe, which went into large-scale production in spring 1944, was very promising (a fact that existing literature has made much of), but it was expensive to build. A new airframe, then, if cheaper would facilitate the Luftwaffe's transition to jet engines by making more jet aircraft available. The Me 262 was expensive in part because it was designed early in the war when the RLM was still emphasizing quality rather than quantity. Its manufacture relied on skilled labor, used scarce aluminum and, because it had been designed when the country's jet engines were still underpowered, took two jet engines, one slung under each wing. In late 1944, the Air Ministry believed that the only way it could get a large number of high-performance aircraft in a short amount of time was to build a small, single-engined fighter airframe that would be an alternative to the costly Me 262 and need fewer jet engines per aircraft. The new design was considered "essential in relation to number."

Unlike the Me 262, the new jet airframe was designed to be produced in Germany's late war economy. The airframe's requirements emphasized cheap and efficient production using unskilled labor, like Germany's jet engines, which were nearing the start of mass production in the Nordwerk factory at the time. The proposed "smallest fighter" ("Kleinstjäger") was to be built primarily from nonstrategic materials, such as wood (two-thirds of Germany's bauxite reserves, necessary for the aluminum production needed to build the Me 262, lay in Hungary, the continued possession of which was increasingly uncertain in late 1944), while its small size would ensure that the total amount of material needed was kept to a minimum. Also crucial was that the smaller airframe would require one less jet engine than the Me 262 and therefore much less fuel to fly. Nevertheless, the Fighter Staff hedged their bets: the program was instructed to use only "spare" production capacity not needed by other weapons programs, including that of the Me 262. Göring observed that with such a definition, even a total failure of the project (if the design performed very poorly, for example) would be guaranteed to be inexpensive.

In order for the new, small jet fighter to be cheap and easy to build, the Fighter Staff accepted familiar trade-offs to facilitate production at the cost of performance. The new airframe's performance specifications were less than those of the Me 262. The staff was willing, for example, to accept a shorter flight range, despite its undesirability, in exchange for the benefits of having a small, cheap airframe.[258] Unlike the Fighter Staff, which was concerned primarily with production, Adolf Galland, commander of the Luftwaffe's fighter force, questioned the new aircraft's possible tactical use. For the general, the small aircraft's shorter flight duration was a disadvantage that rendered the proposed aircraft useless. Even given the shortage of fuel that he expected to affect aircraft deployment in the future, Galland declared in September 1944 that "today I would prefer a single 262 to two of the others." Willy Messerschmitt, the designer of the Me 262, was also unhappy about the new airframe. He challenged the assertion that production of the Me 262 would not be impacted by the new production program. In addition, he questioned the desirability of rushing the mass production of a new, likely inferior airframe that was unlikely to be a successful aircraft instead of choosing to focus all of the nation's aircraft production resources on the Me 262.[259] Still others objected to the small fighter, arguing that the country would be unable to train new pilots for the aircraft. Despite the varied objections, the potential advantages of having a cheap jet airframe available—an airframe so cheap as to be disposable—were judged revealingly by the Fighter Staff to outweigh the recognized risks associated with introducing in the middle of a war an entirely new aircraft type that was not only designed in extraordinary haste but was untested, inferior in speed, range, and weaponry to the more expensive but already operational Me 262.[260]

In favoring the crash program for the small jet airframe, with its clear production advantages, Speer's Fighter Staff made what was by then a typical decision. Saur explained that he could "recommend from experience" the high-risk strategy of betting on a radical new weapon designed for mass production because "it is in fact the same method that we have used for tanks, then a series of other weapons" including "the new u-boats."[261] Although others questioned whether or not those earlier production efforts had produced good weapons, Saur was confident that the use of Speer's ministry's proven methods would guarantee success with yet another attractive but risky project.

In early September 1944, Speer's Fighter Staff thus called for designs for a single-engined jet fighter to be powered by a single BMW 003 engine

(Jumo engines were still earmarked for Me 262s). The new fighter reveal-ingly was christened the "Volksjäger," the people's fighter. From among several submissions, the ministry chose a Heinkel design, which was desig-nated the He 162. Typically, Generaldirektor Kessler was chosen to head the so-called action in light of his earlier success in revolutionizing hand grenade production rather than because he had experience building air-craft. Also typical, the program was expected to move at high speed. The staff's ambitious schedule called for the plans for the airframe to be fin-ished by November 1, 1944. The first prototype was to be completed by December 15. Series production was to commence with fifty aircraft in February 1945, three hundred in March, and 550 in April. To compress the schedule as much as possible, testing was scheduled to proceed in parallel with development and production[262] (an unprofitable practice that aircraft designers had complained to the RLM about before).[263] Series production of the airframe began as the war in Europe was drawing to a close in May 1945, and the first Volksjägers were quickly sent to Luftwaffe units.

Vindicating the planning of the Fighter Staff, the new jet airframe did indeed have significant production advantages over the Me 262. Accord-ing to the respective manufacturing firms, in early 1945, the He 162 took only 1,500 man-hours to produce, whereas an Me 262 airframe at Messer-schmitt required 9,000 man-hours. In total, each Me 262 aircraft (with two jet engines) cost 150,000 Reichsmarks, whereas the He 162 (with one) cost half as much. The He 162 was pursued and embraced primarily for eco-nomic rather than technical reasons. It reached production quickly be-cause it was designed explicitly to fit the coercive production system of the German Reich's last years.[264] Indeed, Speer bragged that the He 162 was a demonstration of what true National Socialist coordination and effort could accomplish.[265] The Volksjäger program, the regime's last major new weapons program, was "a crash programme . . . carried out as a communal work of the participating departments and industrial concerns under the strictest concentration." It was, in Saur's words, "an incredible effort of will" based on a "foundation of absolute faith."[266] So much atten-tion was paid to the aircraft's production requirements that in the end, the Volksjäger made grim economic sense but was rendered tactically non-sensical by its serious limitations in use, especially for the inexperienced pilots then entering the Luftwaffe. The same logic had already shaped the production of Germany's jet engines.

The regime's increasing attachment to jet aircraft became clear as jets rose to preeminence in the country's air-armaments planning. Less than a year after the Me 262 had entered operational service, Hitler appointed Kammler plenipotentiary for jet aircraft on March 27, 1945. Soon after, on March 31, 1945, Kammler ordered the "total cancellation" ("restlos Streichung") of the manufacture of all piston-engined aircraft.[267] This was a desperate and apt climax of the National Socialist government's attempts to change the country's fortunes in the war despite its collapse, and it made practical sense, like moving jet engine production underground, because Germany's jet engines had been designed during the war to be the easiest aero-engines to produce in National Socialist Germany at the time. The economic advantages of producing jet engines helped to attract the political will and resources necessary to produce the new power plants and the airframes to carry them. The Luftwaffe's leadership, for its part, became obsessed with the jet as the solution to their increasingly obvious inferiority in the air war. They came to see the high-performance Me 262 as the aircraft on which the Luftwaffe's (and thus Germany's) fate depended ("die zur Schicksalsmaschine Deutschlands geworden ist")—a view that has colored postwar accounts of the aircraft's importance.[268]

In the context of National Socialist Germany's economy of scarcity, the demand for increased turbojet production was not just a matter of building an extra factory or shifting investment between weapons. It implied engineering and production choices that the Allies neither had to make nor would have made. The extreme nature of jet engine production in the Third Reich repudiates the flimsy universality of a sequential model of weapons production that allows production figures to be used to compare turbojet programs as if they were all alike.

The German program ultimately sacrificed performance, quality, and safety for production. By the last six months of the war, when the output of turbojet engines rose dramatically in Germany, the need for production output had become as fundamental to the National Socialist state's survival as the barbaric cruelty used to achieve it. The jet engine emerged as an aero-engine as important for its production and material advantages over piston engines as for any qualitative advantage over enemy aircraft that it might provide. It is difficult to say when production concerns became the most compelling reason for the regime to adopt jets, but as the war continued, supply issues became increasingly central to the definition and execution of National Socialist policy, infecting almost every area of policy making. The final, official decision to put all of Germany's remaining resources toward the production of a single, new type of aircraft (even if the decision had little practical effect at such a late date) indicated the

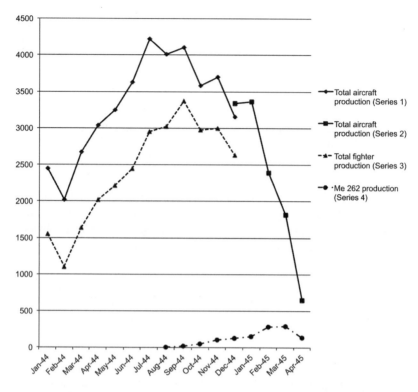

FIGURE I.II Production of Me 262 airframes compared to the production of other German airframes, 1944–45. The peak of production of Me 262 airframes appears in early 1945, well after the peak in total German aircraft production, showing the regieme's increasing preference for jet engines. The data series shown here are taken from different original sources reproduced by Vajda and Dancey. Series 1 and 3 are taken from their table 8-H; series 2 from table 8-L; and series 4 from tables 8-J and 8-L. *Source*: Franz-Antal Vajda and Peter Dancey, *German Aircraft Industry and Production*, 138–43.

jet's importance to the Third Reich, not as a wonder weapon but as the most desirable weapon that its economy could still produce. Germany's jet engines were truly engines of desperation.

Conclusion

Thinking about the requirements and meaning of production completely transforms the standard story of the turbojet. The story that emerged in this chapter was not one of all parties racing headlong to build jet engines.

In fact, it has shown that comparing the three programs using a single measure like production output, as much of the literature does, is deeply misleading. In the extremity of war, turbojet engines took on different roles according to each country's changing military goals, political landscape, material situation, expertise, and expectations about the postwar order. Neither Britain nor the United States needed to deploy jets during the war. Britain therefore turned its production program into one of development. In the United States, the British centrifugal scheme was used primarily to collect the resources needed to learn to design and build jet engines. Instead of creating an essential weapon, the wartime jet work of both Allied countries aimed instead at laying the foundations of postwar industries. Germany, in contrast, decided to produce jet aircraft after both Britain and the United States had made the decision to do so, but the country's program was the most focused of the three on production in a short time. The desperate push to raise the output of German jet engines led the country to produce by far the most jet engines of any nation during the war, but they were also the most unreliable jet engines by far. Choosing production output as the sole index of achievement therefore favors Germany and takes attention away from the development that became the focus of Allied programs and that resulted in engines that dominated the postwar market.

In all three countries, production was buttressed by broad development programs. The engines chosen first for production in each country were not the only engines under development and did not necessarily represent those engines judged most promising in the long run, but rather those easiest to build. This was most clear in Britain, and following Britain, the United States, where centrifugal engines were chosen for speed of development rather than performance potential. Britain's meager wartime production record disguised a development program that ultimately encompassed all of Britain's aero-engine firms. During the war, MAP became increasingly strident in encouraging aero-engine makers to begin work on the new aero-engine type, particularly on axial jet engines. The jet engine designs that British firms developed during the war led the field in the postwar world.

Although it narrowed toward the end of the war when the regime's focus turned to production, Germany's jet engine development program was also established on a broad basis. Despite originally being identified as development rather than as production engines, Jumo and BMW's projects were judged to be closest to production readiness in 1942. Their ultimate form represented the outcome of very particular developmental paths and values.

However profligate the United States' support of General Electric (Britain's production program, in contrast, became more circumspect after the Barnoldswick factory was built), the American company underwent a crash course in aero-engine production during the war. It became familiar with the need for a flexible way of working that combined careful development with the ability to move improved designs into production as fast as possible. The "concurrency" of improvement, manufacture, and use continued to characterize military jet engine development and production as well as the accompanying challenging management problems during the Cold War.[269] By 1945, the United States was poised for accelerated production of General Electric's I-40 engine, and by 1950 the American industry had come to seriously challenge Britain's.

During World War II, jet engines were still new, and the paths different firms followed to develop and produce them were defined by each firm's existing resources and skills. The sheer variety and range of (often experimental) projects that appeared in all three countries, which are discussed in the next chapter, suggest the consequences for early jet engine design of the range of experience brought to bear on the new engineering challenge. Many more firms (including several outside the aero-engine industry) seriously undertook design and development work on jet engines during the war than produced them. Firms differed in their ability to succeed with the new engines, a success that rested on their ability to move a new idea from development to production. When each firm judged that the prospects for its jet engines surpassed those for its piston engines, companies shifted resources toward the development of jet engines. The stories of these many companies as they moved into the unknown demonstrate the many beginnings of jet engine development. They also help explain why some projects succeeded whereas others failed. Rather than alien, revolutionary machines, the first jet engines were much more a problem tackled by companies using their existing skills, resources, and designs, and their existing development and production expertise. Firms did change over time in response to the needs of the new engines, but they tackled them first with what they had.

The Aero-Engine Industry and Turbojet Development

The last chapter showed how production decisions fundamentally shaped the British, German, and American jet engine programs. Although it demonstrated the importance of that approach, it did not, however, show how the firms that produced jet engines during the war learned to make these new machines. This chapter will look at the work of all of the firms that developed jet engines during the war—the firms that produced them as well as those that did not—in order to show how industrial skill was translated into the national capability to innovate. Development work was decisive in determining how the world's first jet engines were constructed, yet the activity is generally understudied if not ignored. Studying development has the further virtue of drawing historical attention to the work of industrial firms, firms that made their most important contributions to the turbojet through development. It also extends our view to include the contributions of huge numbers of people, who are generally wholly overlooked in typical accounts (that focus exclusively on invention or examine only the development work carried out by the inventors).[1]

This chapter is concerned with how new machines were created in the heavy engineering industries of Britain, Germany, and the United States. It is not directly concerned with where the ideas for a new machine came from so much as with why and how these ideas were adopted and translated into a usable design. By the mid-1930s, the idea for a jet engine was widely spread, and many men viewed jet engines as the aero-engines of the future, but no firm had yet built the new aero-engine type. But more than an idea, making new machines requires continuities of expertise, design, and management, continuities that industry can provide. That is the subject of this chapter.

Important continuities between piston engines and jet engines included engineering organizations, methods, and capital goods (along with the skill to operate them) as well as existing designs for parts belonging to other machines. Flexibility and knowing how to innovate were crucial. In both Britain and Germany, the development of jet engines became something of a community effort, making successful development dependent not only on a firm's capability and interest but also on its ability to work with other and with specialist firms.

The story told here, emphasized by the inclusion of firms that developed but did not manufacture jet engines during the war, is not one of inevitable success. Continuities were not always positive. Some firms' established engineering preferences were well suited to mature piston engines, but doomed their attempts to create new types of aero-engines. Nevertheless, even toxic continuities played a crucial role in what occurred and so get attention in the narrative. This chapter will look as carefully at failed engines as successful ones, explaining in each case how and why they were created. Examining development makes the differences between the firms that succeeded with jet engines and those that failed more distinct.

Edward Constant's attempt to reformulate thinking about technical change in his influential book, *Origins of the Turbojet Revolution* (1980), has misled thinking about the jet engine by portraying the turbojet as a revolutionary new type of aero-engine invented by outsiders and forced on an antagonistic and conservative aero-engine industry. Constant approached the jet story looking for radical technological discontinuities with piston engine practice.[2] So although he appreciated the magnitude of what the existing aero-engine industry contributed to the creation of the first jet engines, he nevertheless described existing aero-engine practice as part of the "hostile milieu" that the turbojet revolution had to overcome, emphasizing revolution over continuity.[3] Indeed, Constant writes that the aero-engine industry was "captured" by his turbojet revolutionaries, rather than the reverse. Thus, according to Constant, "the turbojet not only would overthrow the piston engine, but also would capture and exploit the vast pool of technological capabilities, the array of production and quality control techniques that were equally essential to high-powered piston engines or to turbojets. The turbojet would usurp and make its own the very institutional structure that had produced aircraft piston engines."[4] Constant stresses discontinuity, yet he well describes the continuity of aero-engine manufacture in the first part of the twentieth century. The key difference lies in Constant's focus on invention and on presumptive anomaly (according to his definition: a science-based insight that an existing technology will fail under certain conditions in the future) rather than the transfer of an idea into use. Indeed, industry enters his story only after a technological revolution has been completed and therefore hardly features in his account.[5] (Constant is, however, a very good source of what came before the revolution—namely, the predecessors to gas turbines.)[6] Constant praises the aero-engine industry for taking up a "newly established tradition" rather than for helping to create it.

This chapter looks at the coming of the jet engine in a very different way, challenging Constant's argument for the revolutionary novelty of the turbojet. It argues that rather than seeing the aero-engine industry as having been appropriated by the turbojet revolutionaries for their destabilizing ends, we should understand instead how the aero-engine industry used their existing resources and skills to translate the idea of a jet engine into usable machine. It was well-established insiders, not outsiders to the aero-engine industry, who were the most successful at turning the new engine type from idea to a machine. Undoubtedly, the firms under focus here benefited from the work of outsiders. But that is hardly reason to let the work of outsiders eclipse the crucial work done by existing firms. With the important exception of combustion, the turbojet was designed to a significant extent using elements that were already in use in piston engines. Indeed, the continuities between piston engine and turbojet practice are often obvious, and many authors have mentioned them, but none of the existing histories of the jet engine have assembled the evidence of the extent of these continuities or acknowledged the importance of the fact that they were general across all of the successful developing firms in all three countries. This chapter assembles that evidence to show that important features of early turbojet development can be explained by technical continuities between piston and turbojet engines.

In the 1930s, the jet engine was a wholly new means of aero-propulsion, different in important ways from piston engines. Yet to understand the creation of the jet engine in only these terms is very partial and ultimately misleading. The revolutionary account of the jet engine is based on the fact that jet engines ultimately displaced piston engines. A direct consequence of this is that stories ignore firms that took up the jet engine (because this was natural) while branding any firm that didn't immediately begin pursuing jet engines as reactionary and resistant to innovation. Thus

Constant explains the conviction of his turbojet revolutionaries, who were dedicated to turbojets for scientific reasons, yet he never asks why established industrial firms in Britain, Germany, and the United States began to develop jet engines when they did. Looking at the story of the turbojet from the perspective of industry offers an answer. Some firms did resist taking up the new type of aero-engine, but this was often because their piston engine businesses were doing well and/or required all of their resources. Many individuals within established companies were resistant to the new type, but often because it would distract from immediate work on piston engines rather than because it might eventually challenge the piston engine's dominance. Indeed, American policy took this decision out of industrial hands by explicitly preventing the country's aero-engine makers from developing turbojets during the war. Yet despite resistance, turbojet engines found supporters in most aero-engine firms. Conversely, some firms welcomed the jet engine not because they were farsighted, but because they judged that their existing piston aero-engines lacked a future in the civil field. In all cases, industrial firms with commercial motives evaluated not only the technical feasibility and desirability of new engine proposals but also their ability to succeed in the market—both during and after the war.

This chapter has four parts. The first examines Rolls-Royce's determined transition from piston engines to turbojets, which started before the war and laid the groundwork for its emergence as a leader in the turbojet field after the war—a surprisingly underresearched topic given the firm's dominance in the market for large jet engines today. The next part examines all of the remaining firms of the British aero-engine industry, exploring how each aero-engine firm approached the turbojet— with what degree of enthusiasm and with what tools. By the end of the war, every aero-engine company in Britain had begun work on a gas turbine. This chapter explains how and when each company came to work on its first jet engine design and explores the form that those designs took, showing how this was often based on continuities in technical thought. The third part of this chapter extends a similar exploration to the firms in the German aero-engine industry. In contrast to what happened in Britain, in Germany, the entire aero-engine industry was involved in jet engine development before the outbreak of war. Again, an overview of each aero-engine company's competitive situation is combined with a discussion of continuities in personnel and experience to help to explain each firm's work on jet engines. The final part examines American attempts to develop a jet engine during the war. The majority of attention is paid to

General Electric's work, because it was the most successful and extensive strand of development during the war and also because it was an important example of transatlantic technical cooperation, which brought General Electric's centrifugal turbojet team closer to British work than American.

The picture of continuity that emerges through the discussion of development offers a striking testament to the importance of industrial skill to technical change. Aero-engine manufacture was expensive because of its accuracy, emphasis on low weight, costly fixed plant, extensive manufacturing operations, and the complexity of its products. Making jet engines required more than just setting up a new company to make the new product, as Power Jets and the Heinkel Aircraft Company learned to their detriment. Both well known in the history of the turbojet for their roles in invention, neither successfully produced a service jet engine during the war. Their stories are discussed in the following chapter.

Rolls-Royce and the Turbojet

Although Rolls-Royce is an important maker of jet engines today, the history of the firm's earliest involvement with turbojet engines is almost entirely absent from the literature.[7] In company histories, it has been eclipsed by the Merlin piston engine's dominance in the firm's wartime history; in the literature on the turbojet, it is hidden by Frank Whittle's ubiquity. If Rolls-Royce appears at all in works on the turbojet, it is as the beneficiary and savior of the Power Jets W.2B engine.[8] A. A. Griffith's complex axial engine design and Rolls-Royce's early collaboration with Power Jets on the centrifugal WR.1—if mentioned—are both dismissed as failures.[9] The official British history of World War II, in contrast, although it gave few details, stated that Rolls-Royce was an important addition to the national turbojet program as well as "the only firm who on their own initiative had taken active measures to promote gas turbine development."[10] This suggests that there is much more to the story than the literature lets on.

Existing stories need to be challenged on many levels to explain just how it was that Rolls-Royce had become the world's most important designer of jet engines by 1945. By the time that Rolls-Royce took over the production of the W.2B engine in 1943, the firm had already been developing its own turbojet engine for two and a half years. In 1939, the company was developing five piston engines—Merlin, Griffon, Peregrin,

Exe, Vulture—and the (two-stroke) Crecy and had begun the studies that would lead to the company's first experimental axial turbojet. All but the Merlin and Griffon became "casualties of priorities" during the war,[11] as the center of gravity of the firm's work switched from piston to turbojet engines. Thus at the end of 1945, Rolls-Royce was developing only two large piston engines, the Griffon and the Eagle (which would be the company's last piston engine) and no less than six turbojet engines: the centrifugal Derwent, Clyde, Nene, and Dart, and the axial CR.2 and AJ/65.

Rolls-Royce's interest in jet engines may well have been inspired by its need for a new civil aero-engine for the postwar market. The Merlin piston engine, which had been under development at Rolls-Royce since before 1933, became an outstanding military engine during the war. It had poor postwar prospects, however, because the level of routine maintenance by airlines was less than that of the Royal Air Force. After Rolls-Royce's leadership identified the turbojet as the company's future, the firm pursued the new engine type single-mindedly, taking advantage of every opportunity to benefit from the work of other firms and from the resources of the government. The firm succeeded in developing turbojets because it deployed its considerable political power (MAP's respect for the firm was in part a product of the Merlin's success) in pursuit of the new engine type; because it had the resources to support several development and design efforts at once; because it opportunistically learned as much as possible from each such effort that it undertook; and because it successfully used its substantial resources and skills—deriving from its piston engine activities—to develop the new engine type.

The firm's first, long-term axial engine development program began before World War II and was augmented from late 1941 by a second, faster development strand building on Power Jets' work. These two development strands, one based at Rolls-Royce's main works in Derby and the other at the turbojet factory at Barnoldswick, had very different histories. The firm's early work was not all successful, but by the end of the war, Rolls-Royce had positioned itself such that it could rely on the help of government and industry in order to succeed as a maker of jet engines—even when its development efforts went awry, as with the axial compressor of its first commercial axial turbojet engine, the Avon.

Rolls-Royce's First Internal Combustion Turbine

Rolls-Royce's turbojet development program was started in 1939 by the firm's general manager, Ernest Hives. The program, which was seen as a long-term investment, consisted at first only of the work of A. A. Griffith, who had been a senior scientist at the Royal Aircraft Establishment (RAE) before being recruited by Hives as Rolls-Royce's first senior research engineer. Griffith began work at the aero-engine firm on June 1, 1939. At the firm, he continued his theoretical studies of different internal combustion turbine (ICT) and axial compressor layouts. (The phrase "ICT" that Griffith and Rolls-Royce used, emphasized the similarity to piston engines, which were also internal combustion engines.) Although Griffith worked alone, his work was by no means ignored. Hives frequently inquired about it, and his ongoing support demonstrated the manager's early interest in the new engine type as a future product of his company. This interest was reflected in Hives's willingness to support the construction and testing of Griffith's first scheme, known as the CR.1, for many years during the war. A point of pride, in fact, Griffith's CR.1 design was shown to the king and queen when they visited Rolls-Royce in Derby on August 8, 1940.[12]

Griffith's appointment marked a departure in policy for Rolls-Royce; his assignment, according to a colleague, to concentrate on "advanced ideas unconnected to any engine under development was unique within the firm."[13] Indeed, the employment of a scientist was quite a departure for a company that epitomized the British aircraft constructor's "old hostility" to mathematically based science.[14] Yet Griffith was a good choice to lead the firm's new endeavor because he had made the first suggestion—in a paper published in 1926—that a gas turbine could be made efficient enough to be used in an aero-engine, and he had ideas of how to go about designing one. Part of this proposal involved, as Griffith described for the first time, designing ICT compressor blades aerodynamically, like aero-foils (he argued that blades of "current design ... normally run under stalled conditions," in other words they were highly inefficient).[15]

Griffith worked at Duffield Bank House, Rolls-Royce's guest house, about nine miles north of Derby, the building to which all of the firm's nonproduction staff were evacuated when the war broke out. In late 1939, he was assigned a personal assistant and draftsman, Donald Eyre, who was to translate the scientist's elaborate schemes into practical designs that would be comprehensible to the men who would build Griffith's first engine in the firm's experimental shop. In this capacity, Eyre got to know Griffith very well. He saw the scientist's way of working as very distinct from that of Henry Royce, whom Eyre had previously worked for. Whereas Royce had been supremely concerned with the detailed designs arising from his ideas, Eyre recounted that Griffith left the practical details of his schemes to his assistant "with apparent equanimity."[16] Similarly, another colleague remembered that Griffith produced a "varied series of studies that he left to others to test, and he was always throwing out new suggestions," often far ahead of what was then achievable.[17] Indeed, Griffith was known for his "prolific brain" rather than his attachment to any single design; these descriptions don't tally with the accusation that Griffith self-interestedly thwarted Whittle's turbojet design when it was submitted to the British Air Ministry in 1930.

Griffith's style didn't really fit at Rolls-Royce, and he relied on Eyre to translate his ideas into a form that the firm's engineers could understand and appreciate. The novelty of Griffith's advanced ideas meant that Eyre, an experienced piston engine designer, could not rely on a previous fund of design work for guidance but had instead to develop the basic mechanical design scheme for Griffith's first axial unit from scratch.[18] Eyre's work was thus crucial to Griffith's. Stanley Hooker, another Rolls-Royce employee, argued that Griffith's complicated work had received so much support from the firm's (antiscientific) staff because of the "clarity and sophistication" of Eyre's drawings, which, he argued, "sold Griffith's ideas just like Johnnie Walker sold whisky."[19] This allowed Griffith to continue to elaborate new, complex, scientifically inspired schemes with little regard for mechanics in the expectation that, no matter how complicated, his designs would be realized by Rolls-Royce's famed engineering staff. In the case of Griffith's first, complex concepts for a gas turbine, however, his faith proved to be overly sanguine.

Griffith's first axial jet engine was the CR.1. It was actually designed as the high-pressure section of a larger conceptual jet engine (in other words, it was meant to have one or two other compressor/turbine units attached to it in the future). The CR.1 design embodied scientific principles. Although it may have been theoretically perfect, it proved highly impractical.[20] The machine had a "contra-rotating" axial compressor in order to avoid stalling. This called for its fourteen compressor stages to move independently of one another (rather than being mounted on a single central axle), every other one rotating in a direction opposite to the rotors before and after it. Each rotor carried a compressor stage (a circularly arranged set of blades) and, outside and concentric to that, a set of turbine blades. The engine was also "contra-flow": air flowed into the back of the unit, forward through the fourteen inner axial compressor stages, reversed direction in an annular combustion chamber, and then flowed back along the outer row of blades that made up the axial turbine, whence it left the engine.

In early 1940, Eyre's designs for Griffith's CR.1 were released to Derby's Experimental Shop, which was headed at the time by H. P. Smith. Four test units were built before the CR.1 unit itself, mostly from sheet metal, to test the most difficult and novel features of the engine and to provide data for subsequent design work.[21] The testing of these units took place in September 1940.[22] The resources initially devoted to the CR.1 were minimal, but the start of testing brought the firm's first axial ICT into direct competition with other work at the firm—namely, piston engine supercharger development, which used some of the same testing facilities: rigs where different components could be hooked up to a high-pressure airflow.

Hives had assigned a key figure in the development of the centrifugal Merlin supercharger, Stanley Hooker, to work closely with Griffith when Griffith arrived at Rolls-Royce. Hooker had in turn assigned Geoff Wilde to supervise the construction and testing of the CR.1 units on the prewar supercharger test rig at Nightingale Road, Derby. That same test rig was, however, also used to test the firm's centrifugal superchargers, and as work on the Merlin supercharger increased during the first years of the war, time on the test rig became an increasingly scarce commodity. In 1941, the urgent need to increase the performance of the Royal Air Force's Merlin-engined Spitfire over the Luftwaffe's Bf 109 meant that superchargers took precedence. Competition between the two development projects became so keen that at one point, Wilde resorted to physically moving the ICT equipment into the courtyard at night so that the supercharger team could use the test rig. The testing of the CR.1 was ultimately moved to a hangar at nearby Sinfin, where a new test rig was built in which a centrifugal Vulture (piston engine) supercharger powered by a Merlin piston engine produced a stream of compressed air for testing the gas turbine. Wilde, torn between the two projects, decided to devote himself to work on the firm's superchargers and handed the CR.1 over to a new team.[23]

The group formed at Derby to take over development of the CR.1 became known as Rolls-Royce's ICT section. Its first two members, Robin R. Jamison and Douglas Reynolds, were transferred to the new engine project after the decision was made to cancel the Exe, a sleeve-valve piston engine at the firm. In June 1941, the ICT section included four men (as compared to more than 20,000 employed by Rolls-Royce at the time)[24] who had previously worked on piston engines. Under Jamison's leadership it reached about nineteen by 1944, but by then Rolls-Royce had several design teams working on turbojet engines, and the focus of the firm's turbojet work had moved largely away from Derby to Barn-oldswick, where a different team was working on the centrifugal Welland and Derwent turbojet engines.[25]

Although Rolls-Royce funded the CR.1 engine as a private venture, the firm did not hesitate to take advantage of its relationship with the Ministry of Aircraft Production (MAP) to get contracts for work on new turbojet engines. On July 25, 1941, Rolls-Royce requested two contracts for the supply and testing of four new contra-rotating, contra-flow axial turbojet engines based on Griffith's work. The firm proposed to build two each of two types of turbojet engines. The first turbojet type had twenty-two compressor stages and would produce 1,200 pounds of thrust. The second, a turbojet/turboprop engine, would have multiple axial stages (the number was unspecified) and produce 5,000 pounds of thrust. To the second, higher-powered turbojet engine type, the firm proposed adding an additional contra-flow, contra-rotating unit, which would power a contra-rotating airscrew (already used on piston engines: two propellers located one behind the other on a single shaft). Rolls-Royce estimated a cost of £8,000 per turbojet engine and £21,000 for each turbojet/turboprop unit. Aware of Power Jets' progress on their own centrifugal turbojet, Rolls-Royce "confidently" presented its new contra-rotating axial turbojet engine as a "replacement unit" to the Power Jets engine that would offer greater thrust, less weight per pound, and lower fuel consumption. It was even designed to fit into the aircraft Gloster designed for Power Jets. Also revealing the firm's goals and reference points, the turbojet/turboprop engine was advertised as a promising postwar engine that would offer a much higher power to weight ratio than the contemporary Bristol Hercules XI piston engine (Bristol was Rolls-Royce's chief competitor in making piston aero-engines).[26] The plan displayed Rolls-Royce's ambition, its preference for complication, and its confidence that it could master any engineering problem.

The success of Rolls-Royce's requests for contracts that reduced its risk with new engines showed its close relationship with MAP. In early September 1941, after receiving requests from the United States for initial data from their work on the CR.1, Rolls-Royce asked MAP to reimburse it for the cost to date of the development and testing of the fourteen-stage contra-flow turbojet unit before sharing information about it.[27] MAP ultimately issued a contract that included the fourteen-stage CR.1 as well as the four more advanced contra-rotating engines, which had already been approved.[28]

The CR.1 engine was completed in September 1941, but its debut was unimpressive. It was unable to run under its own power. Instead, the ICT section fed it with a stream of compressed air generated by a Power Jets W.2B compressor, which Rolls-Royce had been given to test.[29] (At Rolls-Royce, the W.2B compressor was powered by a 2,000 horsepower Vulture piston engine, with it rotational speed stepped up from 3,000 to 18,000 using two Merlin propeller reduction gears in reverse. The power of Rolls-Royce's piston engines and the skill of its engineers made it possible to test the full performance of the Power Jets compressor for the first time.)[30] Undaunted by the CR.1's poor start, the firm continued to develop the engine until 1944. Part of the further work involved testing numerous different blade profiles in the CR.1 engine.[31]

Subsequent requests Rolls-Royce made for contract cover for additional testing were also granted by MAP. The Ministry approved an appeal by the firm in October 1942 for financial cover for additional combustion chamber development, for example. MAP's director of scientific research, David Pye, hoped that by providing a research contract with accompanying Ministry oversight, MAP could ensure that the firm made use of all available research work into combustion—ensure, in other words, that the firm embraced scientific results (Griffith's work didn't include combustion chamber design).[32] The CR.1 engine finally ran under its own power for the first time in early 1943, after its annular combustion chamber had been replaced with four combustion chambers adapted from another jet engine, the WR.1 (discussed below); the chambers were similar to those used in the W.2B engine and were also designed by the car component company Joseph Lucas Ltd.[33]

The Whittle-Rolls-Royce Engine

Rolls-Royce continued to increase its involvement in gas turbine aeroengines throughout the war. The first of these was the opportunity Power Jets' limitations presented. Hooker, who had been impressed by what he had seen on his first visit to the Power Jets factory in Lutterworth, went to visit it again with Hives in August 1940 in order to see Power Jets' experimental WU unit running. Compared to the CR.1, the Power Jets WU was a vision of simplicity. The hand-built bench unit, then sporting the multiple, smaller (and therefore more manageable experimentally) combustion chambers of its third reconstruction, produced about 800 pounds of thrust. Hooker calculated that the Merlin engine produced the equivalent of about 840 pounds of thrust when propelling a Spitfire at 300 mph. "Compared to the sophisticated design and manufacture of Rolls-Royce, it looked a very crude and outlandish piece of apparatus," Hooker remembered, yet the two men from Rolls-Royce recognized the potential of the Power Jets unit and sought to harness it for their own ends.[34] Hives ensured Rolls-Royce's involvement in the project by volunteering the firm's experimental shop to produce parts for Power Jets engines. (Rolls-Royce served as a subcontractor to Rover, making experimental parts for the Clitheroe factory, from 1941 until January 1942, when it refused Rover's eager request for further collaboration.)[35] Hooker, meanwhile, was tasked with keeping an eye on Power Jets' work, so that Rolls-Royce could expand its involvement in the future.[36]

After Power Jets' success with its W.1 engine, which flew in May 1941, Hives became convinced that jet propulsion was a much shorter-term proposition than he had assumed based on Griffith's CR.1 engine design. As a result, he brought the firm's originally postwar plans for building contra-rotating engines (discussed above) forward in time.[37] Hives continued to believe that Griffith's design was the most efficient possible ICT layout, but he believed that the CR.1 would require several more years of development before it could run (a guess proven through experience to have been not far off the mark). In the meantime, he wanted Rolls-Royce to become actively involved in the design and development of centrifugal turbojet engines, which seemed simpler and therefore faster to develop. So, while Derby's ICT section worked to get the complicated axial CR.1 running, Hives broadened the firm's turbojet program to include a centrifugal engine that was to be designed in collaboration with Power Jets. The new engine was to be like Power Jets' latest engine design, the W.2B. Working with Power Jets, Hives reasoned, was desirable because it would allow Rolls-Royce to gain "the benefit of . . . seven years development work" (Power Jets was established in January 1936).[38]

Accordingly, in January 1942, Rolls-Royce agreed to act as a subcontractor to Power Jets in the design of a new centrifugal turbojet engine with a power output of 2,000 pounds thrust. Power Jets agreed to the proposal but believed that the engine's conservative design could be developed to produce 3,000 pounds thrust.[39] The engine was to be known as the WR.1 (Whittle-Rolls-Royce). MAP gave its consent to the project on the condition that the work not interrupt the "urgent development" of Rolls-Royce's piston engines, especially the Merlin Marks 60 and 61, the Griffon, and the Crecy. Rolls-Royce assented to the Ministry's requirement. Because of the abandonment of the firm's marine Merlin engine project, the firm had a "windfall in personnel" and therefore did not request any additional labor to work on the new engine.[40] The cancellation dictated who would work on the new engine. For Rolls-Royce, the new centrifugal turbojet project thus provided an opportunity to keep on skilled staff members who had been rendered redundant. Hives placed Lionel Haworth, who had headed the canceled marine engine project, in charge of the WR.1 turbojet.[41] Haworth was to become an eminent turbojet designer at Rolls-Royce.

Although it was only in early 1943 that MAP issued a research contract for six WR.1 engines,[42] work on the engine had begun long before. The specifications of the new engine were agreed on in a technical meeting held between representatives of MAP, Power Jets, and Rolls-Royce on February 4, 1942.[43] Rolls-Royce was eager to begin and had already ordered some material for the project.[44] Hives required that Rolls-Royce be made responsible for design, manufacture, and development of the WR.1; the new engine project offered a way for Rolls-Royce to learn about centrifugal jet engines from the pioneers and translate that knowledge into firm expertise. Thus the firm's general manager further insisted that while the first three WR.1 engines were to be built to the specification agreed with Power Jets and the government's Royal Aircraft Establishment, the last three would incorporate modifications by Rolls-Royce based on the performance of the first three engines.[45] Although astutely agreeing to act as a subcontractor to Power Jets, Rolls-Royce was the stronger partner, and in practice, Power Jets became something of a relay between Rolls-Royce and MAP on the project.

The WR.1 used the same layout as the W.2B (see top schematic of figure 1.1): it had a double-sided impeller, ten reverse-flow combustion chambers, and an axial turbine. Its design combined these basic parts with elements from Rolls-Royce's work on centrifugal superchargers.[46] For example, Haworth planned to use a scaled up, doubled-sided version of the Vulture supercharger (also centrifugal) for the engine's compressor, while the rotating guide vanes for the compressor inlet were taken from the Merlin XX supercharger.[47] To these features were added reverse-flow combustion chambers and fuel nozzles developed by the Lucas company for the W.2B, a gearbox designed by Rover, and an exhaust cone designed by the Royal Aircraft Establishment.[48]

Rolls-Royce was an enthusiastic collaborator, eager to learn from other firms, if only to enable the company to later surpass its competition. In this they benefited from the establishment of the "experimental" national Gas Turbine Collaboration Committee (GTCC), an idea that had been discussed by Harold Roxbee Cox, the MAP official responsible for turbojet development throughout the war, during a meeting at Rolls-Royce on August 20, 1941.[49] (Figure 2.3 is a photograph of the committee members at de Havilland's works in Hatfield at the fourth meeting in May 1942.) The committee, which eventually included representatives of all of the firms and government establishments working on gas turbine aero-engines in Britain, was to exchange information on gas turbine research and development.[50] Griffith and Hooker represented Rolls-Royce at the GTCC's first meeting in November 1941. Each member institution submitted monthly progress reports, which were distributed to the other members of the committee. The GTCC met (on average) bimonthly at different firms, where its members saw different gas turbines under test. Problems that faced all of the committee's member firms were addressed in subcommittees, which came to include committees on surging (a particular problem of early jet engines, when flow in the compressor stops completely), combustion, high temperature materials, patents, and production.[51] The committee thus facilitated MAP's expressed desire for the "rapid development and production of gas turbine engines" in Britain.[52] In contrast with the Ministry's competitive piston engine policy, its gas turbine policy began as one of a "mutual and free interchange of ideas" that went beyond the customary level of informal cooperation within industry.[53]

Hives was enthusiastic about the collaboration with Power Jets, but he counseled caution. In late February 1942, he told the Rolls-Royce board that the company should not begin to design a new piston engine until the potential of the ICT was better known. Rolls-Royce's board authorized £25,000 to "obtain an interest in Power Jets" in order to secure full and open collaboration with the firm.[54] This was only a fraction of the more than £275,000 that the Air Ministry had invested in Power Jets by early 1942, but would have significantly exceeded previous private investment in the firm.[55] Nothing came of the suggestion, but the offer illustrated the seriousness of Rolls-Royce's interest as well as the firm's financial clout.

The heavy and relatively low-powered WR.1 engine designed by Rolls-Royce and Power Jets has been cited as proof of Rolls-Royce's inability to design a reasonable centrifugal engine and its reliance on the work of other firms to enter the gas turbine field.[56] That Rolls-Royce's intention was to learn from Power Jets by embarking on the WR.1 project cannot be doubted. In initiating the WR.1 project, Hives was less interested in the

FIGURE 2.1 Image of the members of the Gas Turbine Collaboration Committee gathered at de Havilland in Hatfield, May 1942. *Source:* Rolls-Royce Heritage Trust

actual engine design (and its marketability) than in what the company's employees would learn through the exercise. But equally importantly, the project would serve as a lesson in how Rolls-Royce's methods could transform the prospects of a novel type of engine. Because Rolls-Royce was in charge of design, the design of the WR.1 was intentionally conservative, following the firm's "usual practice" of sacrificing weight and power for mechanical reliability. The engine's performance was disparaged as inferior to that of the W.2B,[57] but Rolls-Royce's engineers were interested in improving on the W.2B's poor reliability before its power. Once the reliability of the new engine had been proven, the firm foresaw turning to improving the engine's performance.[58] That the engine design didn't get that far before being abandoned, however, meant that the firm never worked on improving the WR.1's much criticized performance.

The first WR.1 engine was built, like the CR.1, in Rolls-Royce's experimental shop. It was completed on November 30, 1942,[59] and the engine ran on the bench (as opposed to in a test rig or aircraft) for the second meeting of the GTCC on December 12, 1942. By that time, running was hardly a novelty for the WR.1, for the engine had already passed its first acceptance test. (These type tests, stipulated by the Air Ministry, became increasingly challenging as the capability of the new engine type advanced.)[60] Initial running of the WR.1 produced decent readings, although they were less than the design aims: a thrust of 1,155 pounds with a compressor efficiency calculated at 76 percent. By the end of February

1943, the WR.1 had run 18 hours and 7 minutes, and had reached 1,975 pounds thrust.[61] The engine was reliable, as its designers had hoped, although it suffered from airflow problems that prevented it from being run at full speed. Rolls-Royce's methods had produced an underpowered but mechanically reliable engine, which set the stage for future Rolls-Royce turbojets.

In the end, only two WR.1 engines were built before the program was abandoned in mid-1943. Nevertheless, building and designing the engine undoubtedly benefited the firm's turbojet ambitions. Working on the WR.1, the firm was able to build up its turbojet skills and begin to establish its reputation as a "major influence on gas turbine technique," as Roxbee Cox would describe the firm in 1946.[62] One key technical lesson learned from the engine, for example, had to do with building the high-speed rotating shafts integral to gas turbines. According to Haworth, the project led the firm to learn how to mount an aluminum rotor on a steel shaft in such a way that the differential expansion and rotation strain would not upset its balance when running, making it spin off center. Work on the WR.1 also resulted in a modification of the fir tree turbine blade root that Power Jets had developed (see figure 2.2) to make the part easier to manufacture. Rolls-Royce's new design was used in all of the firm's later engines.[63] Even after Rolls-Royce took over manufacture of the W.2B in early 1943, Haworth's design team, like the ICT section working on the CR.1, remained at Derby, so the main contribution of the engine was to future Derby-designed engines rather than to the engine projects at Barnoldswick.

A crucial social consequence of the WR.1 project for the future of gas turbines in Britain was that it helped to cement Rolls-Royce's positive relations with Power Jets. This was particularly valuable as Power Jets had earned a reputation among industrial firms for its prickliness. During 1941 and 1942, Power Jets came increasingly to see Rolls-Royce (which had uniquely agreed to serve as subcontractors to the upstart company) as its preferred development partner. In early 1942, for example, Power Jets suggested that Rolls-Royce take over flight testing of the W.2B engine. MAP supported the move because it would relieve the overburdened Gloster Aircraft Company and because (demonstrating MAP's faith in Rolls-Royce) it "would certainly get the most work best done in the shortest time."[64] When a Wellington II bomber was allotted by MAP to Rolls-Royce's airfield at Hucknall for flight testing, it was retrofitted to carry a W.2B engine in its tail by the experienced engineer Ray Dorey, who had previously led the testing of Rolls-Royce's famously powerful

FIGURE 2.2 Experimental Power Jets turbine blade showing the fir tree shaped blade root. The blade is 75 mm by 25 mm. *Source*: Collection of the Museum of Applied Arts and Sciences, Sydney. *Photograph*: Debbie Rudder

"R" engine for the 1929 Schneider Trophy race. The adapted bomber became a mainstay of early turbojet flight testing.[65] In late 1942, Power Jets again showed its faith in the larger company by asking Rolls-Royce to produce their newest engine design, the W.2/500. Although Rolls-Royce later declined to produce the engine, preferring to produce its own design instead, the relationship between the two firms that was forged in the early years of the war contributed to Rolls-Royce's assumption of responsibility for the W.2B engine in 1943.

Rolls-Royce at Barnoldswick

By 1943 it had become impossible for Rover's "senior people" at Barnoldswick and Clitheroe to collaborate harmoniously with Power Jets on the W.2B. The companies clashed for technical and personal reasons.[66] Its desperation and its respect for Rolls-Royce led MAP to offer the aero-engine firm the chance in late 1942 to take over the management of the government's two jet engine factories. Thus over the first three months of 1943, Rolls-Royce slowly filtered its managers into the factories at Barnoldswick and Clitheroe in order to avoid disruption. By April 1943, Rolls-Royce was in full control.[67] The company's management brought with it Rolls-Royce standards, expertise, practices, and ambition.

Although working with Power Jets, the firm frequently contrasted its methods with what it portrayed as Power Jets' less professional approach to the new engine. In communications with MAP, for example, Hives assured Linnell that the W.2B engine would quickly reach service status if it were developed by Rolls-Royce's methods, which would treat it like any other aero-engine rather than giving it special treatment.[68] Referring to the W.2B engine's reputation in the aero-engine community (but perhaps equally applicable to Rolls-Royce's own CR.1), Haworth colorfully declared in February 1943 to Hayne Constant, head of turbojet work at the government's RAE, that the "common sense" of Rolls-Royce's experienced engineers would produce a turbojet engine to be "used by ordinary mortals and not by a tribe of scientific dream children."[69] The firm may well have overestimated its own ability to develop the radically new aero-engine type, as future struggles would show, but it nevertheless maintained complete confidence that the approach that it had honed in decades of piston engine development would produce equally good results with the turbojet.

To stamp Rolls-Royce's management onto the government's two jet factories, Hives began by restructuring the organization that Rover had established. Under Rover, Clitheroe had been the design and development center for the W.2B engine. Hives moved Rover's turbojet design team from Clitheroe to Barnoldswick, leaving the former, smaller site to administration. Barnoldswick was thus turned from an engine production factory into a development facility; some of the factory's valuable machine tools were given back to the machine tool control for "reallocation."[70] Hives put Stanley Hooker, the senior engineer who had followed turbojet work at Power Jets, in charge of Barnoldswick's technical team. Hooker and the few key Rolls-Royce personnel he took with him from Derby worked closely with those members of Rover's design team who chose to stay with turbojet development rather than return to Rover's car business. Chief among these was Adrian Lombard, who became a legendary chief engineer at Rolls-Royce after the war. The major-

FIGURE 2.3 Rolls-Royce Welland jet engine. Approximate engine length 158 cm, diameter 109 cm. *Source*: Rolls-Royce Heritage Trust

ity of the factory workforce, which was local, kept working in the Barn-oldswick factory under the new management.[71]

The W.2B engine was still far from service use when Rolls-Royce took over, but the pace of its development accelerated under the new management.[72] The intellectual resources, like the material resources, that the large aero-engine firm dedicated to the new project dwarfed Rover's. Development problems were resolved much more quickly under the new company's control.[73] Rolls-Royce's efforts also benefited greatly from the development of Nimonic 80, a new high-temperature alloy for turbine blades developed by another British company, Mond Nickel. Development of the W.2B engine was officially ended by July 1944 (just a year and a quarter after Rolls-Royce gained full control at Barnoldswick), and scale production of the engine at Barnoldswick finally began. The engine was called the Welland.

Rolls-Royce seemed to deserve much credit for the turnaround of the W.2B engine, but the firm's contribution to the engine was a point of disagreement with Power Jets. Hives insisted that Rolls-Royce had been responsible for the improvement in the engine, whereas Power Jets' managers argued that the engine's improvement was due to suggestions made earlier by Power Jets, which Rover had finally begun to put into practice at the time that Rolls-Royce took over. In an internal company memo, dated May 3, 1943, Power Jets' W. E. P. Johnson recorded that "up to date RR [Rolls-Royce] cannot point to any technical contribution worthy of note." Johnson pointed to the WR.1 scheme as evidence of Rolls-Royce's lack of design ability with respect to centrifugal gas turbines (as discussed above, his reference might not have been apt).[74]

Clearly the two firms took a different approach to the engine. As the first group to exploit the idea of the turbojet, Power Jets felt that it had little to learn from other firms; it persistently characterized production as a derivative exercise rather than a creative one. Rolls-Royce had a different view of aero-engine design and production, however, which rested on that firm's particular industrial and institutional strengths. Its competitiveness as a firm was based on cultivating the ability to design and produce reliable aero-engines (as indeed the WR.1 had been). Rolls-Royce matched its ambition to dominate turbojet production, as it had dominated piston engine production before, with its attempt to learn as much as possible from the work of other firms. This cooperation was not opposed to superiority; indeed, the firm planned to ensure its dominance by one day producing turbojet engines so intricate that no other firm would even be capable of manufacturing them.[75]

The differences between the institutional positions (and hence the strategies) of Power Jets and Rolls-Royce in the high-powered, military aero-engine market soon made themselves felt. As Rolls-Royce became more familiar with the centrifugal turbojet engine, it made less of an effort to work with the smaller Power Jets. In February 1943, after the first WR.1 had passed its acceptance test, Rolls-Royce told MAP that future WR.1-related contracts should be placed directly with Rolls-Royce rather than going through Power Jets (to which it had been a subcontractor). In July 1943, Hives transferred the WR.1 project from Derby to Barnoldswick and requested that MAP cancel altogether the original contract for WR.1 engines that had been placed through Power Jets. Rolls-Royce's decision in April 1943 not to produce Power Jet's W.2/500, as had been agreed in November 1942, severely disappointed the Power Jets firm. Instead, Rolls-Royce chose to embark on a new turbojet project of its own, the B/37, later the Derwent. Although the senior firm benefited from using some Power Jets drawings, the latest Rolls-Royce turbojet engine was de-

FIGURE 2.4 The test bed at Waterloo Mill, December 13, 1946. *Source*: Rolls-Royce Heritage Trust

signed without much consultation with the smaller firm.[76] With the B/37, Rolls-Royce proved that it was capable of designing and producing a service jet engine at a time when Power Jets could design but could not yet produce one (Power Jets' facilities and resources are discussed in more detail in the next chapter).

The B/37 put any doubts about Rolls-Royce's dedication to or ability with turbojets to rest. The engine's design included contributions from other firms in the Gas Turbine Collaboration Committee, but its quick success in moving from design to production demonstrated the power of Rolls-Royce's methods. Eclectic like the WR.1, the B/37's design used the compressor and turbine from the Power Jets W.2/500; the straight-through layout and mechanical design of Rover's production-friendly version of the W.2B, the B/26;[77] a combustion system based on the work of Joseph Lucas Ltd; and Rolls-Royce's production and supercharger expertise. The first B/37 ran in July 1943, and its first flight trials were made in a Gloster F.9/40 in April 1944. A point of some pride for Rolls-Royce, which reveals the firm's emphasis on extensively testing new aero-engines: by October 1944, the firm had clocked up 14,000 hours of turbojet running as compared to the 12,000 hours tallied by all of the other British turbojet firms combined.[78]

Rolls-Royce's design team at Barnoldswick swiftly moved its attention from the B/37 to a new, more powerful centrifugal engine design, the B/41, which would become the Rolls-Royce Nene. The first detail drawings for the Nene went to the shops on May 24, 1944; the first prototype was completed by October 25, 1944; and the unit ran two days later. By October 31, 1944, the first B/41 had passed its acceptance test. On November 9, 1944, the engine reached 4,500 pounds of thrust, confirming it as the most powerful turbojet in the world.[79] After the war, the engine was copied by companies around the globe. Rolls-Royce's rapid success with centrifugal turbojets at Barnoldswick may have led Hives to believe that turbojet development was easy to manage using the firm's trusted methods, but the firm's continuing struggle with the axial turbojet at Derby proved otherwise.[80]

Back at Derby

Rolls-Royce's work on axial engines at the firm's main factory in Derby remained ambitious despite its lack of success. When work on the WR.1 engine was moved to Barnoldswick and then canceled, Haworth began working at Derby on designing an axial engine with three concentric shafts, the Rolls-Royce compound axial engine, or the RCA.3.[81] The compound engine's three shafts each powered a single (high-, intermediate-, or low-pressure) axial compressor from a separate turbine.[82] Because of the failure of the Rolls-Royce-designed annular combustion chamber on the CR.1, the compound axial engine made use from the start of the smaller, Lucas-designed reverse-flow (because air entered from the back side) combustion chambers of the Power Jets engines.[83] David Pye passed the new Rolls-Royce engine proposal to the Royal Aircraft Establishment to be "officially vetted."[84] The Royal Aircraft Establishment objected to the engine's mechanical complexity and suggested that Rolls-Royce build a double compound engine (an engine with two independent central axles, one mounted inside the other) before attempting a triple (three concentric axles). A. G. Elliot, Rolls-Royce's chief engineer, with typical confidence, responded to the Royal Aircraft Establishment's objections with the observation that "we are comforted by the thought that our engines have always been more complicated than others without suffering thereby."[85] Complication was not only something that Rolls-Royce thought it could handle, it was a point of pride and, from the management's point of view, the cornerstone of the firm's competitiveness.[86] The firm's preference for and comfort with complication could be a liability, however, and could lead them astray, particularly in areas of novel development.

Construction of the first compound axial experimental unit began at Derby in 1943. Not unlike the CR.1 engine, Rolls-Royce began by building the compound engine's high-pressure compressor-turbine unit (it would eventually have three). Known as RCA.1, the high-pressure compressor stage fed air into the combustion chambers and so could be built and run independently. The firm planned to subsequently add intermediate- and low-pressure axial units to form the RCA.2 (with two axles) and finally the whole engine, with three axles, the RCA.3.[87] When the program was discontinued in 1946, only the high-pressure stage had been built, and Haworth and his design team, having learned what they could from creating the design of the compound axial engine, had already begun designing a new one.[88]

After sending the axial engine design, RCA.3, to Rolls-Royce's experimental shop, Haworth's team had begun work on the RB/39, the firm's first purpose-built turboprop, known as the Clyde. The firm's first turboprop, the RB/50 or Trent, had been created in 1944 at Barnoldswick by adding a propeller to a Derwent engine (power for the propeller is extracted by a turbine in the gas turbine exhaust). In fact, the Trent had been a successful product of collaboration between the Derby and Barnoldswick turbojet teams that exploited continuities with both piston engine practice and the firm's growing turbojet expertise. For the Trent, the diameter of the Derwent's compressor was reduced, and the energy extracted by the turbine that was no longer needed to turn the (now smaller) compressor was used instead to power a propeller via a gearbox, which was designed at Derby. Only five Trent engines were built, but Rolls-Royce nevertheless took valuable lessons from the engine, which proved useful for later engines such as the RB/39. For example, later practice at the firm was influenced by the perceived importance, demonstrated in the Trent engine, of using helical rather than toothed reduction gears.[89]

Like the Trent, the RB/39 turboprop was also a product of collaboration between Derby and Barnoldswick. Rather than use the unproven axial compressor design of the compound engine, the RCA.3, as Haworth suggested, Hooker recommended that the new engine instead use two known compressors: an axial compressor based on Metropolitan Vickers' axial F.2 (discussed below) and a centrifugal compressor based on the Merlin supercharger.[90] Although it abandoned the compound engine's axial compressor, the turboprop nevertheless borrowed one important feature pioneered in the earlier engine design: the use of compounding (or putting shafts inside one another). In the RB/39 a concentric, double shaft was needed to allow the axle powering the propeller to be threaded through the axle powering the compressor (both shafts ran from the turbine in the back of the engine forward). Haworth's design team at Derby designed the engine's compressors following Hooker's suggestion, while the team led by Lombard at Barnoldswick designed the engine's combustion system and turbine. The design of the RB/39 or Clyde started in March 1944, and the engine was built at Barnoldswick, where J. P. Herriot (a manager transferred from Derby) managed the development of the engine "almost alone."[91] Testing began on August 5, 1945, and development continued until 1949, when the engine was abandoned. No airframe had been foreseen for the engine, and Hives, doubting the engine's suitability for production, turned down a government contract for one hundred Clyde engines.[92] But although the Clyde did not bring Rolls-Royce profits from manufacture, the engine still benefited the company because the firm's designers learned from the experience.

Rolls-Royce did not have an equally good record with all components of the early turbojet engines and struggled especially with its early axial compressor designs. Griffith argued that the company was good at developing known engine types, but too much focused on development to succeed with radically new ones, especially if they required systematic research (as the novel axial compressors did). "For the evolution of new and unfamiliar types," Griffith wrote to Hives in February 1944, defending the CR.1 design, "I believe a leaven of research outlook to be essential."[93] He maintained that the failure of the CR.1 pointed up the limitations of Rolls-Royce's piston engine organization, from which its ICT section had grown. Griffith argued that there were limits to the types of problems to which Rolls-Royce's blunt methods could successfully be applied. In light of his previous career at the RAE, he suggested that developing existing engines and creating entirely novel ones might be mutually opposed.

The requirements of gas turbine design did in fact lead to changes in how Rolls-Royce made new engines, including the introduction of more engineering science to the firm's work, which challenged the traditional hierarchy at the engineering firm. For example, while the firm's first turbojet designers had relied on the firm's performance and stress offices for analysis and calculations—calculations that had been allowed to dictate turbojet engine design to some extent—Haworth, an important designer at the firm, rejected this way of working. Haworth believed in the primacy of design and resented the impingement of any analysis on the independence of the designer. Thus he criticized Griffith's CR.1 as an example of the subjection of design principles to the demands of scientific theory.[94] After the war, Haworth began to recruit graduate engineers into his design office to be "Design Engineers," designers who would be "self-sufficient in terms of routine technical support and who became intelligent customers for the more complex inputs from the Performance and Stress offices."[95] In this way, the firm's employees gradually took on different capacities.

After the CR.1 was abandoned, Griffith prepared a slightly less complex scheme for a new unit, the so-called CR.2, in early 1945. The design of the CR.2 unit gave up the contra-flow aspect of the CR.1 unit, but it nevertheless called for a four-stage ducted fan (a fan mounted in a shroud located before the compressor) and a twenty-row axial, contra-rotating compressor driven by an eight-stage turbine.[96] MAP was as supportive of Griffith's new contra-rotating unit as it had been of the first theoretically inspired engine. It argued similarly that the "CR.2 is a project which, if successfully developed, will be an advance over every other type of gas turbine at present envisaged, and we feel that the work should proceed immediately" so long as development, estimated at four to five years, would not prejudice "the simpler and more immediately needed engines" that the firm was working on.[97] The continued faith maintained by MAP and Rolls-Royce in Griffith's idealistic contra-rotating schemes is remarkable. In fact, the CR.2 unit was given up soon after the war, and Hives demanded that Griffith create a more practical axial scheme that would take advantage of the higher compression ratios possible in axial jet engines.

Griffith's next design, the AJ/65 (Axial Turbojet), later the Avon, was much more practical than any of his earlier contra-rotating schemes. Although the team at Barnoldswick had no experience with axial compressors, the axial engine's development was first undertaken by Hooker's team, then still the center of Rolls-Royce's most productive turbojet development work. In 1946, however, Hives decided to concentrate the firm's turbojet work back at Derby, where the company's two gas turbine design and development (centrifugal and axial) strands were finally united. The axial engine Griffith designed became the focus of Derby's turbojet work immediately after the war.

Several of Rolls-Royce's key turbojet engineers left the firm when Hives combined the two teams because they feared they would not find a place in the new, concentrated turbojet organization.[98] Hooker left Rolls-Royce in September 1948. He eventually became chief engineer at Bristol, where he reversed (with Rolls-Royce's help) that firm's turbojet prospects.[99] Hooker's leadership at Bristol later attracted several experienced

Rolls-Royce engineers to join Bristol, including Jamison, Haworth, and even Lombard, although Lombard was ultimately induced to stay on at Rolls-Royce. The Barnoldswick factory remains in operation today, supporting Derby's work on jet engines, but it no longer develops or produces complete engines. As the company transitioned, the firm's key piston engine personnel took on turbojet projects. A. Cyril Lovesey headed the Avon axial engine development team at Derby from 1947; Lovesey had headed development of the Merlin piston engine during the war.

Rolls-Royce took advantage of the collaboration and resources focused by the British government on turbojets during the war. In so doing, it extended its expertise to the new engine type and shifted its focus irrevocably to turbojet development and production. The firm's early success producing the Welland and the Derwent (as described in chapter 1) was important to both the firm's and Britain's future as a turbojet power. Rolls-Royce's success, however, was not assured, and during the war, the British government also helped and encouraged all of Britain's other aero-engine makers—de Havilland, Armstrong Siddeley, Bristol, and Napier—to design turbojets.

The British Aero-Engine Industry

When the British government decided to support the development of a new type of aero-engine, its primary resource was the firms of its aero-engine industry. In the years before rearmament, when production contracts were sparse, the Air Ministry had split the few contracts up so as to support the four main British aero-engine firms: Rolls-Royce and the Bristol Aircraft Company—its main producers of military aircraft engines—and D. Napier and Son and Armstrong Siddeley Motors (ASM)—which primarily provided engines for trainer aircraft. The Ministry also purchased small Gipsy engines from the engine division of the de Havilland Aircraft Company.[100] Bristol, Armstrong Siddeley, and de Havilland were all associated with airframe companies.

During the war, the Air Ministry and then MAP encouraged each of these five aero-engine supplying firms to pursue jet engines. Although no other firm was as ambitious and successful as Rolls-Royce, each of the British aero-engine firms used their piston engine production machines and testing facilities, designs, and personnel to begin developing turbojets. Before mid-1945, every firm in Britain's protected aero-engine industry had begun developing a turbojet. It is worth telling the story of each firm in detail because it reveals the range of early turbojet designs, the relation of turbojet work to piston engine work at each firm, and the way that each firm's distinct resources and skills were used to develop particular turbojets. Although there was a great deal of collaboration on jet engines, the manner of development justifies that the narrative discuss each company separately.

The Ministry encouraged multiple aero-engine firms to pursue jet engines, as it had earlier with piston engines, because it considered competition essential to producing good results. And indeed, each firm tackled the challenges the turbojet posed in a unique way. When the war ended, each British firm was in a different position with regard to piston engine and turbojet engine development. The (newly founded) de Havilland Engine Company was finishing the development of its first centrifugal turbojet, the Goblin, which produced 3,000 pounds thrust; Armstrong Siddeley was developing the 2,600 pounds thrust axial turbojet that it had built during the war into a turboprop unit; Bristol was developing a turbojet with an axial and a centrifugal compressor of around 2,500 pounds thrust; and Napier was working on a turbojet–diesel engine combination that would lead to a turbojet producing approximately 1,700 pounds thrust. Metropolitan Vickers (Metrovick), a British heavy engineering company that had previously manufactured steam turbines, began to make the transition into the aero-engine industry during the war with the design of its pioneering axial turbojet engines, but never completed (or was allowed to complete) the transition to producing aero-engines. The firm's important work, which influenced the work of the Royal Aircraft Establishment and that of Armstrong Siddeley, is described below in the context of its collaboration with Armstrong Siddeley.[101]

The literature on the British aero-engine industry's wartime turbojet developments is patchy and in rather less detail than is available for the German aero-engine firms that are discussed in the next section. Furthermore, the British literature lags considerably behind the German in terms of accounts of the industry as a whole.[102] Of all the British firms, de Havilland, which produced its first jet engines during the war, has received the most attention from professional and amateur historians. The public visibility of each firm's early turbojet efforts tends to be related to the degree of their success with turbojets during the war; thus the early work on turbojets at Armstrong Siddeley, Bristol, and Napier has attracted much less attention.[103]

Each of Britain's aero-engine firms began work on the new type of engine before the end of the war, although they were not equally successful. For better or for worse, each firm's turbojet designs were connected to their piston engine expertise, built on their existing resources, and were fit in beside wartime work on piston engines. Two key sources of expertise that all British aero-engine firms relied on to augment their own knowledge and skill—both of which have already been mentioned in the preceding section—were the Royal Aircraft Establishment (RAE), which provided virtually unique expertise with axial compressors partly in collaboration with Metropolitan Vickers, and Joseph Lucas Ltd., which produced combustion systems for all of Britain's first jet engines. The story of the circumstances under which each of Britain's remaining aero-engine firms embarked on turbojet design and the designs they pursued is one of flexibility and rigidity, of strengths and weaknesses, of clean breaks and preoccupations, and sometimes of deep government intervention that foreshadowed a company's postwar fate.

De Havilland's New Engine

During the war, the engine division of the de Havilland Aircraft Company went from being a minor manufacturer of small aero-engines to being an independent company and a major developer of large turbojet engines. This dramatic transformation was based on the turbojet engine designs of Frank Halford, a successful piston engine designer. He began the war as de Havilland's aero-engine consultant and the head of its propeller development department, and he ended it as head of the newly formed de Havilland Engine Company.[104]

Halford's designs had been successfully produced in the interwar years by both de Havilland and Napier, but the links between Halford's consultancy and de Havilland deepened during the war. In 1937, the designer's fifty-two-strong consulting company moved into a set of offices built for them in de Havilland's Stag Lane Works, near Edgware, London— less than fifteen miles away from de Havilland's main works in Hatfield. The establishment on February 1, 1944, of the de Havilland Engine Company at Stag Lane formalized the intimate link between Halford's consultancy and the aircraft company. The new engine company incorporated Halford's existing engine design team, including his chief engineer, E. S. Moult. Halford became its technical director and chairman of its board of directors.[105]

FIGURE 2.5 Testing the first H.1 engine prototype at Hatfield, April 13, 1942. *Source:* Rolls-Royce Heritage Trust

At the outbreak of World War II, Halford's consultancy put on hold its work on small, civil aero-engines. This meant that the firm had spare capacity to take on a new project in January 1941, when Sir Henry Tizard, scientific advisor to MAP, proposed to Halford that he design a jet engine. In collaboration with de Havilland's aircraft designers, Halford decided on a design goal of 3,000 pounds thrust. The chosen figure would make the new engine powerful enough to make a single-engined jet aircraft competitive with existing piston engine fighters. This was almost twice the thrust that the W.2B was designed to produce; the W.2B engine was intended for an airframe with two engines. When Halford's team completed its design for the H.1 turbojet, de Havilland's engine division was the obvious place to build it. (Napier was busy with piston engine development problems— see below.) By August 1941, the first H.1 drawings had been released to the de Havilland engine shops. The first prototype was built largely using machines used in the production of de Havilland's small Gipsy piston engines, which were not in great demand during the war. The first H.1 engine prototype was put on test at Hatfield on April 13, 1942, in a converted test tunnel originally used for testing Gipsy piston engines, and the jet engine reached its design thrust of 3,000 pounds for a brief moment.[106] This success seemed to be proof of Halford's design skill, but in fact, the development of the engine would throw up many new challenges for the firm before the engine would reliably produce that same thrust.

Although he was one of the first to ever design a jet engine, Halford was confident in his own design experience and technical choices. He had the benefit of knowing about Power Jets' earlier centrifugal turbojet design (a copy of which he received from the RAE after May 15, 1941, several months after starting work on his own engine), but his design diverged from it in several respects, each of which Halford had to defend against the RAE's objections.[107] Like Whittle, Halford chose to use a centrifugal compressor rather than an axial one because it was a familiar component. But instead of Power Jets' double-sided compressor, which Power Jets had devised in order to increase airflow through the engine while keeping the engine's radius small, Halford decided to use a single-sided centrifugal compressor that could move a similar amount of air. Halford's compressor therefore had a much larger diameter than that of the W.2B engine. Despite the challenges of making (and mounting) a large centrifugal compressor, Halford opted for a single-sided compressor in order to gain the maximum advantage from the fact that the aircraft's forward speed would compress the incoming air, thereby increasing the engine's power (this would be more true in a single-sided than in a doubled-sided compressor, in which half of the incoming air would have to be led around to the back of the compressor). A single-sided compressor was also advantageous because Halford could base its design on one of his previous designs that had already been built: the supercharger of the Napier Sabre piston engine. Although the Sabre was still being feverishly developed at Napier, high-speed tests were carried out on a Sabre supercharger at Napier to aid the adaptation of the Sabre design to a turbojet impeller. Napier later lent equipment for H.1 combustion tests, for like Power Jets, Halford too began developing a new combustion system for his engine (it too would ultimately be abandoned).[108] De Havilland reasoned that a single-sided impeller would be simpler to produce (although the large size brought manufacturing and aerodynamic problems of its own) and wouldn't need a complicated casing like the one required to route air through a balance chamber for the two intakes (forward and back) on the W.2B. Symmetrically too, the single-sided impeller was appealing because the axial thrust from the compressor would be balanced by the axial thrust the turbine produced.[109] All of the firm's spare capacity contributed to the jet engine project: even de Havilland's propel-

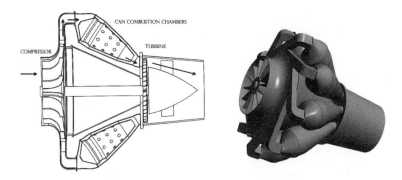

FIGURE 2.6 The de Havilland H.1 or Goblin jet engine (designed by Frank Halford). The engine had a one-sided impeller and used straight through combustion. Approximate length 270 cm, approximate diameter 130 cm. *Source*: Models by Denie Technology

ler division contributed to the propeller-less engine by developing new test methods to determine the vibration characteristics of the H.1's giant impeller.[110] The H.1 compressor moved about 60 pounds of air per second; at that rate it would empty all of the air from a room of 48,000 cubic feet in just over a minute.[111]

Another major difference between the two designs was that Halford's engine used a straight-through combustion system rather than the reverse-flow combustion system designed by Power Jets (see table 2.1 for a discussion of the two systems). Power Jets developed its compact system in order to keep the engine, and thus its main axle, as short as possible and thus avoid the problematic flexing of a long, thin shaft joining compressor and turbine. This system (which remained a part of Power Jets' designs throughout the war) reduced overall engine length but complicated the engine's air path. Halford, in contrast, with his ample experience designing crankshafts for piston engines, was undaunted by the prospect of a long compressor-turbine assembly. He kept the compressor-turbine shaft in his turbojet design rigid (avoiding flexing) by increasing its diameter.[112] The combination of Halford's experience and de Havilland's resources thus produced different design decisions than were made by Power Jets—a very different organization.[113]

In those areas in which Halford was knowledgeable, he frequently departed from Power Jets' practice. In other areas, when he could, Halford relied on the expertise of others (also a skill learned from experience). The final H.1 design replaced the engine's initial components with an RAE designed turbine and a combustion system that was the product of

TABLE 2.1 **Comparison between reverse-flow and straight-through combustion on double-sided impeller turbojet engines.**

	Reverse-Flow Combustion System			Straight-Through Combustion System
1.		+	−	Requires a longer shaft, involving a third bearing and a flexible coupling.
2.	Permits automatic compensation of expansion without special joints.	+	−	Requires expansion joints to allow differential expansion.
3.	Is such that the turbine stator and rotor blades are shielded from direct high-temperature radiation from the flame.	+	−	Subjects turbine stator and rotor blades to direct high temperature radiation from the flame with possibly higher metal temperatures.
4.	Involves two 180° bends in the path of the gas.	−	+	Involves no bends and so has the lower pressure loss.
5.		−	+	Allows the greater cross-sectional area for combustion within a given overall diameter.
6.		−	+	Is the simpler manufacturing proposition.
7.	Allows easier location and inspection of burners and easier assembly and removal of combustion chambers.	+	−	
8.		−	+	Should provide the more even air distribution.

Reproduction of table 1 from Roxbee Cox's lecture on British aircraft gas turbines given in December 1945, showing one example of the design trade-offs inherent in choosing any engine component. The weighting of advantages and disadvantages depends on a company's expertise and changes over time.

According to Roxbee Cox, if the points in the table "were of equal weight, the merits of the two systems would be evenly balanced. Item 5 . . . however represents a preponderating design advantage in the larger engines, and the disadvantages noted against the straight-through system have with development, shrunk in importance." Frank Halford's 1941 H.1 engine used straight-through combustion and a single-sided impeller. All of the engine designs developed by Power Jets during the war used reverse-flow combustion.

Source: Harold Roxbee Cox, "British Aircraft Gas Turbines," 66–67.

work by member firms of the Gas Turbine Collaboration Committee—it used a scaled-down version of the chamber-type combustion chambers used for the Power Jets W.2B. Halford, like Power Jets, preferred using individual combustion chambers because they were easier to test than a single annular chamber, since they required a fraction of the amount of compressed air needed to test a larger axial combustor.[114]

Using existing design elements brought advantages but also presented new challenges. Scaling up the Sabre supercharger by about three times and running it at a much higher speed than in the piston engine required serious additional work. For example, because of the greater accuracy necessary for the H.1 turbojet's compressor vane contours (which impacted the engine's efficiency), the firm's normal practice of bending the vanes of centrifugal compressors into shape could not be used for the new engine. As a result, the firm had to spend time and resources developing new production methods for the new, larger compressor. To machine the H.1 engine's giant impeller, de Havilland called on the expertise and relatively rare American Keller copying machines of the nearby United Glass Bottle Company.[115]

Reusing the Sabre design was also problematic because it brought Halford into collision with MAP's policy of open sharing on turbojet engines. Halford was reluctant to reveal the design—a product of the private interwar piston engine industry—to the Royal Aircraft Establishment or to other commercial members of the Gas Turbine Collaboration Committee, which Halford had reluctantly joined when it was established in 1941. Halford's fear was that de Havilland's competitors on the committee would use information gained through turbojet exchange to covertly improve their piston engines. He also mistrusted the motives of the Royal Aircraft Establishment, which Halford wrongly assumed to be commercial. It wasn't until November 1942 that William Farren, the Royal Aircraft Establishment's director, managed to convince Halford of the nature of the research establishment's interest in the new engines.[116]

In the end, despite several disagreements, being a member of the Gas Turbine Collaboration Committee and the national turbojet program brought de Havilland important benefits. The H.1 engine benefited from the work done by other firms and the Royal Aircraft Establishment as well as the open discussion of development problems among members of the committee. The Gas Turbine Collaboration Committee also coordinated the use of powerful state-owned facilities for gas turbine testing, which made it possible to test the powerful H.1 engine for the first time. For the H.1 engine, the test facilities the committee found included a powerful compressor requisitioned from the construction of the Dartford tunnel under the Thames, which was halted during the war—the only available compressor that could provide a sufficient volume of compressed air for the H.1 engine. The committee also allotted the steam turbine of the Northampton electric power station to H.1 testing because it could provide enough shaft power to test the H.1 engine's large impeller.[117]

During the first years of de Havilland's development of the H.1 engine, the Air Ministry adopted a hands-off policy, leaving the company to

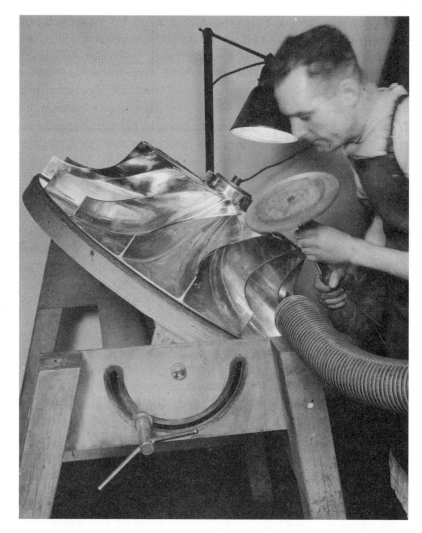

FIGURE 2.7 Man working on an impeller for a de Havilland Goblin (H.1) engine compressor. The engine's diameter was about 127 cm. *Source*: Smithsonian National Air and Space Museum (NASM 9A08292)

its own devices. De Havilland was a trusted firm, with "wide engineering experience and with a sound production background," and the H.1 prototype had quickly achieved promising results. Furthermore, Halford had the confidence of George Bulman, MAP's director of engine development.[118] Bulman's warm relations with Halford dated from their time as colleagues during World War I, and his enthusiasm for the de Havilland company, which unlike Power Jets and Rover had experience developing aero-engines, was in stark contrast to his more cautious approach to Power Jets' work and his strained relations with Power Jets' company's chief engineer, Frank Whittle. Nevertheless, MAP became increasingly wary as H.1 development fell further behind de Havilland's promises and the work of other companies.

The engine that had seemed such a promising alternative to the troubled W.2B engine ran into its own development problems. In July 1944, after Bulman's departure from MAP's directorate of engine development on February 4, 1944, to be replaced by F. R. Banks, MAP began to take a more interventionist approach to de Havilland's turbojet work.[119] According to MAP's Wing Commander Watt (who had served as Roxbee Cox's deputy and was made head of the deputy directorate established for jet engines by the end of 1944), the firm's experience had brought them a good distance, but "unless we make de Havilland help themselves they will not be able to compete with firms like Rolls Royce for very much longer." Watt was certain that "the firm are capable of turning out very good work and most of their engineers and technical people are keen. I think it only requires a little pressure from us on the people who are directing their technical policy."[120] The wing commander's primary worry was de Havilland's method of development, which he believed was responsible for the turbojet's slow progress. Following their own ideas, de Havilland had been slowly developing the H.1 engine to incrementally higher powers, relying on flight rather than bench testing. Watt thought that this method, while useful in developing small piston engines of the sort that de Havilland had worked on before, was ill-suited to larger engines like the H.1 turbojet. He advocated the use of extensive bench running to determine as soon as possible whether the de Havilland turbojet would be able to reach its design parameters.[121] Watt argued that MAP should intervene to make sure that the H.1 engine passed a 100-hour reliability test at its rated 2,700 pounds thrust (the "minimum acceptable thrust" under the company's pilot production contract and thus to make the engine appealing to MAP)[122] as soon as possible. In addition, he wanted to encourage the firm to use components from other turbojet engines to improve the H.1 engine, replacing in particular the H.1's poor combustion system with that of the B/37 (Derwent; first produced November 1944), which had benefited from the thorough research done by the combustion panel of the Gas Turbine Collaboration Committee.

Whether or not Watt was right, after MAP's intervention, de Havilland's development work on the H.1 engine began to move faster. By November 1944, the engine had completed 100 hours of running at 2,500 pounds thrust followed by 100 hours at 2,700 pounds of thrust,[123] and in January 1945, it passed the Air Ministry turbojet type test at 2,700 pounds. The engine would not be produced during the war. Armstrong Siddeley, which volunteered to make parts for the H.1 engine, was never asked to produce H.1 engines, although it was requested in late August 1942 to lend a few of its engineers to de Havilland to help de Havilland build up its aero-engine production organization.[124]

By November 23, 1944, Halford had already designed a new version of the engine, the H.2 (later the Ghost), which would to reach 4,500 pounds thrust. The Ghost dominated the firm's work after the war.[125] So the "almost inhuman pressure" of de Havilland's wartime work didn't let up in 1945.[126] The de Havilland Engine Company, which was established on February 1, 1944, produced engines for the first generation of postwar British jet fighters and for the world's first civil jet aircraft, the Comet. The Comet first flew with de Havilland Goblin jet engines. It was later outfitted with de Havilland Ghost engines and then the Rolls-Royce Avon.

Even with two turbojet designs under development, MAP wasn't satisfied. As the record of the H.1 engine makes clear, this was justified. Promising prototypes could disappear from a production program because of development problems—especially when production and development were going on at the same time. Spreading its bets, MAP encouraged other British aero-engine firms to develop turbojets too. The next that it guided into turbojet design was Armstrong Siddeley Motors.

Armstrong Siddeley Motors Changes Course

Armstrong Siddeley Motors, a second-tier British aero-engine firm, gave up its piston engine development business and began working on a turbojet in 1941. The firm's reputation for producing unreliable, low-powered, large aero-engines made the Air Ministry skeptical of the firm's development competence; the firm had been struggling with its 1,115 horsepower Deerhound piston engine since 1935. Convinced of the need to "guide" the firm's development policy, MAP counseled Armstrong Siddeley Motors to focus its efforts instead on a single new turbojet engine in order to "catch up to its rivals." In October 1941, Armstrong Siddeley finally agreed to abandon the Deerhound and its successor, the Wolfhound, a 3,000 horsepower piston engine that was on the drawing board, in favor of a turbojet. Armstrong Siddeley had experience with gas turbines, albeit not with a turbojet aero-engine. It had previously worked with the Royal Navy on the construction of a Swedish-designed gas turbine engine for a torpedo. Furthermore, in 1939, the firm had constructed a nine-stage, contra-rotating compressor with independent rotors for the Royal Aircraft Establishment.[127] To gain experience building a complete turbojet aero-engine, MAP suggested that Armstrong Siddeley join an existing turbojet project as a collaborator, and Armstrong Siddeley was invited to join the Gas Turbine Collaboration Committee in November 1941.[128] After the proposed collaboration failed, however, the Royal Aircraft Establishment helped the firm to design a simple axial turbojet and turbo-prop engine of its own, which it completed developing before the end of the war. Although the engine was not a great commercial success, it nevertheless paved the way for Armstrong Siddeley's subsequent involvement in the very successful Sapphire turbojet engine.

To start Armstrong Siddeley's work on turbojets, MAP suggested that the firm collaborate with Metrovick, a steam turbine manufacturer that was working on an axial turbojet. Metrovick had designed a turbojet engine, the F.2, in collaboration with the Royal Aircraft Establishment. The research establishment had begun working with Metrovick in 1937,[129] long before the F.2, when the Royal Aircraft Establishment's engine department was first given permission to place contracts for experimental apparatus with outside firms.[130] So when the RAE designed its first complete turbojet flight engine, Scheme F, in 1940, it looked once again to outside contractors for help. (The design departed from the Establishment's earlier work, which had been aimed at producing research units rather than functional engines.) After being turned down by Power Jets, which was under great pressure to do other work,[131] the Establishment approached Metrovick, which in July 1940 agreed to build the F.1 engine—albeit as considerably more than a subcontractor, as it had become accustomed.[132]

Since the outbreak of war, Metrovick's turbojet team had proceeded without waiting "for approval from RAE for every step," as they had done before.[133] Together, the two parties decided that the axial flight engine, now known as the F.2, would use a nine-stage axial compressor designed by the Royal Aircraft Establishment; Metrovick would design the engine's combustion chamber, turbine, and nozzle.[134] An F.2 prototype ran for the first time in December 1941.[135] But the unit was much too heavy to be an aero-engine. The combination of steam turbine experience, aerodynamic research, mathematical analysis, and skillful design that characterized the fruitful partnership between Metrovick and the Royal Aircraft Establishment created a first axial gas turbine bench unit that was powerful but heavy—closer to steam turbine practice than aero-engine design.

It was for this reason that Armstrong Siddeley was brought in to the help with the F.2, particularly with an eye to production. Not only did MAP and its advisors at the Royal Aircraft Establishment feel that the aero-engine firm could make an important contribution to the F.2 engine, but they also hoped that the project offered a good first turbojet for Armstrong Siddeley, which would by developing it become familiar with the particularities of turbojet engines. The collaboration was thus seen as a first step toward Armstrong Siddeley designing its own turbojet engine.[136] However, quite contrary to MAP's hopes, Metrovick refused to let Armstrong Siddeley "undertake any redesign or modification work in connection with the actual unit"[137] and insisted that it retain all technical responsibility for the engine. This neither allowed Armstrong Siddeley to give Metrovick the benefit of its aero-engine expertise nor to learn about turbojet design. In December 1941, the manager of Armstrong Siddeley Motors, Sir Frank Spriggs, despaired that Metrovick "still fail to realise that while we all recognise their experience and ability in what they term the 'rotating plant' they do not realise how much they do not know about aero-engines."[138]

The testy relationship between the two firms remained unchanged despite Armstrong Siddeley's complaints to MAP, and the F.2 engine remained far from achieving what the Ministry described as "aero engine standard."[139] The engine reached an impressive 2,160 pounds thrust in bench testing by the start of May 1942, well ahead of the performance achieved by either the W.2B or the H.1,[140] yet it slowly began to fall further behind its competitors. In late 1941, MAP had considered the F.2 to be a promising alternative to the languishing W.2B, a "second string" engine for the F.9/40.[141] Indeed, the F.2 engine actually powered an F.9/40 prototype only on November 13, 1943, some eight months after an F.9/40 airframe had flown with H.1 engines. Whereas in March 1942 it was still unclear "whether the de Havilland H.1. or a cleaned up Metropolitan Vickers F.2. is likely to be the first to reach that combination of reliability, weight and performance to make it suitable as an operational engine,"[142] by autumn 1942 the F.2 had dropped behind the H.1 in the Ministry's priority ranking.[143] The refusal of Metrovick to collaborate with Armstrong Siddeley likely led to the failure of the F.2 unit as an aero-engine.

It also led the aero-engine firm to look for another way to start work on a turbojet engine: hiring an experienced turbojet engineer, Fritz Albert Max Heppner.

Hiring Heppner was a direct affront to the firm's chief engineer, Stuart Tresilian, who had been grudgingly installed as the chief engineer of Armstrong Siddeley Motors at MAP's request in 1939. Tresilian had inaugurated speculative work on axial ducted fan gas turbines engines under T. P. de Paravicini and his able assistant Brian Slatter, who later moved to the Royal Aircraft Establishment, and was critical of Heppner's work.[144] Yet in late 1941, the board of Armstrong Siddeley Motors, particularly the two directors, Frank Spriggs and Thomas Sopwith, hired Heppner, a German refugee who brought an ambitious plan for a contra-rotating, axial turbojet engine with him.[145] Like Griffith's, Heppner's contra-rotating engine was made up of two interleaved sets of rotors, which rotated in opposing directions; unlike Griffith's, however, the first set was mounted on a central shaft, while the second was moved by a rotating external casing. There was opposition to Heppner's complicated plans from within and without the firm.

Although the board was enthusiastic about Heppner's scheme, the British government was wary of its complexity. In fact, in November 1939, well before the invitation from Armstrong Siddeley, Heppner had submitted a patent specification for a turbojet to the Air Ministry's Directorate of Scientific Research that had been turned down because it was judged overly complex.[146] In June 1941, the RAE reported similarly unfavorably on the short-term prospects of a new Heppner proposal for a contra-flow, contra-rotating engine, which was judged, like other contra-rotating schemes, to have great potential but to be "a more remote development than the engines at present under construction."[147] The Ministry thought that Armstrong Siddeley should begin with a simpler design that would be easier to develop quickly. It was perhaps to be expected that when Armstrong Siddeley dispatched Heppner's December 1941 proposal, which had the same structure as his June proposal but was now under the aero-engine firm's name, to the RAE for comment, it was once again rejected as impractically complicated, especially for the company's first turbojet. The RAE was skeptical of Heppner's performance assumptions in particular, which far exceeded anything achieved in contemporary practice in Britain. Heppner claimed that his contra-rotating compressor would be able to achieve a pressure ratio of 8:1, whereas the highest that a British engine had yet reached was a ratio of 3.5:1.[148]

Throughout 1942, chafing at Metrovick's refusal to let the firm contribute to the F.2 engine, Armstrong Siddeley continued to try to start an engine project of its own. Criticism of Heppner's proposal by Tresilian was brought to an end when the company board used his lack of support as a pretext to fire him on January 19, 1942. Tresilian then took up a position as the engine coordinator between the US Army Air Forces and the Royal Air Force, which he held from 1942 to 1945. He was not the only ex-chief engineer in Britain to be given a government post; Frank Whittle and Roy Fedden both benefited from similar treatment. Tresilian was replaced by Dr. H. S. Rowell, Armstrong Siddeley Motors' general manager and a strong supporter of Heppner's plan.[149] The way was then clear for the firm to act on Heppner's schemes; it subsequently submitted a complicated turbojet engine proposal to MAP, which included three sets of compressors and a contra-rotating turbine. The new proposal was rejected by both the Royal Aircraft Establishment and MAP in May 1942.[150] MAP worried that the Heppner scheme "appear[ed] to have all the earmarks of another Deerhound"—excessive complexity that would lead to interminable development.[151]

By August 1942, Armstrong Siddeley had still not managed to start its own turbojet project. On August 8, 1942, Spriggs wrote to MAP that the firm's "development team" was eager to start on an engine; "they have the facilities, knowledge and enthusiasm, and could have something tangible for test within 6 to 9 months." He also volunteered Armstrong Siddeley's production capacity for the H.1 engine or any other turbojet that was ready for production.[152] Worried by the complication of Heppner's schemes, MAP had urged the aero-engine firm throughout 1942 to begin with a simple turbojet engine design that would be relatively quick to develop. By mid-August 1942, Armstrong Siddeley finally agreed with MAP to begin developing a single-axle axial engine that would complement the greater number of centrifugal engines already under development in Britain. Having given up the prospect of collaboration, the design would also offer competition to Metrovick's F.2 engine, "consistent with a policy of competition similar to that existing in the reciprocating [piston] engine field," although neither axial engine was expected to reach production during the war.[153]

Still worried about the firm's capability, however, MAP continued to rely on the Royal Aircraft Establishment to supervise Armstrong Siddeley's first steps into the new field. The Establishment volunteered to help Armstrong Siddeley Motors to design an engine similar to Metrovick's F.2 engine, which was also designed with input from the Royal Aircraft Establishment.[154] (The Royal Aircraft Establishment's turbine section was founded in mid-1941.) As it had with Metrovick, the Royal Aircraft Establishment provided Armstrong Siddeley with a basic aerodynamic design based on the research establishment's unique expertise with axial compressors. The aero-engine firm was responsible, however, for the layout of the actual engine, which became known as the ASX.[155] The ASX engine incorporated the Royal Aircraft Establishment's continuing research on axial compressors. It had a fourteen-stage axial flow compressor, which consisted of the same nine-stage high-pressure axial compressor as the F.2 engine plus five additional low-pressure stages.[156] Designed to produce 2,500 pounds thrust—about 1,000 pounds more than the F.2 engine, the ASX had eleven combustion chambers designed and manufactured by Joseph Lucas. The firm ultimately used a unique system of vaporizing fuel burners designed by the firm's Sidney Allen (other British companies used atomizing fuel burners).[157] To reduce the engine's length, the engine used a reverse-flow layout: air flowed through the compressor from back to front, where the combustion chambers were located—an arrangement that later caused serious aerodynamic losses in the engine[158] (see figure 2.8 below).

In October 1942, the Royal Aircraft Establishment approved Armstrong Siddeley's ASX engine design, perhaps after lobbying from the firm's turbojet engineers, including Pat Lindsey.[159] The Establishment suggested that MAP place an order for six jet engines and spares from Armstrong Siddeley, but (still not entirely trusting the firm) only on the condition that "technical supervision of the detail design by RAE should be stipulated."[160] The contract was granted on November 7, 1942.[161] The first ASX engine ran on a converted Deerhound test bed on April 22, 1943, and soon produced more than 2,000 pounds thrust. The firm's construction record for the ASX engine was as impressive as the better-known 248 days between the first detailed drawing of de Havilland's H.1 engine to its first run on a test bed.[162]

Satisfied that Armstrong Siddeley was capable of making a turbojet, in late summer 1943, MAP gave Armstrong Siddeley a second contract for research on a Heppner-inspired aero-engine with many "untried features": the ASH. Armstrong Siddeley, however, soon gave up the project.[163] Instead of pursuing a completely new turbojet engine design after building a single ASX engine, the firm decided to use the ASX turbojet as the basis for a turboprop engine, the ASP. By April 20, 1945,

FIGURE 2.8 Photographs of (above) the Armstrong Siddeley ASX axial jet engine and (below) the Armstrong Siddeley Python turboprop engine.

Both engines pictured are located at the Rolls-Royce Heritage Trust in Derby. The ASX is cut away to reveal the engine's axial compressor and uniquely long, reverse-flow combustion chambers. The Armstrong Siddeley Python was based on the ASX.

Photographs: Hermione Giffard

the company had completed twenty-two hours of test running on the new engine.[164] The ASP engine was developed into the Python under Ministry contract, and in October 1945, it recorded the highest power of any turboprop in Britain (and far ahead of any piston engine): 3,656 shaft horsepower plus an additional 1,130 pounds jet thrust.[165]

Neither the ASX nor the ASP engine was produced during the war, but unlike the Metrovick F.2 engine, the Python was eventually put into limited military service after the war—a striking success for a firm that had virtually no experience with axial gas turbine engine design before 1942.[166] Although denied the chance to contribute to the F.2 engine, in the end, Armstrong Siddeley did directly benefit from Metrovick's development work on its axial turbojets. When Metrovick was forced to give up its turbojet ambitions in the late 1940s,[167] Armstrong Siddeley took over Metrovick's next axial engine design, the F.9, which it developed into the successful Sapphire turbojet engine. In the end, MAP thus created a successful turbojet design firm out of Armstrong Siddeley, but it faced a different challenge with the Bristol Aircraft Company. MAP didn't have to convince Bristol to work on turbojets, which the firm had already begun on its own, but rather to devote resources to turbojet development.

Bristol Aircraft Company Is of Two Minds

In 1939, the engine division of the Bristol Aircraft Company was, alongside Rolls-Royce, one of Britain's two most important and accomplished aero-engine firms.[168] Unlike the Derby-based company, however, Bristol was reluctant to give up its work on piston engines during the war and devoted few resources to the new turbojet engines. This reflected the firm's judgment of its engine's prospects rather than any fear of the new type. (Rolls-Royce, in contrast, had judged the prospects of its piston engines in civil aviation to be poor and had therefore embraced the turbojet.) In fact, Bristol was already familiar with gas turbines well before 1941, when Roy Fedden, the firm's chief engineer, and Frank Owner, the head of its project office, visited Power Jets' design office at Sir Henry Tizard's suggestion.[169] As a student at Manchester University, Owner had designed a gas turbine before joining Bristol in 1922.[170] A couple of years later, in 1924, Fedden had charged him with carrying out an investigation of the potential of a turboprop engine.[171] At the time, Owner argued that the turboprop was much too inefficient to replace a piston engine. When Flying Officer Frank Whittle approached Bristol for support in February

1931, it was Owner who evaluated Whittle's gas turbine aero-engine proposal. He judged it sound in principle but unlikely to become practical for
another decade—a prediction that proved to be remarkably accurate.[172]
Although the firm thus did not support Whittle's first scheme, Bristol continued its familiarity with gas turbines by collaborating with the Royal
Aircraft Establishment in the first years of the war to test a six-stage high-
speed axial compressor that had been built for the establishment by Fra-
ser and Chalmers, a British steam turbine company.[173]

Roy Fedden was interested from early on in the new engine type but
was also wary of its limitations. After seeing Power Jets' work in 1941—
an engine that he considered relatively low-powered, he proposed pairing
a turbojet with a piston engine so that the combination would generate
enough power to propel an airplane. In a letter dated October 20, 1941,
Fedden proposed to MAP that the W.1A (Power Jets' most advanced
gas turbine prototype at the time, which produced around 1,400 pounds
thrust) serve as an accelerator or additional power plant for Bristol's most
advanced sleeve-valve engine, which was then under development, the
Centaurus (designed to produce more than 2,500 pounds thrust at 300
mph). Bristol submitted a brochure for the accelerator project on December 4, 1941, but the opportunity to develop it was never to arise as
war work kept the company firmly focused on piston engines.[174] Indeed,
the shift of MAP's interest from fighters to bombers after Germany in-
vaded Russia (Operation Barbarossa began on June 22, 1941) made Bristol's bomber-powering piston-engines particularly important to the nation's war effort.

Although the firm put its combination proposal aside, Fedden nevertheless asked MAP whether the head of his project office, Frank Owner,
could join the Ministry's newly established Gas Turbine Collaboration
Committee. Owner duly began attending from the second meeting of the
committee on December 13, 1941, providing another strong link between
Bristol and the turbojet development work going on in Britain.[175] In fact,
Owner had already requested in mid-1941 that the firm's project office
begin preparing a proposal for a 4,000 horsepower turboprop with fuel
consumption comparable to that of a piston engine at 300 mph and 20,000
feet (turbojet thrust varies with speed and altitude). In 1942, this became
Bristol's first turboprop engine scheme.[176] Owner's requirements were important because of the fact that he used contemporary piston-engine performance as a point of reference and also because they reveal his interest
in improving on the high fuel consumption of the early turbojets, which
meant the engines were ill-suited to longer-distance transport or bomber

flights. The project office's turbojet proposal was completed in May 1942, and the engine design was finished by October 1942. The project was however suspended when Fedden was forced to leave his position as chief engineer in October 1942.[177] It wasn't until April 7, 1943, that Owner finally presented a "definite proposal" for a Bristol turboprop to MAP.[178]

The turboprop proposal that Bristol finally submitted to MAP was scaled down from Owner's ambitious original goal to a more reasonable 2,000 horsepower. It included several novel features, which appealed to MAP's desire to promote variety in the national program. The engine had both an axial and a centrifugal compressor. Unlike all of the existing projects at that time, the firm decided to pursue a turboprop rather than a pure jet because it was particularly interested in making engines with high fuel efficiency for long-distance flights (pure jets were best suited to high-speed, short flights and notoriously fuel inefficient). But although the features of Bristol's engine were novel within the national jet engine program, none of them was completely unknown. The design of the turboprop's centrifugal compressor and the engine's turbines drew on the firm's experience in developing a turbo supercharger for the Hercules piston engine; Owner had led the project in 1937 and 1938.[179] The firm consulted the Royal Aircraft Establishment to design the engine's axial compressor and Joseph Lucas to design the engine's fuel system.[180]

The centerpiece of Bristol's turboprop proposal, which was unique among all of Britain's turbojet projects, was its heat exchanger. By recycling heat from the engine's exhaust stream that would otherwise be lost, the heat exchanger offered a solution to the problem of the low fuel efficiency of turbojets.[181] Including a heat exchanger allowed the firm to aim to make a turboprop with characteristics similar to its best piston engines and follow its existing design preferences. The use of a heat exchanger followed Bristol's established preference for keeping an engine's internal stresses minimal— the heat exchanger limited the unit's compression ratio to 5:1 because above this pressure the compressed-air stream would be too hot for efficient heat exchanger operation.[182] The decision to include a heat exchanger in the design dictated much of the engine's overall layout. In order to allow the heat exchanger to be placed between the streams of combustion and exhaust gases (low and high energy streams respectively), for example, the engine's air path made two 180-degree bends.

Roxbee Cox was particularly excited about the potential of the engine's unique heat exchanger, and five days after Bristol submitted a brochure for their turboprop engine, he promised the firm a research contract for experimental work on the element. Because of its novelty and intricate mechanical design, the heat exchanger was expected to require a lot of preliminary development work.[183] In early August 1943, Bristol decided to accept not a research contract but only a manufacturing contract from MAP, which paid for material for three turboprops.[184]

With Fedden's departure from Bristol in late 1942, Owner became the main force behind the firm's turbojet work. His strong advocacy led the company board at a meeting on July 10, 1943, to support the firm's involvement in "the design and manufacture of gas turbines" (two years after the Rolls-Royce board had made a similar decision).[185] Despite this high level backing, however, Owner struggled to assemble people and resources for his turboprop project at the firm. Several key members of Bristol's leadership were reluctant to support gas turbine work.[186] According to Owner, the Engine Division's managing director and general manager in charge of production opposed a switch in emphasis to gas turbine engines.[187] Undoubtedly, both the firm's commitment to long-haul piston engines for the war and the conviction that the same engines would play an important role in the postwar period contributed to the management's reluctance to reduce the resources invested in the improvement and production of piston engines.

Seeing Bristol's work on turbojets thus held up and wanting to enrich the nation's turbojet program, MAP sought to counter the power of intra-firm resistance and facilitate the firm's gas turbine work. In January 1944, Roxbee Cox urged MAP that "a great firm like" Bristol must be allowed to "gain the necessary experience of gas turbines if it is to continue to fill its important place in the aircraft engine industry."[188] Not quite as keen perhaps as Roxbee Cox, in early 1944, the minister of aircraft production, Stafford Cripps, gave Bristol explicit permission to build its turboprop, known as the Theseus. His permission was only given, however, on the condition that the project did not interfere with work on the firm's piston engines, which were still needed for the war. Bristol honored the undertaking to Cripps so "conscientiously" that the staff doing the detail design for the Theseus engine was made up only of six apprentices and trainees borrowed from the main drawing offices.[189]

By the end of July 1944, fifteen months after its initial proposal to MAP, Bristol was finally in the midst of manufacturing its first turboprop unit and had begun testing a section equivalent to one-eighth of its heat exchanger.[190] The engine ran for the first time without its heat exchanger on July 18, 1945.[191] A complete Theseus engine outfitted with a heat exchanger ran in December 1945, but the complex device produced a fuel savings of only 8 percent and actually reduced the engine's overall performance. Despite Bristol's high hopes, the heat exchanger was subsequently abandoned. In September 1944, while the Theseus design was being completed, Owner diverted an engineer from his already small turbine staff, Charles Marchant, to begin work on a new private venture that would become the Proteus turboprop. Once again, low specific fuel consumption and high efficiency were the engine's key features; the engine did not include a heat exchanger and instead relied on a higher-pressure ratio to increase its power output.[192]

Despite MAP's attempts to shift the balance of work at the firm, Bristol's management remained firmly focused on the large Centaurus piston engine. This despite the fact that in February 1945, with the Luftwaffe all but defeated in Europe, MAP had decided that it would allow Bristol to drop a large portion of its remaining piston engine work so that the firm could more aggressively pursue a gas turbine engine design.[193] Further supporting the firm's work on its turboprop, in mid-1945, the Ministry approved almost half a million pounds for gas turbine facilities at Bristol.[194] Yet even this could not overcome the firm's resistance. In July 1945, MAP observed that personnel, whom the firm had promised to switch from piston engine to gas turbine work, had either arrived late or were not transferred at all.[195]

At the end of the war, much of the firm's meager gains in turbojet design were lost as a significant part of Bristol's turboprop staff left the company. With their departure, Owner lost much of the design expertise that he had managed to build up during the war. He complained that the new staff members who replaced the departing designers had no experience with turbines of any sort.[196] In part because the firm's leadership would not commit resources to developing the new engines, Bristol's early record of designing and building gas turbines was poor. The firm's reputation for aero-engine design, which had declined with Fedden's departure, was only rebuilt in gas turbines in the 1950s, after assistance from MAP, Rolls-Royce, and Stanley Hooker, who joined the firm from Rolls-Royce in 1949. The last British aero-engine firm, Napier, was even more caught up with piston engine work than Bristol. Yet before the end of the war, MAP had involved Napier too in its turbojet program.

D. Napier and Son

Napier was the last of Britain's aero-engine companies to enter the gas turbine field. Although its jet engine story is short, it is included here because its experience with turbojets is characteristic of how firms reused resources and balanced work on the new type with piston engine development. Napier's story is also important because it shows MAP's eagerness to broaden their program to include all of Britain's established aero-engine designers, and it demonstrates how many different ideas were tried in early gas turbines, even those that were ultimately unsuccessful. Napier was, with Armstrong Siddeley Motors, one of Britain's second-rank aero-engine firms, and it started the war bogged down with the development and production of the Sabre piston engine, which had been designed by Frank Halford before the war. Commitment to the Sabre prevented the firm from diverting resources to any other project, and in December 1942, MAP encouraged English Electric to take over the disorganized firm, which it did.[197] It was only during 1944, when development of the Sabre had made steps forward, that MAP encouraged Napier to enter the gas turbine field. Looking more than ever to the postwar period, this was part of the Ministry's effort to rationalize the aero-engine industry's remaining work on piston engines and to make sure that all of Britain's aero-engine firms made a "contribution to British gas turbine progress."[198]

MAP issued a specification for Napier's engine on August 21, 1944.[199] The proposal for a turbojet unit that Napier ultimately submitted was not for a pure jet unit, which it felt would be more suited to war than peacetime markets. The firm's first proposal was for a power plant that would combine a gas turbine engine with very high-powered, two-stroke diesel piston engine. The use of a two-stroke diesel was well suited to Napier's strengths; the Sabre was a two-stroke piston engine, and English Electric specialized in diesel engines.[200] The engine was to have a centrifugal supercharger and was to drive a contra-rotating propeller, both of which the firm had been developing with the Sabre.[201]

Known within the firm as the E.124 (the E.125 was a smaller engine considered initially as a "phase in the development of the large E.124 engine"),[202] the engine was projected to reach an impressive 6,000 horsepower (for comparison, the Sabre piston engine, which Napier had been working on, produced a maximum of 3,000 horsepower).[203] The design goals of the E.124 reflected the need for a very high-powered aero-engine that had been identified by the country's Brabazon Committee, formed on December 23, 1942, to plan for the postwar needs of British civil aviation.[204] The E.124's low projected fuel consumption recommended it to MAP as an engine for long-range aircraft, both military and civil.[205] The Ministry declared that the engine was a "most attractive" addition to the national gamut of aero-engine development and would allow the firm to remain competitive with other aero-engine firms. In addition, the Ministry hoped that the compound idea might provide a way for the nation's highly developed, high-powered piston engines to compete with pure turbojets in the future.[206] (The dual-engine was ultimately abandoned because it could not do so.)

Napier began serious work on its E.124 design in early 1945, before the end of the war. It began by researching two-stroke compression ignition cycles with both sleeve and poppet valves (the Sabre piston engine had sleeve valves) while preparing designs for the constituent parts of the engine's gas turbine half. Because Napier's engine design included an axial gas turbine, with which the firm had no previous experience, the firm requested and was assured access to all available knowledge on gas turbines and axial flow compressors in Britain. By 1944, the provision of such information through the Royal Aircraft Establishment and the Gas Turbine Collaboration Committee had become a matter of course. Napier's representatives were included for the first time at the fifteenth meeting of the Gas Turbine Collaboration Committee on November 11, 1944.[207] After the war ended in Europe, Napier also requested information on Germany's axial gas turbines as soon it became available.[208]

To help Napier get started with gas turbines, the government paid for extensive facilities for gas turbine research and development at Napier's Liverpool shadow factory (which had been paid for by the government, but was run by Napier). Early cost estimates for the new facilities ranged from half a million to one million pounds.[209] Despite this support, the E.124 was never realized. Instead, the design evolved into two postwar engines: a high-economy, two-stroke piston engine with an axial turbo supercharger known as the Nomad and an axial turboprop, known as the Naiad. Neither was very successful commercially, although they both contributed to Napier's experience with turbojets; the firm produced its first competitive gas turbine engine, the Eland, in the 1950s, after about a decade of work.[210]

The British aero-engine landscape at the end of the war thus fulfilled two of MAP's ideas. Firstly, it exemplified what the Ministry considered to be productive competition in turbojets, and secondly, it signaled MAP's view of aero-engine development as a national capacity that would benefit everyone if maintained in the best way. Needless to say, the British program contrasted with that in Germany, where turbojet development followed a rather different path.

The German Aero-Engine Industry

German Air Ministry (Reichsluftfahrtministerium, RLM) policy dictated that the country's aero-engine industry should be the main site for the development of turbojets.[211] So although German piston engine makers made less powerful piston engines than their British counterparts in the late 1930s,[212] the German aero-engine industry had already embarked on developing turbojets by 1938, before the outbreak of war. In Britain, as has been shown, aero-engine firms moved gradually to work on the new engine type as piston engine development wound down, whereas in Germany the whole aero-engine industry began developing the new engines at the same time as carrying on piston engine development. The chief exception to the RLM's policy was the privately funded work of the Heinkel Aircraft Company, which will be dealt with in the next chapter. The distinction between development and production is clearest in the case of Germany's aero-engine makers; the two were related but not absolutely.

All of Germany's major aero-engine firms ultimately decided to develop axial turbojets, which were expected to produce higher engine powers than centrifugal engines. There was nothing inevitable about the choice of axial turbojets, no government fiat to that effect, and several German firms did design centrifugal gas turbines during their research.[213] Nevertheless, this decision resulted in a deceptive uniformity in German turbojet design. The preference for axial compressors in Germany is important because while many British firms made use of their own experience with centrifugal compressors, German firms could not exploit as many design continuities with piston engine work. Indeed, the decision for axial compressors was also based, in part, on German aero-engine firms' lack of experience with centrifugal turbo superchargers, which therefore gave them little expectation of advantage.

The decision to develop axial compressors made all of Germany's aero-engine firms reliant on the expertise of the aerodynamic research institute (Aerodynamische Versuchsanstalt, AVA) in Göttingen. The availability of the establishment's expertise doubtless encouraged the development of axial compressors in industry. The Aerodynamische Versuchsanstalt was reorganized under RLM auspices in 1937, and it carried out extensive research for the aeronautical industry during the war. Its work on turbo machinery primarily included axial compressors for superchargers and gas turbines.[214] During the war, the Aerodynamische Versuchsanstalt became Germany's leading source of axial turbine expertise.

Understanding development at each of Germany's aero-engine firms as part of a single, national effort changes our interpretation of that story radically; the different firms' approaches, far from being a judgment on the firm in question, complemented one another. Thus the work of the most well known producers, Jumo and BMW, supplemented each other. Jumo chose a conservative design that intentionally exploited continuities in order to be quick to develop, whereas BMW decided to build up new experience in house. The quicker Jumo engine allowed BMW to take longer to develop an ambitious production turbojet engine. BMW's program, furthermore, owed much to the enterprise of the turbojet design team of the Brandenburgische Motorenwerke (Bramo), which like the firm, was incorporated into BMW near the start of the war. The connection between the firms has been noted by historians before now, but none has emphasized the importance of the ethos of Bramo's program in shaping the turbojet program of the piston-engine-centered BMW.[215] The literature generally dismisses Daimler-Benz's "impossibly slow progress" with turbojets[216] as a failure, but that firm's work too needs to be seriously revisited as part of a national turbojet program as well as in light of Daimler-Benz's commitment to piston engines.

Junkers Motorenwerke

The first aero-engine firm to begin the series production of turbojet engines in Germany was Junkers Motoren (Jumo). That Jumo produced the first turbojets in Germany was not surprising given the firm's place in the national development program. The engine firm was made independent from the Junkers Aircraft Company, both originally founded by Hugo Junkers, after the National Socialists seized power. When the government demanded that the firm work on turbojets in the late 1930s, Otto Mader, Jumo's head of development until 1944, was reluctant to begin work on a turbojet, fearing that the work would detract from the company's important piston engine work. Nevertheless, the firm started a small turbojet development program at its main works in Dessau (the early work on jet propulsion done at the firm's Magdeburg branch in 1937 had been unenthusiastic and made little progress), and Mader put Dr. Anselm Franz in charge of it.[217]

Franz was a good choice to head Jumo's turbojet development. At the time, he was the head of the firm's small department for aerodynamics and supercharger development. In addition, he had worked on turbo machinery before joining Jumo in 1936. When he joined Jumo, the firm was doing research on so-called free piston engines (in which a piston engine was used not to produce shaft power but to produce a high-energy exhaust stream from which power was extracted), but the scheme was soon given up as too difficult to make reasonably light and efficient.[218] Franz spearheaded the firm's subsequent efforts to improve piston engine performance by ducting engine exhaust to produce useful thrust (a common idea at the time, as it was known that directing the exhaust of a piston engine through carefully designed nozzles could produce a significant amount of thrust from the waste stream). In 1940, he was awarded a PhD by the technical university in Berlin for a thesis on exhaust jet development.[219] Franz is one of the few people who has defended the RLM's policy of placing turbojet contracts preferentially with aero-engine firms.[220]

Franz's work was not the first turbojet project at Junkers, although it was the first on the aero-engine side. Before 1938, Professor Herbert Wagner and his assistant Diplom-Ingenieur (equivalent to a master's in engineering) Max Adolf Müller developed several gas turbine engines as a private venture at the Junkers airframe company. In keeping with the RLM's policy to only support turbojet development work at aero-engine firms, the work never received government funding. In fact, it was halted in the summer of 1939.[221] A widespread assumption that Franz went out of his way to contradict after the war was that Wagner's work on an axial turbojet had served as the basis for Jumo's axial turbojet work.[222] In fact, Franz did not use Wagner's designs, although he did attempt to convince the members of Müller's group to join Jumo (experienced engineers were more valuable to him than the design). Nevertheless, only the team's skilled materials specialist, Dr. Heinrich Adenstedt, did so.[223] Nevertheless, before the team left Junkers (for Heinkel, as it turned out), Franz engaged in a four-week exchange of information with Müller's group that was "intensive and without reservation." But Franz thereafter started his own design rather than take over the previous group's engine design, which was developed further at Heinkel.[224]

Franz, his own experience augmented by the things learned from Müller's team, created a new axial turbojet design that reflected his own decisions and priorities as well as the needs of the national program. The resulting engine was assigned the government number 109-004.[225] Franz chose to use an axial compressor for his gas turbine engine because he considered a smaller frontal area essential for high-speed flight (a centrifugal turbojet, having a greater frontal area, would suffer much more drag). Yet in order to get the engine to the point of production as soon as possible, he settled for performance goals lower than what might ultimately be possible.[226]

To ease development, the 004 engine design was intentionally conservative, staying close to existing expertise. Franz enlisted the help of others to design every crucial part of the engine except the combustion system, for which he could find no ready expertise. Jumo had earlier developed an axial flow supercharger (with an axial compressor) with the design help of Walter Encke of the Aerodynamische Versuchsanstalt, so Franz enlisted Encke's help to design the new turbojet engine's compressor. The eight-stage axial compressor of the 004 was the first multistage (multiple axial compressor rows) axial compressor Encke designed.[227] Franz's team developed its own combustion system. He opted to build six individual combustion chambers, which because of their smaller size would be easier to test than the single, large annular chamber that Franz might otherwise have preferred. In designing the turbine for the Jumo 004, Franz couldn't call on prior experience with turbo superchargers, as many firms in Britain did, because Jumo had only developed geared superchargers (in which the compressor is powered directly by the piston engine rather than via a turbine in the exhaust stream) in the interwar period. He therefore called on the help of the steam turbine builders at the Allgemeine Elektricitäts-Gesellschaft (AEG) to design the 004's turbine. The engine's elaborate jet nozzle was a product of Franz's earlier, detailed studies of piston engine exhaust jet propulsion. Uniquely in Germany and Britain, the 004's nozzle had a moving center cone that could be used to vary the nozzle's cross-sectional area.[228]

Because Jumo did not have the necessary plant to produce the power needed to test the compressor or turbine separate from the engine, it tested its engine (except the combustion chambers) only as a complete unit. Franz attempted to carry out scale tests of a model compressor, but this failed spectacularly.[229] The engine's conservative design paid off, however, as the first prototype ran in November 1940. Nevertheless, the engine was not sufficiently reliable to power the Me 262 airframe until July 18, 1942.

Franz's preference for calling on the expertise of outside firms might have been in part because Jumo's first turbojet team was small. In the

late 1930s, the firm was overwhelmingly committed to piston engine development, and Franz struggled to find people at the company to work
on his new project. The core of his early turbojet design group were a few
people from his aerodynamics department plus some recent graduates,
who were recruited from universities and colleges.[230] As in Britain, young
recruits were not necessarily favored because they were better trained
or less committed to piston aero-engines, but because they were available for work on the new project. Franz later praised his youthful team
for their work but insisted on the importance of his group's "engineering
talent and scientific knowledge," both products of extensive experience.[231]
Although he had to fight to get new team members, Franz's work did receive support from the company's experimental shop in building its early
turbojet prototypes.[232]

This early situation changed, however, as Jumo's commitment to the
turbojet grew dramatically during the war. From virtually nothing, Franz's
team at Dessau grew to about 500 engineers, scientists, and mechanics in
early 1943 (including staff in the experimental shop).[233] Lucrative production contracts from the RLM radically changed the company's view of the
turbojet. Franz recalled that by September 1944, when the firm's turbojet
was finally being manufactured, the opposition that he had initially faced
at Jumo had entirely disappeared.[234]

The company continued to invest more resources in turbojet development. At the end of 1944, the firm's turbojet development section in the
Otto Mader Works (where development rather than production took
place) reported having 580 people on "pure" turbojet development, 134
who were busy on development relating to turbojet preproduction, and
an additional seventy engaged in development concerned with large-
series production of the engines. For comparison, at the same time, piston engine development engaged 815 men on "pure" development, 186
on preproduction development and twenty-five on large series production. A total of 1,026 workers were developing piston engines as compared with 794 on jet work. These figures include foreign workers employed by Jumo, mostly workers from Belarus ("white Russians"). The
use of foreign workers was approximately 28 percent in piston engines
and 21 percent in jets.[235]

These numbers must be treated with some skepticism because they
were officially reported figures, but they nevertheless indicate a marked
shift in the firm's development focus toward turbojets. Design represented 4 percent of the piston engine development work force at Jumo in

late 1944, experimental work 34 percent and the workshop 42 percent. Jet engine work, in contrast, was about 11 percent design, 34 percent experimental, with only 29 percent in the workshop. The different weighting of design and workshop work for the two engine types suggests the greater maturity of the reciprocating engines and the greater ease of construction and relatively greater amount of design work necessary for the new engines. According to company statistics, the firm's development activity increased during the war: the number of people working in the development section at Dessau increased dramatically from around 757 in January 1938 to 1,600 in early 1942, to 2,760 at the end of 1944.[236] So although the firm's turbojet organization emerged from a long-standing piston engine organization, it reflected the unique problems of the new engines.

Sacrificing performance for quick development by relying as much as possible on existing expertise (whether in house or belonging to subcontractors) was a successful strategy, and it assured the place of the Jumo 004 in the RLM's production planning. Franz was extremely proud that the Jumo 004 was the first German engine to reach production.[237] Indeed, by far more 004 engines were produced than any other engine during the war, as detailed in the first chapter. The engine's design was never seriously changed throughout its life, although its risk-averse design meant that the engine's axial compressor needed more stages (rows of axial blades) than other engines to reach its compression ratio of 3.14.[238]

As the company's production version of its turbojet engine, the 109-004B, began to be manufactured in 1943, the Otto Mader Works began to develop a range of different versions of the 004 turbojet engine, experimenting with fuel injection, after-burning, and more intense combustion. Designers from finished projects also began work on a more powerful engine, the 109-012, and a corresponding turboprop, the 109-022, although neither was ever built.[239] Work on both continued for some time after the war in the Soviet Union. Jumo's program made sense as part of a larger German effort as the firm's engine design was matched by BMW's more ambitious engine design, which owed a great deal to BMW's acquisition of Bramo—and its turbojet expertise—in 1939.

Bramo and BMW: A Good Team

The second German aero-engine company to successfully design and produce a turbojet during the war was the Bayerische Motoren Werke (BMW), but the firm approached turbojets in a very different way than

Jumo. The two large aero-engine firms shared a focus on piston engines, but BMW had more resources than Jumo, so its turbojet program was commensurately bigger. The firm's wartime turbojet program was formed from two earlier turbojet programs, that of BMW and that of the Brandenburgische Motorenwerke (Bramo). BMW designed and produced radial aero-engines, and it was encouraged by the government to take over the country's second radial aero-engine manufacturer, Bramo, in September 1939.[240] Before the merger with Bramo, BMW's development section was already the largest of the major aero-engine firms in Germany. After taking over Bramo, the number of workers in the firm's development department was more than twice that at either of its remaining competitors, Jumo or Daimler-Benz.[241]

The merger took place after (rather different) jet engine programs had been started at both companies, and they were melded into a single program at the new BMW. Bramo's earlier turbojet program was much more extensive and more ambitious than that at BMW. Unlike the larger aero-engine firm, Bramo had eagerly accepted a turbojet development contract from the RLM in August 1938, after the Ministry had threatened to stop funding Bramo's piston engine development earlier that year.[242] Bramo's turbojet program was headed by Dr. Hermann Oestrich, who had earlier worked on both high altitude piston engines and jet propulsion.[243] In fact, Oestrich began working on jet propulsion aero-engines rather than gas turbines. His group first built a motorjet, in which a Bramo piston engine powered a compressor, which pumped compressed air into a combustion chamber where it was ignited to form an exhaust stream that produced thrust. Demonstrating the speed with which such a combination of existing parts could be built, the Bramo motorjet was test flown in a biplane at Rechlin before the end of 1938.[244]

Bramo abandoned its motorjet project early in 1939 and focused its efforts instead on an axial turbojet, which it had calculated would give the best results in terms of weight and size.[245] The idea for the engine came from the independent engineer Helmuth Weinrich, with whom the RLM had put the firm in touch. Weinrich had in 1936 submitted a proposal for a counter-rotating turboprop engine to the RLM, and Bramo's first turbojet design was an ambitious, contra-rotating axial engine to Weinrich's design.[246] It received the government number 109-002 in December 1938.[247] To test features of the 002 design, Bramo initiated work on a simpler axial turbojet. Like Jumo, Bramo relied on the Aerodynamische Versuchsanstalt for its axial expertise, as Bramo's piston engines too used

FIGURE 2.9 The BMW 132 piston engine, front (left) and back (right) sides. The engine pictured is located at the Deutsches Museum's Flugwerft Schleissheim. Several pistons are missing from the engine; they would be arrayed circularly around the edge of the engine. The supercharger on the back side of the engine demonstrates that piston engine components held quite visibly relevant lessons for jet engines. *Photographs*: Hermione Giffard

centrifugal, geared superchargers. The Aerodynamische Versuchsanstalt designed the axial 002's six-stage compressor and its axial turbine.[248] For the 002 engine, the firm used the unique annular combustion chamber that it had been developing from scratch in 1938. Bramo and then BMW continued to develop the combustion chamber throughout the war, and it eventually featured in the BMW 003 axial turbojet as the only annular combustion chamber of any wartime jet engine.[249]

BMW, for its part, also began a turbojet program at the RLM's request in 1938, but it was far more reluctant to divert resources to turbojet development. The firm's managing director, Franz Josef Popp, made sure that the firm's focus remained firmly on improving its piston engines. BMW's early turbojet work was led by Kurt Löhner, who was the head of the firm's research department. In contrast to Oestrich at Bramo, Löhner was responsible for the firm's studies of both new piston engines and new gas turbine engines. In order to quickly assemble a turbojet (thereby diverting as few resources to the new engine as possible), Löhner decided to create one from components that BMW was already familiar with.[250]

His first plan, known as BMW P.3303, reused main components of BMW's earlier supercharger; the turbojet had two centrifugal compressors in series. In addition, it had an axial, hollow-blade, air-cooled turbine that was based on the work of the firm's independent turbine consultant since 1937, Dr. Alfred Müller.[251]

At the end of September 1939, BMW purchased Bramo and with it, its turbojet program.[252] This had important consequences for BMW's turbojet program. Popp wanted to concentrate development of the important BMW 801 piston engine at BMW's main works at Munich-Allach, so BMW's small turbojet program was transferred to join Bramo's group at Spandau (near Berlin), and Oestrich was put in charge of the joint turbojet program.[253] The single jet program thus forged included both the contra-rotating 109-002 (BMW P.3304) and a simple axial engine, soon given the Ministry number 109-003. The 003 engine design incorporated elements of Bramo's earlier axial design as well as elements from BMW's first turbojet; it combined the axial compressor designed by the Aerodynamische Versuchsanstalt, Bramo's annular combustion system, and BMW's air-cooled axial turbine.[254]

In mid-1940, after poor initial results, Oestrich's team at Spandau decided to entirely redesign its simple axial turbojet.[255] In designing its new axial turbojet, Spandau both expanded its own capabilities and called on additional expertise to augment its own. The combined resources of BMW and Bramo meant that the group could afford to build up in-house expertise in order to pursue a more ambitious design. The decision to do so put the firm's turbojet a year and a half behind Jumo's, however, the existence of the 004 made the delay acceptable for the national program. After relying on the Aerodynamische Versuchsanstalt for the design of their first axial compressor, BMW had worked to build up its own calculation section, which allowed it to design the new engine's axial compressor itself using aerodynamic profiles published by the American National Advisory Committee for Aeronautics with corrections according to the Aerodynamische Versuchsanstalt.[256] The new BMW compressor had a higher-pressure ratio than the compressor designed by the Aerodynamische Versuchsanstalt that it replaced, and it added approximately 300 pounds of thrust to the axial engine.[257] Perhaps reflecting the firm's perception of its work, one member of the firm claimed in a postwar interrogation that in designing the 003, BMW's team had received "no significant help from national research organisations whose work in this field was considered ineffective."[258]

BMW developed the fuel burners for its annular combustion chamber at Spandau (a design taken over from Bramo) with the help of the Henschel company and the Luftkriegsakademie in Gatow (near Spandau), which provided the firm with information about, among other things, the fuel-air mixing system developed by Jumo.[259] In the interest of hurrying the 003 engine into production, the RLM later urged BMW to use the superior Jumo fuel regulator rather than its own partially developed system—again having an alternative design in the national program was crucial for reducing the risk of BMW's novel development.[260] By the end of the war, the combustion systems of Jumo and BMW's turbojet engines were similarly efficient, although the service lifetime of Jumo's system of individual combustion chambers was much less than that of BMW's single, annular chamber.

By mid-1942, the 109-003 A-o preproduction design was complete, and BMW began developing a production version of the engine (which became the 109-003 A-1 and A-2);[261] it had earlier been instructed to give up the 109-002 engine to focus its resources on the simple axial 003 engine.[262] Government demand for production of the 003 virtually halted the firm's work on longer-term designs and kept its attention instead on "immediate development problems."[263] To avoid bombing, the turbojet development section was moved in late 1943 from Spandau to Wittringen (Saarbrücken) and then, after the liberation of France in August 1944, to Neu Staßfurt (code name Kalag).[264] Only once the 109-003 design was reasonably complete did BMW's turbojet team turn its design efforts to the new, more powerful engines that the RLM had asked BMW to design, the 109-018 turbojet and 109-028 turboprop.[265] Neither engine got very far before the end of the war.

By 1945, BMW had the largest turbojet development group in Germany. The merger with Bramo was thus beneficial for both BMW and Bramo's turbojet teams. In 1939, about forty people ("designers and engineers") had been engaged entirely on turbojet work at Spandau, sharing the factory's experimental shops and test facilities with piston-engine development; perhaps half of these were working on the 109-002 and half on Bramo's simple axial engine, which would become the 003.[266] By 1942, the RLM had canceled the 002 engine, and the turbojet staff working on the BMW 003 had grown to about one hundred.[267] Under BMW, the dedicated turbojet engineering and design team grew to about four hundred by 1945. At the same time, the number of people working on turbojets in the machine shop increased from about seven hundred in 1942, when piston engine work finally ceased at Spandau, to one thousand in 1945. This exceptional focus on turbojets was in part due to government intervention: piston engine work was ended entirely at Spandau when the RLM took over the firm's development program in 1941 in the wake of a technical crisis brought on by Popp's handling of the firm's piston engine development.[268] Because there were few external sources of trained engineers, the section's growth came largely from recruiting engineers from the firm's piston engine department.[269] The more technically sophisticated and more powerful 003 turbojet lagged behind Jumo's conservative alternative, the 004, but it did reach production before the end of the war. BMW thus joined Jumo as a successful turbojet developer, unlike Germany's other major aero-engine firm, Daimler-Benz, which was denied the chance to learn how to develop turbojets.

Daimler-Benz Makes Time for Turbojets

Daimler-Benz was one of Germany's most successful aero-engine firms in the interwar period, developing piston engines for land, sea, and air applications. Despite the extent of its work, however, the firm's development capacity was much smaller than that of any other major aero-engine maker in Germany at the time. In 1938, Daimler-Benz's development section at Stuttgart employed less than two-thirds as many workers on development as Jumo and less than half the number employed by BMW (which corresponded to only a quarter of the joined Bramo-BMW development staff).[270] Nevertheless, Daimler-Benz too began researching turbojets after being asked to by the RLM in autumn 1938. After a survey of the possibilities including ramjets and turbojets, Daimler-Benz agreed with the RLM that a high-powered bypass turbojet (Zweikreis-Turbine-Luftstrahl Triebwerk) would be the most appealing addition to the national turbojet program.[271] Daimler-Benz's proposed bypass turbojet engine was to produce 3,000 pounds thrust, and it aimed to achieve lower fuel consumption than the other axial engines, bringing the new engine type closer to the standards achieved by piston engines in this regard. It was more complex than the simple axial turbojets under development at Jumo and BMW—both of the 2,000 pound thrust class. Working on a higher-powered engine made sense for Daimler-Benz because it would give the busy firm a longer time to develop the more challenging engine; the first generation turbojets in the national program, the 003 and 004, were expected to be completed and tested first.

Daimler-Benz entrusted the design of its engine to Karl Leist, who had recently joined the firm from the Deutsche Versuchsanstalt für Luftfahrt (DVL), the RLM's research center in Berlin-Adlershof, which did research used directly by the country's aviation industry.[272] At the DVL, Leist had done research on turbine cooling as part of the institution's work on centrifugal exhaust turbo superchargers. Turbine cooling was particularly important to Daimler-Benz's turbojet because the turbine in the bypass engine would operate at higher temperatures than either of the simple axial engines. Researchers at the DVL had tried two systems for turbine cooling. The first system provided cooling air through hollow turbine blades—as used by BMW and later Jumo. The second was a system of partial impingement pioneered by Leist, whereby the turbine blades were only exposed to high-temperature exhaust gases coming from the combustion chamber(s) for a fraction of each rotation. When they were not in the hot gases, the blades were cooled by passing through a stream of cold air.[273]

Leist chose to use his system of partial impingement in Daimler-Benz's turbojet engine, and this heavily influenced the engine's overall design. The cooling system Leist chose required gaps in the combustion chamber arrangement so that the turbine was only subject to hot gas from the combustion chambers (built by Henschel), during 70 percent of its rotation.[274] During the other 30 percent of its rotation, cooling air for the turbine was provided by the stream of air that bypassed the combustion system. To increase the engine's power (in part because only 70 percent of the turbine was extracting heat from the hot gas flow at any given time) and to decrease the engine's fuel consumption, Leist chose a high compression ratio of 8:1 for the engine's compressor. This ratio was only achievable with an axial compressor. Thus Daimler-Benz too consulted the Aerodynamische Versuchsanstalt, as Jumo and Bramo had done, on the design of the engine's axial compressor. Its representatives met with members of the Göttingen institute for the first time on October 4, 1939.[275] At the meeting, it was agreed that a seventeen-stage contra-rotating axial compressor would be required in order to achieve the stated high-pressure ratio in a short distance. The contra-rotating compressor chosen for the engine was, like Weinrich's, made up of two counter-rotating drums. The outer cylinder, geared so as to rotate at half of the speed of the inner one, carried half of the compressor, the compressor stages facing inward and the ducted fan's blades facing outward.[276] The engine's complicated structure is the factor most frequently cited as the reason for its cancellation.[277]

Daimler-Benz submitted a scheme to the RLM for its bypass engine on January 29, 1940, noting that the provision of labor for design and construction would be decisive to the continuation of the work in the firm's small development department. On March 5, 1940, the RLM provided Daimler-Benz with an intent to purchase ("Vorbescheid") for five ducted-fan turbojet engines, given the number 109-007. This was confirmed with a contract on July 29, 1940.[278]

On June 14, 1940, Daimler-Benz gave the Aerodynamische Versuchsanstalt a contract to build and test the engine's contra-rotating compressor. The aero-engine firm undertook to provide the required construction expertise, such as that needed for high-speed gears, which the Aerodynamische Versuchsanstalt lacked.[279] However, because the Aerodynamische Versuchsanstalt was already overburdened with work for other firms and was continually losing workers to conscription (a complaint heard throughout the German aviation industry), it made little progress with Daimler-Benz's compressor.[280] Further exacerbating the problems caused by lack of labor, the complicated contra-rotating compressor that the design called for was hugely difficult to manufacture. Each of the compressor's seventeen stages had different blade profiles and between twenty-four and thirty blades each; it was estimated that the entire compressor would require 3,400 hours of milling time to complete.[281]

Despairing that the Aerodynamische Versuchsanstalt would never finish the compressor, on March 26, 1941, Daimler-Benz gave J. M. Voith of Heidenheim, a nearby mechanical engineering company, a subcontract for the development of an alternate compressor for the firm's ducted fan engine.[282] Voith designed the compressor but was slow to build it, allegedly because of the delayed delivery of a special blade-milling machine from yet another company in Germany.[283] Nevertheless, Daimler-Benz continued to work on the remainder of the engine, while pushing its subcontractors to deliver a compressor for the engine.

The first prototype of the 109-007 was completed on March 26, 1943, well after the Jumo and BMW engines. The engine ran for the first time, with external assistance, on April 1, 1943. Throughout 1943, the firm actively tested and developed the engine, focusing initially on mechanical soundness. On October 2, 1943, the company's first prototype ran self-sustaining for the first time, and by October 19, 1943, the company had finished the remaining four prototypes initially contracted for—all but their compressors.[284] The spurt in turbojet development activity at Daimler-Benz in 1943 was the result of a significant increase in the size of the firm's turbojet workshop in early 1943, which corresponded to the launching of the firm's 605 piston engine into production freeing up many workers for the turbojet.[285]

Daimler-Benz's turbojet engine project was part of the long-term planning of the RLM's Office for Special Power Plants. Helmut Schelp, the head of the office, prized Daimler-Benz's development prowess, and he was willing to accept delays in engine design in order to benefit from it and include the firm in the national program. Showing his continued interest in Daimler-Benz's turbojet engine despite its slow progress, he ordered four additional 109-007 bypass engine prototypes on February 16, 1942.[286] The RLM's Office for Special Power Plants had also granted Jumo and BMW additional turbojet development contracts to help their work along. Schelp may indeed have underestimated the difficulty of designing and developing a turbojet;[287] the effect of the continual demands made by the Office for Special Power Plants was that each aero-engine firm had to decide about how to split their resources among many potential projects.

While the 109-007 bypass engine was held up by its lack of compressors, Daimler-Benz moved on to work on an even more powerful turbojet. The 109-016 engine was to be one of a family of three geometrically similar, nine-stage axial jet engines ("Einheitstriebwerke"), which were, according to Schelp's ambitious plans, to have thrusts of about 4,400, 14,330, and 28,660 pounds respectively. The project was a good demonstration of the office's goal of reaching higher engine powers quickly through the design of similar, increasingly large units rather than entirely new engines.[288] In September 1943, Fritz Nallinger, Daimler-Benz's head of engine development, ranked the 109-016 as more important to the firm than the production of another batch of 109-007 bypass engines (a decision that proved prophetic when the 109-007 engine was canceled the following month), and Daimler-Benz soon ordered parts to be made for the compressor of the first experimental 109-016 prototype.[289] This time, the firm avoided the overloaded Aerodynamische Versuchsanstalt in designing the 109-016 engine's axial compressor and chose instead to consult a different axial compressor expert: Bruno Eckert of the Forschungsinstitut für Kraftfahrwesen und Fahrzeugmotoren (Research Institute for Motor Vehicles and Engines, FKFS) at the technical university in Stuttgart, who also advised other turbojet teams. Eckert worked on a compressor for Daimler-Benz's large 109-016 turbojet engine and may have been consulted on the layout of Voith's compressor for the 109-007 bypass engine.[290]

In mid-October 1943, just at the time that Daimler-Benz's work on turbojets was picking up speed, State Secretary Milch canceled the 109-007 bypass engine, arguing that despite four years of development no "positive results" had been achieved by the firm. Milch was interested in immediate production, and as he observed, Jumo and BMW's turbo-jets were ready to go into production. Milch ordered that all personnel, workshop, machines, and test stands that formed part of Daimler-Benz's jet-engine development capacity be transferred immediately from work on the 109-007 bypass engine to support Heinkel-Hirth's development of the second-generation (so more powerful) 109-011 axial engine, which the RLM had high hopes for, having been promised by Heinkel-Hirth that the engine was nearly ready for production. Heinkel-Hirth's claim proved to be wildly optimistic, however; the engine did not reach series pro-duction during the war, although a few engines were made afterward.[291] Milch further undermined Daimler-Benz's work by ordering, in response to repeated requests from the technical university in Braunschweig, that Leist was to be transferred to Braunschweig.[292] Leist's movements, like those of Oestrich and Wagner, demonstrate the high mobility of person-nel between and within the German aviation industry, research establish-ments, and academic institutes.

In accordance with the RLM's instructions, Daimler-Benz began collaborating with Heinkel-Hirth. The two firms agreed that Daimler-Benz would help produce the 109-011 turbojet engine, and in June 1944, Daimler-Benz officially took out a license for production of the Heinkel-Hirth engine.[293] Daimler-Benz's involvement with the new engine was limited however by Heinkel-Hirth's desire to remain in control of the work and to keep its own production facilities and staff busy.[294] So despite the necessary preparations, the early 1944 plan for Daimler-Benz to pro-duce 150 of the 109-011 engines was canceled when Heinkel-Hirth de-cided to retain production entirely at its own works. In the end, Daimler-Benz built about twelve 109-011 axial engines.[295]

Daimler-Benz's efforts were thus frustrated at every turn, but the firm nevertheless continued to extend its work on the development of turbojet engines. The firm persisted in testing and learning from its 109-007 by-pass engine prototype, for example, even after the engine's cancellation, and by March 1944, the engine had been run for some 152 hours. The firm's deliberate attempt to look out for its own future prospects sup-port's Gregor's argument for the firm's active independence in pursuing its own interests during the war.[296]

Nallinger came to see the cancellation of the 007 and the firm's subor-dinate role to Heinkel-Hirth as an unfortunate curtailment of the firm's activity in a new and important field of engine design. In January 1944, he argued that while the 4,000 horsepower piston engine already under contract from the RLM should be pursued, the firm had to work as "intensively as possible" on gas turbine engines.[297] He appealed to Schelp to extend Daimler-Benz's remit on the 109-011 axial engine to include design and development in addition to license production.[298] In autumn 1944, Schelp agreed that the engine firm should be allowed to use the expertise gained from its work on the 109-007 bypass engine to "more or less independently" develop the 109-011 axial turbojet. Schelp also gave Daimler-Benz permission to work on a turboprop version of the 109-011 engine, the 109-021, which was to be constructed by adding a third turbine stage and propeller to the 011 turbojet unit. Twenty 109-021 turboprop engines were ordered from Daimler-Benz, although none was ever built.[299]

Nallinger's forward-looking insistence that Daimler-Benz should become more deeply involved in turbojet design and development marked a change in strategic focus at the firm and led to an increasing commitment of resources to turbojet engines, even at the cost of piston engine work, in order to develop in-house design expertise. In September 1944, Nallinger appealed to the chairman of Daimler-Benz's board, Wilhelm Haspel, for more resources for turbojet work. He argued that the "failure" of the 109-007 bypass engine was due to a lack of labor, and it would be unbefitting a leading aero-engine firm to make the same mistake again. If the company failed to invest in the new field, he argued, "we will not be a party to these developments for many years to come."[300]

Indeed, Daimler-Benz remained active in turbojet design and production. Like other German firms, it rebuilt its company in the first decades after the war. In the late 1960s, it joined in the formation of the Motoren- und Turbinen-Union München GmbH (MTU), which fused BMW and Daimler-Benz's remaining aero-engine businesses. Later, in 1989, Daimler-Benz created the Deutsche Aerospace AG, which included MTU and Messerschmitt-Bölkow-Blohm (MBB).[301] Messerschmitt-Bölkow-Blohm was created in 1969 from the merger of Bölkow GmbH, the Hamburger Flugzeugbau GmbH, and Messerschmitt Flugzeug-Union Süd, which had bought Junkers Flugzeug and Motoren in 1967.[302] (In 1980, Messerschmitt-Bölkow-Blohm assimilated the remainder of Heinkel; its aero-engine business is described in the next chapter.) So despite failing to succeed as turbojet engine developers during World War II, Daimler-Benz's aero-engine business ultimately came to incorporate the descendants of all of Germany's wartime aero-engine programs. Daimler-Benz's aerospace subsidiary later fed into a European aerospace identity when it became part of the European Aeronautic Defence and Space Company (EADS) in 2000; Daimler-Benz owned 30 percent of the shares. As a leading representative of Germany's aero-engine firms after the war, Daimler Benz's chief competitors and collaborators were American firms, including particularly General Electric, which emerged onto the turbojet engine development scene during the war.

Development in the United States

The United States' jet engine program was transformed by the British offer of the design for the W.2B, or the Whittle jet engine, as American officials called the W.2B engine. Getting the foreign design made it conceivable for the first time that an American firm might produce a jet engine quickly enough for wartime deployment. This possibility was crucial in leading the United States to apply to Britain in June 1941 for complete information about the W.2B engine and in September 1941 for manufacturing rights to the engine. While strongly pushing this effort, General Henry "Hap" Arnold also saw copying the British engine as offering a quick route to a competitive American-designed jet engine. In this, Arnold's focus was on competition in the postwar military market. The realization of his ambition relied on the development skill of the American companies chosen to manufacture the British engine.

The story of the development of jet engines in the United States needs to be no less thoroughly rethought than the story of American jet engine production discussed in the last chapter. The standard story dwells on the American failure to invent the turbojet and its reliance on British designs rather than highlighting, for example, the fact that an American-built version of the W.2B jet engine flew before a W.2B engine had flown in Britain.[303] In focusing on invention and discounting development, the existing narrative of failure betrays a familiar rigidity of thought regarding how new machines are made. In insinuating that the W.2B engine was copied because it was the best solution to building a turbojet (which American firms were supposedly uninterested in or unable to solve on their own), the narrative ignores the serious ramifications for the American firm that developed and produced the engine of importing the British design, thereby locking its efforts into a particular developmental path. Furthermore, the conventional story of inventive failure strongly implies a similar story of developmental failure, yet the expertise of American companies was undoubtedly essential for assimilating and further developing British designs.

That American companies were very successful after the war with jet engines that were not based on British designs suggests that the standard story leaves much unexplained. The British design alone did not lead directly to American jet engine success, nor does the fact that the design was taken over by General Electric imply any lack of creative ability by American firms. By the end of 1941, in addition to General Electric's work on the British W.2B design, there were at least seven indigenous gas turbine engines being developed by American companies: turboprops were being developed by Northrop (Turbodyne), Pratt and Whitney (PT-1 free gas turbine), and General Electric's steam turbine division in Schenectady, New York (TG-100); turbojets were being developed at Lockheed (L-1000, later the J37), Westinghouse (19A), and Turbo Engineering Corporation (booster-sized turbojet); a turbine-driven ducted fan was being developed by Allis-Chalmers.[304] The following discussion of development in the United States gives the most attention to the most contentious case, that of the relationship of General Electric's supercharger division, which produced a W.2B engine derivative, to Britain's aero-engine industry. By putting the decision to import a British engine design back into its development context, this section will demonstrate what the consequences of this decision really were—and what they were not. In so doing, it seeks to establish how General Electric positioned itself to succeed in the postwar market for large jet engines.

The NACA Starts Work

The historiography of the American jet engine almost universally blames the 1922 report by Edgar Buckingham of the Bureau of Standards, titled "Jet Propulsion for Airplanes," for discouraging research into jet engines in the United States. The report was commissioned by the Engineering Division of the United States Army Air Service and published by the National Advisory Committee for Aeronautics (NACA), which did not have its own aero-engine laboratory at the time.[305] In fact, Buckingham's paper wasn't about gas turbine, jet propulsion aero-engines at all. As might be expected from the early date, it examined the potential efficiency of a motorjet (in which the compressor was driven by a piston engine not a turbine). Although he was not optimistic about it, Buckingham left open the possibility that "other schemes for producing the jet" that might be lighter or simpler than a piston engine might be found.[306]

The standard negative interpretation of Buckingham's report by historians like Alex Roland relies on a confusion of gas turbine (a mechanical object) and jet propulsion (a means of propulsion), a projection backward of later practice.[307] In his history of the National Advisory Committee for Aeronautics, Roland wrote that Buckingham accepted "the common fallacy that a turbojet would weigh too much to be practical." According to Roland, Buckingham "and most others who considered the application of turbines to aircraft in the 1920s and 1930s, assumed that such turbines would resemble the heavy industrial turbines then being used in blast furnaces and boilers."[308] Yet the engine modeled in Buckingham's paper included no turbine at all. Far from advising against jet propulsion, Buckingham actually recommended further research into thrust augmentation (producing thrust from an exhaust stream), which he saw as the most promising way to make jet propulsion more economical than existing piston engine–propeller engines. This, as he saw it, was the necessary first step before researching new ways of producing a jet.[309] It is hard to fault Buckingham's conclusions. If Buckingham's report can be accused of anything, it is of perpetuating the separation between studying gas turbines and studying the production of jet streams, a division that remained a feature (uniquely) of the National Advisory Committee's work in the following decades.[310]

Eastman Jacobs, an exceptional aerodynamicist who joined the National Advisory Committee for Aeronautics in 1925, pushed forward work on both gas turbines and jet propulsion (separately) at the organization's Power Plants Division at Langley Memorial Aeronautical Laboratory. In 1938, Jacobs began the design of an axial compressor with Eugene Wasielewski, who had previously been employed by Allis-Chalmers as a power plant expert.[311] Their axial compressor work sought to use information on airfoils (a topic that Jacobs had been researching previously) to predict the performance of axial compressors. (The National Advisory Committee for Aeronautics' report on the results of their study makes no reference to Griffith's earlier work, dating from the mid-1920s, on the same topic.)[312] Far from being stymied by Buckingham's conclusions about jet propulsion, in January 1939, Jacobs encouraged his coworkers to revisit the report and to recalculate its findings based on higher aircraft speeds.[313] Meanwhile, Jacobs began work on "a jet-propulsion system applicable to flight," which led to a wartime research project, a motorjet unit that became known as Jake's Jeep. Jacobs decided not to use a gas turbine in Jake's Jeep in order to avoid unnecessarily complicating things by introducing two experimental elements at once (jet propulsion and a gas turbine).[314] The motorjet research was ultimately unsuccessful, but the expertise that the National Advisory Committee for Aeronautics gained thereby was nevertheless an important building block for postwar work, particularly on axial compressors.[315]

Unlike in Britain or Germany, American groups pursued jet propulsion and gas turbines independently at first. Whereas the British Air Ministry's aerodynamic research establishment, the Royal Aircraft Establishment, was more interested in gas turbines than jet propulsion (it favored gas turbines powering a propeller rather than producing a jet because of the turboprop's potentially higher efficiency), the National Advisory Committee for Aeronautics worked more extensively on jet propulsion than on gas turbines.[316] General Electric's supercharger division in Lynn, Massachusetts, which would manufacture W.2B-like engines, was focused first only on gas turbines for superchargers and only later became interested in jet propulsion.[317] The ultimate success of gas turbine jet propulsion engines has virtually erased the distinction between the study of jet propulsion (a principle of propulsion) from the gas turbine or internal combustion turbine (as a power plant) in the literature. American companies and the National Advisory Committee for Aeronautics are criticized for not combining the two, a fact often explained by the country's greater need for large, long distance transport aircraft rather than small, short distance, high performance fighters.[318] Yet the National Advisory Committee's gradual approach to the union of the gas turbine and jet propulsion—the great insight of the European engine turbojet programs—does not demonstrate that conservatism slowed technological progress,[319] but rather how gradual and disparate developments could still lead to dramatically novel technologies—even easing their adoption by making them more plausible. As Edward Constant argued, somewhat confusingly given the focus of his study of the turbojet on complete engines, technological revolutions can occur at certain design levels without necessarily revolutionizing hierarchically higher levels.[320] Nevertheless, the armed forces expressed ample interest in a new, more powerful aero-engine as soon as they became aware of it.

The Air Force Pushes Development

General Arnold, commander in chief of the United States Army Air Forces (USAAF), first became aware of the practicality of turbojet engines through the hints of the Tizard Mission in late 1940 and through his own visit to Britain in April 1941. As a result, Arnold then requested that Vannevar Bush, who was then chairman of both the newly established National Defense Research Committee and the National Advisory Committee for Aeronautics, organize a group of scientists to investigate the new aero-propulsion system.[321] Turning to the scientific community for weapons-related research was not a new departure for the United States Army Air Forces, which had supported the research into jet-assisted take-off led by Theodore von Kármán at the California Institute of Technology's Guggenheim Aeronautical Laboratory since 1939. (In 1943, the institute was renamed the "Jet Propulsion Laboratory." It was to be operated by the California Institute of Technology for the army.) Bush decided that the problem of aircraft propulsion belonged with the National Advisory Committee for Aeronautics. After considering placing it with an existing committee on auxiliary jet propulsion, he opted instead to create a special committee on jet propulsion to be headed by Dr. William F. Durand. Durand, an eminent retired aerodynamics professor, served as the head of the committee until 1946. Bush remained involved with the jet propulsion committee until Jerome Hunsaker took over as chairman of the National Advisory Committee for Aeronautics in October 1941.

The Durand committee reflected the nature of the United States Army Air Forces' turbojet interests: it included politicians, scientists, researchers from the National Advisory Committee, representatives of the army and navy, and—following General Arnold's request—the country's leading makers of steam turbines: Allis-Chalmers, Westinghouse, and General Electric (Schenectady).[322] Each of the steam turbine firms submitted schemes for gas turbine jet propulsion aero-engines to the committee, which finally agreed with each company on the design that each would pursue. In 1941, Allis-Chalmers embarked on designing an axial ducted fan, Westinghouse began work on an axial turbojet, and General Electric's steam turbine division set about making an axial turboprop.[323] Each company's interest in axial compressors came out of steam turbine practice; notably, none of them suggested designs that included a centrifugal compressor.[324] The Durand committee also supported Jacobs's work on jet propulsion at the National Advisory Committee for Aeronautics, encouraging the research establishment to go beyond research on a test rig and construct an actual aircraft for ground testing. The first ground tests of the aircraft propelled by Jacobs's motorjet were made in July 1942.[325] The National Advisory Committee for Aeronautics continued to support the Jacobs project until April 15, 1943, when the jet propulsion committee voted to drop it.

The American government's application to the British government for information about the W.2B engine took place after the start of the three indigenous jet engine programs at America's leading steam turbine companies. Importing an actual design for a machine represented a very different approach to the problem than that which the Durand committee had adopted. When permission was received for manufacture of the W.2B engine in the United States, Arnold assigned the job to General Electric's supercharger division in Lynn, Massachusetts (rather than its steam-turbine division in Schenectady). The British supported the company's selection "in view of their [General Electric's] standing and their work on turbo-superchargers."[326] The arrival of the W.2B engine design in New York during the autumn of 1941 remained secret, and did not lead to the abandonment of the projects the Durand committee had chosen. When the W.2B engine proved unready for large-scale production, it became an additional, if advanced, strand of development in the American program. General Electric was ready and willing to embark on the development necessary to prepare the W.2B engine for manufacture and to increase the aero-engine's power output. It was only in 1944, with General Electric struggling with its many development commitments, that the American air force would enlist the help of a "main [piston aero-] engine manufacturer," as suggested by the British in 1941, in order to speed up jet engine production.[327]

In addition to the possibility of producing jet engines for deployment during World War II, Arnold and the United States Army Air Forces wanted to use the British engine plans to jumpstart American turbojet design work. On September 20, 1941, an enthusiastic army observer in London wrote to General Arnold that the jet had great potential as an American product: "England does not have the money to put into adequate Jet Propulsion and Gas Turbine laboratories. We have." Thus, Arnold's correspondent argued, "The production of the Whittle [W.2B] engine should be considered as purely a stop gap measure. Development programs should be initiated with the industry on a scale commensurate with our present program for the conventional type of engine development."[328] With America's greater resources (including significant development expertise), he felt that the country's companies could quickly build a serious turbojet program that would soon surpass Britain's. He recommended that the United States Army Air Forces support this by undertaking jet engine development on a significant scale—indeed, a scale equivalent to that for piston engines. Arnold agreed. He replied, "I gave instructions that the Whittle [W.2B] Engine would be completed in every detail so that we would have something that would work, and after they completed that article, they could go ahead at will in an endeavor to improve upon it and secure better design, better construction methods or 'what have you.'"[329] Because the design for the W.2B engine was not, in fact, usable as received, General Electric undertook significant development work of its own on the engine. As the possibility of quickly producing the W.2B engine receded, the goal of the American jet program shifted its emphasis from supporting the British war effort to building an American turbojet industry for the postwar period. Indeed, it would be less the design of the W.2B engine than the logistics of its development and production that would make General Electric a world leader in the field.

The usefulness of the W.2B engine design to American turbojet makers was limited. Because the English requirement that the engine be kept a military secret was conscientiously honored by the United States Army Air Forces, its existence remained known to only a small community before 1943. In fact, in the United States, the W.2B engine project had a higher secrecy rating ("supersecret") than the indigenous American jet engine projects, which were all "secret."[330] The National Advisory Committee for Aeronautics didn't know about the jet airframe that Bell was designing for General Electric's engine until a rumor reached them in March 1942.[331] The airframe was disguised, when necessary, as a piston-engined aircraft by attaching a dummy propeller to its jet engines.[332] Even within General Electric, which had two different divisions developing turbojets (they were in different cities), the two teams proceeded in ignorance of each other's work.[333] Knowledge of the W.2B engine project was severely limited until it was downgraded from "supersecret" to "secret" in summer 1943.[334] For General Electric, the most direct consequence of the high level of secrecy in the United States was that the firm's development team became much more a part of Britain's collaborative jet engine community than of an incipient American jet engine community, which instead remained compartmentalized and competitive.

General Electric

General Electric's first knowledge of the British jet engine came in 1941. In September, Arnold requested from Britain the "complete production drawings" for the W.2B engine, the loan of an already completed engine, one of Gloster's E.28/39 airframes, and "such key engineering personnel as may be spared" to help start the project in the United States.[335] That is not exactly what General Electric got, however. Because Britain had no W.2B engine to spare, General Electric received instead the Power Jets W.1X engine. The W.1X was a unique, non-flight-worthy experimental version of the W.1 engine, Britain's first flight jet engine, which had served as the model for the W.2B engine but was not identical to it. MAP refused to release a Gloster E.28/39 airframe for fear that damages might delay their own testing program, although they later shipped a prototype of Gloster's jet fighter, the F.9/40, to the United States.[336] The British turbojet engineers who were unhappily transported across the Atlantic in the bomb bay of a B-17 beside the disassembled W.1X engine were crucial in reassembling the engine so that General Electric's engineers could test it, but the group did not include the engineer Frank Whittle, who did, however, visit the United States the following year from June to August 1942. The drawings that General Electric received and translated into American measurements and materials came from Rover, which was then still struggling to build its first W.2B development engines.

By working on the development of the W.2B engine, General Electric (Lynn) became a contributing member of the British jet engine development community. Although General Electric was new to aero-engine design and development, the company was nevertheless able to contribute directly due to its previous experience designing turbo superchargers. One area of particular expertise at General Electric, for example, was in turbine blade production and metal alloys that could withstand high temperature and high stress environments.[337]

A key area of collaboration between British firms and General Electric was on turbine blades. Throughout 1942, breaking blades caused British firms headaches, and these failures were exacerbated by constant engine design changes and production problems. On February 14, 1942, the Gas Turbine Collaboration Committee's subcommittee on high temperature materials recommended the formation of a group to focus on blade production. The high temperature materials subcommittee met on March 13, 1942, to hear a report from Bristol's representative in the United States on American manufacturing methods.[338] With British firms struggling to develop and to consistently produce new high-temperature alloys like the promising Nimonic 80 alloy (nickel, titanium, and chromium) from Mond Nickel,[339] General Electric's blades appeared to hold many lessons for British manufactures.[340] British firms were particularly interested in General Electric's Hastelloy B alloy (nickel, manganese).[341] Turbine blade samples were exchanged between General Electric and

Britain's firms and establishments to test the quality of different metal alloys and the effectiveness of different manufacturing processes, which were crucial for they changed the alloys' structure and mechanical properties. In May 1942, a set of British Nimonic 80 blades was sent to the United States for test, and a set of Hastelloy B blades arrived in Britain.[342] Sets of Vitallium blades from American superchargers were also used during development in a W.2B engine and in a Rolls-Royce turbine wheel.[343] British government establishments carried out creep and fatigue tests on both British and American alloys.[344] To compare manufacturing techniques, Power Jets tested Nimonic 80 blades forged by both British and American producers.[345]

Although each country's firms ended up using metals made by national manufacturers (and therefore locally available), the open exchange of information during development remained constant and useful. When Rover had trouble with the high stresses on their turbine blades, for example, the Royal Aircraft Establishment observed in March 1942 that "if it is assumed that Rex 78 material [the British alloy] is not strong enough in the present design ... the use of a better material as a temporary measure, pending a change in the design, must be considered and samples of American Hastelloy B blades have been sent to you for early test."[346] By March 1943, Rover had discovered that locally produced blades performed better than the American Hastelloy B blades it had tried—not necessarily because of the alloy's properties, but because of the "greater accuracy achieved in blade profile."[347] Nevertheless, the exchange proved fruitful.

Starting in September 1942, the American contribution to British development was formalized, when representatives from the Army Air Forces (Colonel Kiern) and United States Navy (Commander Kauffman) joined the British Gas Turbine Collaboration Committee. Work at General Electric was regularly reported to the members of the committee via progress reports. Through information MAP provided, General Electric in turn knew about all of the engines being developed in Britain's various firms. British firms meanwhile learned about developments and problems in all of America's jet engine programs through the combination of the progress reports issued by the Gas Turbine Collaboration Committee, American representatives, and English engineers who visited American plants.[348] British firms were thus familiar with progress on the American version of the W.2B engine as well as all the other government-sponsored projects overseen by the special committee of the

National Advisory Committee for Aeronautics, including General Electric's indigenous project in its steam turbine division in Schenectady. So unlike American firms (including General Electric), British firms knew about both projects at General Electric.[349] The Gas Turbine Collaboration Committee continued to report on American progress to its members up until its nineteenth meeting, on February 1, 1946.[350]

Collaboration across the Atlantic on the W.2B engine became unusually "intimate," in Frank Whittle's words, because the engine type was so new.[351] Every turbojet engine developer faced a legion of unknowns that sharing information and experience might help resolve. Information, engines, engineers, airframes, and turbojet engine parts all crossed the Atlantic between 1942 and 1945. As General Electric developed the W.2B engine design, the American firm became more confident and began to introduce its own mechanical elements into the engine design. Drawings of and data from General Electric's I-16, which was based on the W.2B engine but included General Electric's own improvements, were sent back to Britain in 1943, around the same time that plans for Rolls-Royce's Derwent—Rolls-Royce's improvement on the W.2B engine, came the other way.[352] Machines made and used by both Allies were tested on both sides of the Atlantic. In December 1943, testing of a Bell XP-59A airframe with General Electric Type I-A engines began at the Royal Aircraft Establishment in Farnborough in Britain.[353] The first production Meteor Mk I with Welland engines (which first flew in Britain on January 12, 1944) was sent to the United States in exchange. The Meteor first flew at Muroc Dry Lake in California on April 15, 1944.[354]

The importance of technological collaboration in forging Anglo-American solidarity during World War II is notable, if rarely commented on, and aviation was an important area in which this cooperation took place. The key contribution of the Tizard Mission of autumn 1940 to the jet engine was not primarily in delivering a finished design to the United States (the head of the mission, Sir Henry Tizard, may only have known about the jet on the most general level)[355] but rather in establishing the groundwork for constructive technical exchange between the two Allies and to overcome technical chauvinism on both sides.[356] The wide-ranging and unprecedented technical exchange that took place across the Atlantic during the war, including collaboration on anti-aircraft gunnery, radar, proximity fuses, and atomic bombs, was founded on favorable diplomatic relations and common military interests, but also reinforced these. Sharing the plans for Britain's first jet engine projects built on the tradition of openness begun in late 1940. Like other collaborative developments the Tizard mission stimulated, there was very close cooperation between British and American development teams on turbojet aero-engines, even when secrecy kept that information compartmentalized in the United States.

General Electric undoubtedly benefited from its W.2B engine development program. One key thing it gained was extensive experience with the machining accuracy required for aero-engine development and manufacture. The historian Robert Schlaifer argued in 1950 that General Electric's expertise in supercharger development and production was the reason the firm was the only company outside the aero-engine industry to successfully become a jet engine manufacturer after the war.[357] General Electric's earlier experience in manufacturing superchargers surely helped it to make the transition to aero-engines more quickly than other firms, like Westinghouse (which left the field by 1960),[358] that were wholly new to the aero-engine business. Nevertheless, General Electric still had to learn how to manage the complexity and unique challenges of the development of large turbojet engines, and thus its wartime experience was extremely valuable.

Starting with the W.2B engine design offered a way for General Electric to reduce the risks of a novel aero-engine development, although pursuing the W.2B engine also meant that the firm followed a particular path of development rather than other potential paths. General Electric was fortunate in that it did not have to choose between its centrifugal and axial developments early on, so the company benefited from developing both. After the war, General Electric stopped its centrifugal engine line, deciding that axial engines had the potential to be smaller, lighter, and more powerful than centrifugals. When Allison took over production of General Electric's centrifugal jet engines after the war (Allison also began developing turbojets for the navy, suggesting that it was difficult to produce the engines without developing them too—or conversely, that production was a good route to development experience), General Electric was left to devote its efforts to development alone. When General Electric combined its two turbojet projects to form a single division, the lessons learned from developing the W.2B engine were passed on even if the design wasn't.[359]

One should not underestimate the importance to General Electric of the intimate transatlantic exchange that grew up around jet engines, but nor should one assume that it was all one way. Indeed, the American firm's contributions to development in Britain make nonsense of the argument that General Electric passively copied the British engine. Developing the W.2B engine in collaboration with Britain's aero-engine industry gave General Electric valuable experience, but General Electric certainly did not rely on British sources for all of its turbojet aero-engine designs.

It is to General Electric's credit that its company culture and its skills in development served it well through the many years in which it struggled to develop and produce the United States' first service turbojet engines. While General Electric's contributions to the British jet engine program demonstrated that the American firm was good at development, the firm's wartime record of disorganization and missing deadlines demonstrated how much it still had to learn. As the pressure to deploy an American jet fighter during the war receded, General Electric was given the time and resources necessary to undertake the "slow and expensive" business of creating an efficient aero-engine development and production organization that came to rival that of Britain's Rolls-Royce, ultimately the chief beneficiary of Britain's wartime work.[360]

Conclusion

The turbojet did not create a new industry, although it presented a chance to alter existing relations within the aero-engine industry, as existing firms' expertise, resources, and personnel were applied to gas turbines. This can be contrasted with the emergence of the American automobile industry from the bicycle and carriage industries, in which the bicycle and carriage industries provided elements, skills, and concepts that shaped a new transportation industry that consisted overwhelmingly of newly founded firms.[361] As has been shown in this chapter, the development of the turbojet took place within and was achieved by an existing industry.

By the end of the war, all of Britain and Germany's prewar aero-engine companies had begun designing turbojets. Those that succeeded used their expertise in and resources for aero-engine development effectively to develop the new type of aero-engine. All of the aero-engine firms had a stake in being good at innovation; flexible firms who could act with "confidence under conditions of uncertainty"[362] were the most successful. Although in many cases, as with axial compressors or combustion systems, firms had to acquire new expertise in order to develop turbojet engines, firms' more general skills and methods provided a scaffold for the new.

Dealing with the unavoidable uncertainty surrounding turbojet development was crucial to producing a good engine quickly.

It was the firms of the aero-engine industry in Britain and Germany and General Electric in the United States that controlled when and how the turbojet came into use. Firms had a range of incentives to begin developing the new engine type, but all of these depended on the firm's positive judgment of the engine's plausibility and competitiveness. A crucial part of each firm's decision to develop turbojets was the future commercial prospects of its piston engines, rather than any theoretical allegiance to them. It was not when the turbojet was proved to be possible, but when piston engine production was well in hand (and thus no longer demanded development priority), that the aero-engine industry embraced the new engines. For no firm was the changeover to jet engines without some financial risk, but firms with stronger incentives (including government contracts) to make the switch took up the new engine earlier and with more dedication than others.

Being a successful aero-engine firm was not a guarantee of success in developing turbojet aero-engines, but it brought significant advantages because of the intricacy of aero-engines and the accuracy of their manufacture. (The only successful firm outside the aero-engine industry, General Electric, had however experience making aviation products.) Most aero-engine firms had among their staff some design experience that touched on aspects of gas turbine design, whether compressors, turbines, or exhaust nozzles. In addition, the industrial know-how of these firms, which was much more than a collection of machines and designs, provided method, self-confidence, and a network of contacts across an industry dedicated to making new things. These connections, which included people, resources, designs, machine tools, and factories, explain why insiders were so much better placed than outsiders to make engines that could actually be manufactured, even in the case of entirely unknown machines like axial turbojets. As this chapter has shown, it was from the at times overconfident aero-engine industry that the most ambitious and elaborate gas turbine schemes emerged, and these projects—whether or not they were developed further—represented an immense creative effort. Restoring "failed" early turbojets to the record rather than just focusing on the successes gives us an understanding of this larger innovative picture.

Yet being an established firm could also be a liability. Continuities could be toxic if they led firms to developmental dead ends. The commitment to a single element, like Bristol's heat exchanger or Daimler-Benz's partial impingement system, could impact the direction of an entire development. Each firm's specific resources, both physical and mental, not only limited the number of projects that a firm could be working on at a given time but also crucially shaped how a firm made engineering compromises. The many different designs developed early on testified to the fact that different design teams, with different reserves of experience and different engineering aesthetics, were unlikely to make the same design decisions. The trade-offs that a given team made tended to favor elements or adaptations of elements that its members were familiar with, adopting the simplest solutions in areas in which the team was least confident, even when that created more complicated problems elsewhere.

Because the gas turbine was and is seen as centered on the turbine, is has become accepted that the steam turbine industry should have played an important part in its development. Yet the role played by steam turbine companies in the design and production of turbojets was very different in Britain, Germany, and the United States, and most such firms were ultimately unsuccessful, as the advantages of their experience with developing steam turbines were outweighed by the disadvantages of tending to design heavy, immovable plant. The record of the steam turbine firms that worked on gas turbines suggests that, as Robert Schlaifer argued, the aero-engine industry was crucial to the early development of turbojet engines because it was able to develop the turbojet aero-engine more rapidly than any other type of firm.[363] In contrast to accounts that stress invention as the key activity in the creation of new machines, Schlaifer argued that "experience in the development of conventional aircraft engines gives a very great advantage in the development of even a completely novel type of aircraft engine." He argued that "brilliant theoretical engineering will not replace either experience or adequate experimental facilities."[364] Aero-engine firms competed on more than just price and novelty. As Philip Scranton argued in a paper in *Business History*, price was just "one aspect of an elaborate set of socio-technical relations ... within the world of enterprise."[365] Innovation should be understood as one among many skills and strategies of competitive firms. Strength in the development of complex machines is a crucial national capability.

It was no accident that aero-engine firms made the world's first turbojet engines. Turbojets were entirely new machines, using a wholly different principle to propel aircraft than the propeller then in universal use. Yet in important respects, particularly to do with development as discussed above, turbojet aero-engines were not that different from piston aero-engines, just as companies that made turbojets were not that different from companies that made piston engines. Further underscoring this argument is the history of those institutions that invented turbojets but did not manage to produce them because they never successfully developed production models, Power Jets and the Heinkel Aircraft Company, which will be told in the next chapter.

CHAPTER THREE

Inventive Institutions

The standard story of the jet engine identifies two companies as the pioneers of the turbojet: Power Jets Limited and the Ernst Heinkel Aircraft Company. The two companies are famous because they were the places that the most prominent turbojet inventors, Frank Whittle and Hans von Ohain, worked. Neither firm has been discussed so far in this book because neither was directly involved in the large-scale production of turbojets. These two firms specialized in producing something different from service engines (and something of great cultural value): invention.

Inventive institutions, or companies or parts of companies, that remain competitive through invention specialize in deriving commercial advantage from novelty rather than production. The objective of both Power Jets and the Ernst Heinkel Aircraft Company was inventing a turbojet—neither was in a position to produce turbojet engines or to create a design that was particularly easy to manufacture. Nevertheless, although both firms were new to aero-engine design, the turbojet took a central place in the identity of both Power Jets and the Ernst Heinkel Aircraft Company.

Firsts were a key claim to commercial importance, so it is no accident that Power Jets and the Ernst Heinkel Aircraft Company encouraged and promoted the work of Whittle and von Ohain rhetorically as well as technically. Largely because of this promotion, we know much more about the work of the unsuccessful turbojet innovators Power Jets and the Ernst Heinkel Aircraft Company than about the many companies discussed in the previous two chapters. Although the two pioneering firms both had a strong identity as pioneering turbojet firms, both went out of business before 1960. This seems to support the argument that the institutional requirements for producing firsts was opposed to those needed for successful development and production; it was only by becoming another type of company that the firms could hope to succeed as manufacturers. Neither did so.

This chapter gives the first account of Power Jets and the Ernst Heinkel Aircraft Company as inventive institutions, an approach that changes not only how we understand them as organizations, but also how we evaluate the work of the inventors associated with them. It places the work of the well-known inventors into a broader context and looks at these institutions throughout the entire war. This makes sense because neither company was reducible to one man, no matter how important, nor did the complex concerns of the firms and their inventors exactly align. Thus rather than focusing on individuals, this chapter will examine each firms' activity also after the main early forces behind turbojet development—Whittle and Heinkel (not von Ohain)—had become less central. Filling in blind spots in the standard accounts of both of these institutions reveals that the trajectories of the work of the two most famous turbojet inventors was very different from what the standard story implies.

The successes attributed to Whittle and von Ohain in the literature contrast with the failure of their work to impact turbojet practice in the long run. The key to this riddle lies in the institutions where they worked. This is clear in the distance between the familiar heroic account of the inventors as independents and the account given in this chapter culminating in the failure of their institutions. Indeed, the most famous work associated with these individuals was fundamentally shaped by their firms: each company's concerns and capabilities, its institutional security, and its access to resources for turbojet work. The new focus this chapter adopts is crucial if we want to recover what Power Jets and the Ernst Heinkel Aircraft Company achieved or failed to achieve as inventive institutions associated with the turbojet, which has been entirely overshadowed by the claims made for the inventors that worked for them.

In taking inventive institutions as its theme, this chapter asks us to reconsider the importance generally attributed to technical firsts as markers of technical achievement. Earlier histories of the jet have made much of firsts, whether first flights, first production, or first deployment. The assumption has often been made that there were obvious and close connections between being first to fly and first to produce service engines. The timing of Germany's first turbojet flight, first jet deployment, and its large wartime jet production numbers all seem to support the argument for German superiority. Yet, as this chapter will discuss, the first turbojet to fly in Germany was of an entirely different type than the first to be produced in that country; von Ohain's engines never powered any Luftwaffe squadrons. In Britain, the first turbojet to fly was of the same type as the first to be produced in quantity, but the centrifugal design Whittle pioneered did not prevail (in Britain or the United States) long after World War II.

This chapter seeks to explain the fact that the two institutions associated with inventing the turbojet failed to get their engines into the air in substantial numbers. Indeed, this chapter argues that the nature of the inventive institutions where each inventor worked was of crucial importance to the future of the inventors' ideas. Power Jets' failure to become a commercial enterprise during or after the war contributed to the speed of the abandonment of Whittle's ideas by the British aero-engine industry, whereas the often-celebrated nature of the Ernst Heinkel Aircraft Company was the major reason that von Ohain's wartime work had no lasting impact on turbojet design. Yet we know primarily about these two turbojet inventors; why the apparent contradiction?

Inventors

The elements or ideas embodied in a new machine may exist long before the machine is conceived or practically pursued. This fact certainly doesn't detract from the work of the individuals that ultimately promote them.[1] Conversely, and equally importantly, the fact of such individuals doesn't detract from the importance of the work necessary to make the idea into a real, practical machine. Although this chapter is not concerned with the question of who the first man to conceive of a turbojet was, a brief summary of the claims made for the key figures that feature in this chapter are necessary in order to show just how different the conclusions of this chapter, with its novel focus, are.

The prominence and number of accounts of Whittle written by historians, enthusiasts, and aviation writers speak to his status as the British inventor of the turbojet.[2] Whittle proposed the desirability of developing an air (rather than steam) turbine for aircraft propulsion in 1928. He did so in his thesis, "Future Developments in Aircraft Design," which he wrote at age twenty-one while a flight cadet at Cranwell. Based on a mathematically based proof that an "air turbine" could be more efficient than an internal combustion engine, his thesis argued that increasing aircraft range would be achieved only by "careful streamlining, and better structural design and also more efficient prime movers." Whittle suggested that achieving faster flight could best be done by flying at higher altitudes (where there was less air resistance), at which the air turbine was the best-suited prime mover.[3]

After graduating, Whittle continued to pursue both invention and flying. From Cranwell, he joined a flight squadron before moving to the Royal Air Force's Central Flying School in Wittering, where he studied to become a flying instructor. One of his flight instructors at Wittering, W. E. P. Johnson, particularly encouraged Whittle to pursue his idea for a gas turbine jet propulsion aero-engine. In late 1929, through Johnson's intervention, Whittle was able to present his idea to the Air Ministry's top scientists. When it was turned down at the advice of a senior government scientist, A. A. Griffith, Johnson, himself trained as a patent lawyer, urged Whittle file a patent. Whittle applied for a patent on "Improvements relating to the propulsion of aircraft and other vehicles" on January 16, 1930. The patent was granted on April 16, 1931.[4] Whittle applied for a second patent on October 16, 1930, for "Improvements relating to centrifugal compressors and pumps." It was granted on May 7, 1931.[5] These were the first of the thirty-odd patents that were assigned to Whittle throughout his career. In 1931, he took his ideas (backed by patents) to two commercial firms, the British Thomson-Houston Company and Bristol Aircraft Company, but neither was interested or willing during the depression to contribute the required finances to develop the ideas, estimated at £30,000.

Meanwhile, Whittle's air force career continued to advance. Commissioned a pilot officer on leaving Cranwell in July 1928, he was promoted to flying officer in 1930. During 1931, Whittle was posted from Wittering to the Royal Marine Experimental Aircraft Establishment in Felixstowe, where he earned a reputation for brilliant and risky flying but by no means gave up inventing.[6] In 1932, he joined the Royal Air Force Officer's Engineering Course at Henlow. Very successful there, he was promoted again to flight lieutenant in 1934. In July, the young flight lieutenant and promising engineer was sent by the Royal Air Force to Cambridge to pursue his training as an engineer, despite the fact that the scheme had officially been discontinued. At Peterhouse College, Whittle continued his formal training as an engineer. While there, he also joined the University Air Squadron, which allowed him to maintain his pilot's license. Whittle's studies at Cambridge culminated in first class honors in the Mechanical Sciences Tripos of 1936, and the Royal Air Force agreed that Whittle could stay on at the university for a postgraduate research year. Whittle spent this year working in large part on his turbojet engine design. His supervisor at Cambridge was Professor Melville Jones, who was head of the university's recently established aeronautics department.[7] When Whittle left Cambridge as a qualified engineer in mid-1937, he was posted not to a squadron, but to a fledgling private company, Power Jets Limited. As the company's chief engineer, still being paid as a Royal Air Force officer, Whittle devoted all of his time to designing and developing a turbojet engine. Despite having worked at Power Jets, he was promoted in 1938 to the rank of squadron leader. Whittle later thanked the Royal Air Force for giving him "the training which made possible the jet engine."[8] Many contemporaries praised Whittle's ability, his enthusiasm, and his energy—so it's not surprising that the Royal Air Force sought to develop these.[9]

Claims to Whittle's importance as an inventor rest on three arguments: that Whittle's proposal for a gas turbine jet propulsion engine was the first practical proposal for a pure turbojet (and was supported by a patent); that Whittle inspired work on turbojets in Britain; and that Whittle was the first in the world to run a centrifugal turbojet aero-engine that burned kerosene (as later turbojet engines did), which occurred at Power Jets on April 12, 1937. As discussed below, claims to the significance and fundamental nature of Whittle's work were central to the policies and practices of Power Jets, the company founded to develop his ideas.

Hans von Ohain, due largely to American advocacy after the war (as discussed in the next chapter), joined Whittle as coinventor in the standard story: the second man who independently invented the turbojet aero-engine and achieved the idea's practical execution. The son of an aristocratic family, von Ohain had the idea for a jet engine while pursuing a PhD in physics and aerodynamics at the University of Göttingen. Unlike Whittle, he had no personal experience as a pilot or any interest in the turbojet except as a "device" that was an "elegant" solution to the limitations and discomfort of civil piston engine flight (which he had the privilege of having experienced as a passenger at a rather early date). Von Ohain got his first experience of the importance of patents to new ideas when he applied to patent the results of his PhD work on an optical microphone, in June 1934. He later sold the work to Siemens. After finishing his doctoral studies, von Ohain pursued his jet engine idea on his own, paying a local car mechanic to build an experimental model from sheet metal. This first model never ran self-sustaining, but was nevertheless conceptually important. The professor, whom von Ohain was working for at Göttingen, Robert Pohl, encouraged the young physicist to work further on his jet engine design, even allowing von Ohain to test his prototype engine (loud and dangerous) in the Physics Institute's courtyard.

In order to make it possible for von Ohain to pursue his idea for a turbojet further, Pohl offered to write the young physicist a letter of introduction to an industrialist. Von Ohain selected Ernst Heinkel, who immediately hired him. At Heinkel's company, von Ohain built a prototype radial turbojet unit, the HeS1 (HeS meaning Heinkel Strahltriebwerk or Heinkel Jet Engine), which was the first turbojet engine to run self-sustaining in the world. It ran in March 1937, although it burned hydrogen gas rather than a liquid fuel like Whittle's kerosene.[10] (The fuel choice undoubtedly reflected each man's training.) Von Ohain, now working on his turbojet engine in a commercial environment, patented his work throughout.[11] The two inventors' early engines were notably different—not only in terms of fuel but also in their layout: von Ohain's engine had a centrifugal compressor (like Whittle's) but also a centrifugal (rather than axial) turbine and so looked roughly symmetrical.

In terms of putting engines into the air, both Whittle and von Ohain were pioneers, although von Ohain's work, coupled to a commercial company and Ernst Heinkel's drive, moved ahead faster. Von Ohain's engine powered the first turbojet aircraft in the world, the experimental Heinkel He178, on August 27, 1939. A later engine designed by von Ohain, the HeS8, powered the prototype of Heinkel's He280 jet fighter on March 31, 1941. Whittle's first flight engine, the Power Jets W.1, powered the experimental Gloster E.28/39 airframe in the first jet flight in Britain on May 15, 1941.

Whittle and von Ohain were initially a few years ahead of the competition, but by 1940, they were competing with commercial aero-engine designers. In Germany, the first axial turbojets ran during 1940: a precursor to the BMW 003 ran in spring 1940 and the Jumo 004 ran for the first time in October 1940.[12] The first axial turbojet in Britain, the Metrovick F.2, ran in December 1941.[13] Although Herbert Wagner of the Junkers Aircraft Company was the first to develop a scheme for an axial turbojet in 1935–36, the engine built based on his designs did not run until 1942 (at the Ernst Heinkel Aircraft Company). Within three years of the experimental operation of the very first Whittle and von Ohain turbojets they thus had many competitors, a point central to the histories of the institutions this chapter discusses.

Power Jets

Because of the fame of its chief engineer, Frank Whittle, Power Jets is the best-known British institution associated with the turbojet. As is clear from its leading role in accounts like that of Postan, Kay, and Nahum, the firm is considered to have been central to turbojet development in Britain. Despite the surfeit of accounts that discuss Power Jets, however, we know very little about the company apart from Whittle. Indeed, the firm is often treated as an extension of the officer, whose work it was founded to realize. Yet the two stories are not identical. This becomes especially clear after 1942, when Whittle's involvement in the company began to lessen, while the company's turbojet activity continued to grow.

There are two major reasons that we know so much about the short-lived company and in much greater detail than about any other company that made gas turbines during the war. The first is because Whittle's own account of the firm in his memoir (published in 1953) was so detailed and so widely known. It is also an important source in this account. The second is because the company's end in 1946 allowed its senior figures to discuss the intimate details of the firm's finances with the business historian Robert Schlaifer a couple of years later, and he included a detailed financial history of the private company in his history of the turbojet published in 1950—the only such section in his book.[14] Other historians, like Postan, Constant, and Pavelec, have treated the firm's financial circumstances in less detail, but all have emphasized the company's private status and how hard it had to work to attract funding.[15] All of the major historical accounts of the turbojet and some others by enthusiasts lay stress on Power Jets' antagonistic relationships with the British Ministry of Aircraft Production (MAP) and with the first firms chosen to produce its engines, the British Thomson-Houston Company and Rover.[16] All of these facts change complexion, however, when seen in a larger context.

So preoccupied are most accounts with Whittle that they tend to be vague about the years when Whittle became less central to the firm. For example, because of the standard story's desire to emphasize Power Jets' demise as a personal attack on Whittle, the standard narrative basically jumps from the time when Power Jets gave up on collaboration with Rover in 1942 to the moment of the firm's nationalization in 1944. The nationalization, a key event in the firm's short history, is frequently confused with the point when the nationalized company was turned into a government research establishment in 1946.[17] This emphasis corresponds roughly to Whittle's visibility in the story, but little else. The new story told here makes clear not only the meaning and relation of these events, but also what the company was doing at those times when it slips out of the historiographical limelight.

Antagonism notwithstanding, government support was crucial to Power Jets' existence, and the firm benefited greatly from its relationships to other firms, which MAP encouraged. It was only from 1940 onward, with government funding, that the firm was able to build up a huge staff and significant facilities, reaching almost one thousand employees in 1943. Over time, the company increasingly found itself in an atypical position with regard to the aero-engine industry and the Air Ministry's research establishment, which were both becoming increasingly involved with the new engine type. This fact was intensified by the firm's failure to get its second engine design, the W.2/500, produced by another firm. In a significant recasting of the nationalization story, this account will argue that nationalization was advantageous for the firm because it served to preserve the company as an operational unit. In fact, nationalization allowed the firm briefly to expand its capabilities and stretch the boundaries of what an inventive institution could be, even if the postwar aero-engine industry then united against it. The story of Power Jets has too much been seen through Whittle's eyes, with the assumption that the firm deserved special treatment because of its role as a pioneer. This section will show that Whittle and Power Jets' moral claims for their right, as inventors, to be free from all oversight—even from their paymasters, ultimately alienated both the government and the commercial aero-engine industry, which enjoyed no such latitude. Power Jets' uncertain institutional position and its clumsy attempts to advance its commercial interests were the firm's undoing in a context where it was very far from being the only institution able to design and develop a turbojet in Britain.

Small Beginnings

Power Jets was established on January 27, 1936, by an agreement between Flight Lieutenant Frank Whittle ("The Inventor"); the president of the Air Council; Rolf Dudley Williams and James Collingwood Burdett Tinling ("The Vendors," Whittle's representatives); and the financier Oswald Toynbee Falk and Partners ("The Firm").[18] The Air Ministry was a party to the agreement because Whittle was a serving officer of the Royal Air Force, and for that reason the Air Ministry retained crown user rights to all of Whittle's patents, which made them useless as commercial tools during the war. (If any of Whittle's patents had been declared secret when they were granted, the Air Ministry would have retained all rights; since they were not, the government retained only the right of free use. Power Jets was granted the commercial and foreign rights on Whittle's patents, for which the firm was required to pay a fraction of royalties received to the government.)[19] Because of the instrumental role played by Williams and Tinling in helping Whittle to find the funding required to exploit his ideas, Whittle granted each a 22.5 percent interest in his patents, which became the new firm's sole assets. In order to establish the company's intellectual property, given that Whittle's initial patent had already lapsed, Whittle filed for several new patents during 1935 and early 1936, while he was still studying at Cambridge.[20] O. T. Falk and Partners, a branch of the investment-banking firm O. T. Falk and Company, which made high-risk investments in new companies, was responsible for the firm's finances. It pledged an initial £2,000 to the firm and undertook to raise more funds for Power Jets from its own capital and from interested investors.[21]

On March 19, 1936, when Power Jets was officially registered, the company consisted of only a board of four directors and one alternate director. At the company's first meeting on March 26, 1936, Whittle was named its honorary chief engineer and technical consultant (seconded from the Royal Air Force, which continued to pay his salary).[22] The firm's board of directors consisted of two members of Falk and Partners, Lancelot Law Whyte and Maurice Bonham Carter, and the neophyte business associates, Williams and Tinling, who were both nominated by Whittle. M. L. Bramson, a pilot, engineer, and aeronautical consultant, who had put Williams and Tinling in touch with O. T. Falk and Partners, was the firm's technical advisor. Bramson authored the independent engineer's report on Whittle's ideas in October 1935 that was crucial in getting Falk's support,[23] and until January 25, 1940, he served as Bonham Carter's alternate on the board.[24] Bonham Carter, a Partner at O. T. Falk, held a number of business directorships in addition to that of Power Jets, including several in the aviation industry.[25] Along with Whyte, who became the first managing director of Power Jets, Bonham Carter was also a director of an equally promising company, Scophony Limited, a similar firm founded in the early 1930s with O. T. Falk's help to develop a mechanical television based on the patents of George William Walton.

At first, Power Jets was reliant on the private capital attracted by Falk and Partners.[26] Between 1935 and 1939, the firm's shareholders came to include friends, family, and Power Jets employees as well as experienced investors, industrialists, and other private companies. Early in 1937, Whyte appealed to Sir Henry Tizard, "the highest opinion outside the Government," for support in attracting finance. The letter that Whyte received from Tizard in June 1937 stating the well-known scientist's support for the development of a gas turbine aero-engine and his faith in Whittle's abilities proved decisive in Whyte's further efforts to raise funds.[27] The firm's largest shareholders were William and James Weir, the second generation of Weir brothers in Scottish industry, who invested £5,000 between November 1937 and spring 1939. According to the history of the Weir group, the investment was not primarily meant as a commercial one. Instead, "the investment was made, like others of Weirs' investments, without regard to the possibility of profit but simply as a contribution to technical progress."[28] So short of funds was the young firm that shares were accepted in lieu of cash payment by the firm's managers and by its key subcontractors. By May 1939, the British Thomson-Houston Company, Power Jet's first major subcontractor, held in this way £4,500 worth of shares, making the later maligned company Power Jets' second-largest shareholder.[29] In July 1939, Power Jets' total income (including shares granted for services rendered) since March 1936 came to £19,880. Of this, £16,640 was from private sources, while 16 percent of the firm's income had come from Air Ministry contracts paid in 1938 and 1939. With these contracts, the Air Ministry increased the support that the Royal Air Force had already given the firm by sponsoring Whittle's training as an engineer and agreeing to his secondment to the company.

Power Jets raised an additional £3,830 privately after July 1, 1939, primarily through additional purchases by its original investors because its ability to fund-raise was limited, after the Air Ministry began taking an official interest in the project, by the strictures of the Official Secrets Act. Funding from the Air Ministry quickly overtook the amount of private capital invested in the firm.[30] Schlaifer argued that this couldn't have been a surprise to O. T. Falk and Partners and its investors, who should have understood that "the entire undertaking would ultimately have to be a partnership with the state."[31] Despite later resistance from shareholders, the extent of indirect government support from early on suggests, as Andrew Nahum has argued, that the company really was "a surrogate official venture" from the start.[32]

In order to build its first prototype, Power Jets, which had no workshop of its own, enlisted the British Thomson-Houston Company, a major electrical engineering firm and steam turbine maker based in nearby Rugby, as a subcontractor. The tiny Power Jets concern selected the British Thomson-Houston Company for fear of losing control by involving an established aero-engine manufacturer in its work.[33] The electrical firm undertook the design drawings and construction of Power Jets' first experimental turbojet engine, known as the Whittle Unit or WU. (Whittle and Johnson had approached the firm previously in 1930, when it had declared itself interested but unable to offer financial support.) In a contract signed on March 4, 1936, Power Jets promised the Rugby company manufacturing rights to its first production engine, implying not only confidence in quick success but also the expectation that Power Jets would likely not build the engines itself.[34]

A year into its existence, Power Jets had let a subcontract but still had neither employees nor facilities. Indeed, Whittle, Power Jets' only technical member, designed the company's first experimental unit from an office at the British Thomson-Houston Company. The tiny Power Jets continued to rely on the Rugby firm's engineering and drawing staff until well into 1940.[35] Nevertheless, Whittle retained close control of the work done by Power Jets' subcontractor; he required all of the design drawings to have his signature on them before being sent to the shop.[36] Undoubtedly due to Whittle's influence, the experimental gas turbine built for Power Jets under Whittle's supervision was much lighter and more precise than anything the British Thomson-Houston Company was accustomed to building in the course of steam turbine manufacture.[37] It ran for the first time at Rugby on April 12, 1937.[38]

In August 1937, after the WU's first tests, the director of the British Thomson-Houston Company, H. N. Sporborg, decided that the jet unit could no longer be safely tested directly outside the firm's steam turbine factory.[39] Power Jets consequently negotiated to rent part of the British Thomson-Houston Company's disused foundry, the Ladywood Ironworks in Lutterworth. It was there, in Lutterworth, that Power Jets began in early 1938 to establish its first factory.[40] A month later, in September 1937, expenditure on the firm's first significant items of equipment and testing apparatus were approved by its board at a cost of about £156. In December 1937, the firm hired its first employee, Victor E. Crompton.[41] The Air Ministry gave the company its first payment of £1,000 in May 1938.[42] By December 1938, Power Jets had five employees, none of whom, however, were engineers.[43]

Through the end of the 1930s, Power Jets continued to rely on the expertise and facilities of other companies. In addition to working with the British Thomson-Houston Company, Whittle consulted numerous leading British firms on aspects of each of the first unit's main components. He spoke to the Hoffman Company in Chelmsford about bearings, Alfred Herbert of Coventry about machining the WU's rotor shaft, High Duty Alloys in Slough about the unit's compressor impeller, and Firth-Vickers of Sheffield about forging a turbine with integral blades, disc, and shaft from its Stayblade alloy. Laidlaw, Drew and Company of Edinburgh designed the WU's single, elongated combustion chamber.[44] Whittle also frequently traveled to London to consult with Bramson and Power Jets' directors.[45]

Whittle's contacts outside of industry were equally important to his work. The compressor rotor of the WU was designed with the help of Arnold Alexander Hall, a colleague of Whittle's at Cambridge.[46] In late 1936 or early 1937, Whittle initiated contact with Hayne Constant and A. A. Griffith at the Royal Aircraft Establishment, whose support was crucial in securing government funding for Power Jets in 1937.[47] Because one of the research establishment's chief functions was to assist industry with aeronautical design work, Power Jets' contacts with the Royal Aircraft Establishment were standard. Although it was busily designing axial gas turbines for the Air Ministry, the Royal Aircraft Establishment had a reservoir of knowledge relevant to the design of a centrifugal gas turbine aero-engine from its earlier work on centrifugal exhaust gas turbines and combustion. According to Constant, who later oversaw the establishment of a turbine section at the Royal Aircraft Establishment in 1941, the research team was engaged in research on "the same problems as Whittle was attacking from the design point of view."[48] Despite the positive relations that it had with many organisations, as Power Jets began building up its own skills and capabilities, it became increasingly dismissive of the competence of its earlier collaborators.

In August 1939, Power Jets received a government contract for a flight engine, the W.1, and the firm again subcontracted the construction to the British Thomson-Houston Company. The next year, in June 1940, Power Jets designed a test engine from un-flight-worthy W.1 engine parts, which was called the W.1X. The engine was also built by the British Thomson-Houston Company. Power Jets' growing staff designed and built a range of test rigs, but until the W.1X engine arrived from the British Thomson-Houston Company on December 11, 1940, the firm had only one complete engine (the WU) to use for testing. After the arrival of the W.1X, the firm used the engine for experiments, including installation in Gloster's experimental E.28/39 airframe.[49] Remarkably, the firm continued to modify and test the original WU until February 22, 1941. It was in February 1941 that the W.1 was delivered to Power Jets by the British Thomson-Houston Company as a group of subassemblies. The company did not assemble it, however, until they were confident that they had learned enough from the experimental W.1X engine. The W.1 engine ran for the first time in April 1941.[50]

When the Air Ministry decided to put a Whittle designed turbojet into production, after seeing the reconstructed WU unit running at Lutterworth, Power Jets had a staff of not many more than twenty-five people and had never entirely manufactured even a single engine. Whittle, no matter how gifted, was young and inexperienced, and in the middle of a war, he could not be given the time to learn the skills needed to run a production organization. For the same reason, it was impossible to put him in a position of authority over an established engineering firm. If Power Jets' early success meant that the engine was "a production job," as Tizard announced in January 1940, it meant that production would have to be undertaken by an experienced company.[51]

The Air Ministry's very optimism about the aero-engine's prospects dictated that it should turn to industry to place an order for the engine's manufacture, while it continued to employ Power Jets on "research work."[52] When Power Jets realized that it would not be given its own production organization, it put forward Rover as a likely collaborator,[53] suggesting that a joint manufacturing company be established. Power Jets urged the Ministry to grant contracts to itself rather than to Rover. It wanted to subcontract work to the production firm in an arrangement similar to its relationship with the British Thomson-Houston Company, thereby allowing the small firm to retain control over its aero-engine design. The Air Ministry decided however to place contracts for production engines directly with Rover, and Power Jets continued to rely on Rover and the British Thomson-Houston Company for the provision of experimental engines and parts. From late 1940, because of its own interest in the gas turbine, Rolls-Royce contributed to the work by producing turbine blades, gear cases, and other components for Power Jets free of charge.[54]

The next prototype built by the British Thomson-Houston Company for Power Jets was the W.1A engine, which was meant to test the novel features of the planned W.2 engine design, the first engine designed by

Power Jets for series production.[55] The first W.2 engine was built by Rover and delivered to Power Jets in February 1941.[56] By then, however, Power Jets had already redesigned the engine when additional calculations made possible by an increase in its staff revealed problems that would occur under some flight conditions.[57] Power Jets' revised engine design, the W.2B, was later chosen to become the country's first production turbojet, and the drawings were given to Rover in March 1941.[58] Rover prepared its own drawings based on those received from Power Jets, introducing in some places Rover designed components, which the car company claimed would ease manufacturing. These changes would later become a point of contention with Power Jets.[59]

The assurance of increasing government support after 1938 allowed Power Jets to grow dramatically. The firm employed just five people at the end of 1938, fifteen at the end of 1939, 134 at the end of 1940, 329 at the end of 1941, 562 at the end of 1942, and 983 at the end of 1943 (see figure 3.1). In August 1940, the Power Jets factory at Lutterworth had contained only one machine tool, a lathe, and the firm's experimental shop was mostly staffed by skilled sheet-metal workers making combustion chambers, the chief problem plaguing the WU unit in 1939–40.[60] By September 1940, the company had a well-defined organization, including a drawing office, a design office, an experimental workshop, combustion test and engine test sections, an inspection department, and a purchasing department. Nevertheless, with seventy staff members, the firm was still working in a subdivision of the Ladywood Works, Lutterworth—the British Thomson-Houston Company having reoccupied the rest of the foundry with war work—and had no facilities for producing engine parts.

Power Jets' breakneck expansion was guided by a stringent recruitment policy that placed a high value on academic qualifications, which were at that time being more widely pursued by young engineers. In 1939, seven of the company's fifteen employees had scientific or technical qualifications (almost half). Applicants to the company were interviewed by Whittle and by Whyte, the most technically accomplished of the firm's managers, who held a physics degree from Cambridge.[61] Those employed were a select group; Whittle recalled that Daniel N. Walker, a senior engineer hired during 1939, was the only candidate accepted from 140 applicants. Soon Power Jets began inviting recent graduates to join the firm, as wartime made the labor position (especially for experienced, skilled workers) more difficult. Nevertheless, a degree in engineering was the most common qualification among the firm's highly qualified staff. Power Jets also took on metallur-

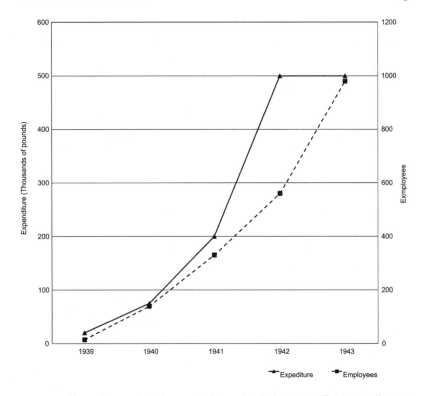

FIGURE 3.1 Expenditure and employment at Power Jets Ltd., 1939–43. Total expenditure at the company grew even faster than its workforce, more than doubling each year between 1939 and 1942. *Source*: Robert Schlaifer and S. D. Heron, *Development of Aircraft Engines*, 346; Lancelot Law Whyte, *Focus and Diversions*, 145.

gists, chemists, physicists, and mathematicians in addition to the personnel who supported the firm's engineers: general staff, machinists, draftsmen, and managers.[62] Through selective recruitment, the company built up a loyal core engineering team of young men, who frequently had first class degrees in mechanical science from Cambridge or London and who often came straight from academia and thus had no other commercial experience. The team was united by their shared academic background, their familiarity with mathematical analysis,[63] their awe of Whittle, and their dedication to the firm's goals. The disposition of the core engineering team imbued the firm with its "definite character": that of a firm devoted to the development of a working turbojet engine design.[64]

Despite the firm's rapid growth, Power Jets was still highly dependent on external expertise. This was illustrated by the problem of combustion, that part of the new engine that Whittle was least comfortable with. A reliable combustion system was ultimately created through collaboration with the Shell Company, which MAP brought into its turbojet program in response to an appeal for help from Power Jets. Shell's Dr. Lubbock developed his so-called atomizing fuel burner in October 1940, which solved the problem of fuel dispersal for the time being.[65] Lubbock subsequently worked with all of the British turbojet firms.[66] During 1940, the Royal Aircraft Establishment seconded three qualified men to Power Jets to work on combustion: Dr. William R. Hawthorne, R. F. Darling, and J. B. Bennett-Powell.[67] Hawthorne, who had a first class degree from Cambridge, actually led Power Jets' combustion group for a time.[68] MAP also helped by making the powerful compressor used for the excavation of the Dartford tunnel available to the firm for combustion tests.[69] Thus Whittle was connected to and supported by a network of Britain's ablest researchers, engineers, and industrialists.

The Contradiction of Power Jets

At the time of its founding, Power Jets' only assets were patents. The firm's management thus constructed a business model in which patents, which they saw as their only commercial defense, were central to their success as an independent firm. Power Jets' aims were to build a turbojet aero-engine and to enter the aero-engine manufacturing industry. Because the war broke out before Power Jets had established itself, however, these two "inseparable" aims came into conflict.[70] Whittle, who came to view the company as his service command during the war, was dedicated to its survival as an independent company.[71] Yet he wanted more than anything to make a contribution to Britain's war effort. Sharing the firm's work widely with other firms hastened wartime development (and was therefore demanded by MAP) but contradicted Power Jets' desire to protect its intellectual property, which was central to its ability to become commercially successful.

Nevertheless, Power Jets tried to defend what it saw as its intellectual property. The firm took a hard line on independent patenting by its collaborators and insisted on the paramount value of invention—its primary output—over design and development. While Power Jets took out joint patents early on with Laidlaw, Drew and Company on combustion,[72] it contended that similar equality did not exist with the British Thomson-Houston Company and Rover. In fact, Power Jets strongly objected to patent applications by British Thomson-Houston Company and Rover, which it asserted could only be based on "information supplied by Power Jets." Such patents, the firm argued, should therefore properly be assigned to Power Jets, with the British Thomson-Houston Company and Rover retaining nonexclusive licenses. Although Whittle admitted that it was often difficult to identify "the true origin of any invention" when a problem was solved in conference, Power Jets consistently held that its claims to patents were stronger than those of its collaborators.[73] Not only did Power Jets think the British Thomson-Houston Company and Rover incapable of making useful contributions to the turbojet,[74] but the firm also believed that its pioneering position entitled it to "dominate design and development of the engines which the other two firms were to make"—in other words, to retain "engineering control" over the W.2B engine.[75]

Power Jets' obsession with patents aroused suspicion in the Air Ministry, which insisted during the war that proprietary interests not hinder the development of new weapons.[76] Months after war broke out in 1939, the Air Ministry had deferred the question of rewards for patents until after the war, hoping that by promising to establish a mechanism for postwar settlement it could "remove any reluctance which [claimants] might have to giving their fullest cooperation ... without a prior settlement of the terms for the use of their designs and patents."[77] As Air Marshal Linnell, controller of research and development in MAP, remarked with respect to an argument between Rover and Power Jets in late 1942, "questions of patent rights and petty jealousies are not to be allowed to cause one day's delay in the supply of the best possible engine to the Royal Air Force."[78] The government held rights to use Power Jets' patents, thus MAP could dictate the best way to proceed.

Enabling cooperation between British turbojet firms was the central aim of the Gas Turbine Collaboration Committee—of which Power Jets was a key member, and the committee early on considered the question of patenting work "arising out of joint discussions."[79] Harold Roxbee Cox, the committee's chairman, opened the discussion of patents at the first meeting by emphasizing MAP's position that "patents and commercial obligations" were less important than ensuring "that there should be no obstacle to the free interchange of ideas."[80]

Harry Ricardo's consultancy collaborated, like Power Jets, with all of the British firms developing turbojets (the Ricardo barostat was used on all wartime British jet engines), but that firm decided to abstain from taking out patents on work done for members of the Gas Turbine Collaboration Committee. The firm insisted that it did not see its consulting agreements with Rolls-Royce, Napier, and de Havilland as compensation for "renunciation of patent rights"; instead, its decision was meant as a contribution to the war effort. Roxbee Cox offered nevertheless to "tactfully suggest" that Rover, Power Jets, and Metrovick take out such consulting agreements.[81]

Despite extensive discussion, neither the Gas Turbine Collaboration Committee nor its patents subcommittee, to which the question was deferred, reached a consensus about how to deal with the early jet patents. The question was made less pressing by the decision that firms should take out provisional patents to protect any new ideas before sharing them with other firms.[82] The lack of conflict over the issue during the war suggests that this was protection enough or that the collaboration of the committee did not necessarily lead directly to patentable results. Whittle remembered that meetings of the Gas Turbine Collaboration Committee were relatively formal and "people were a bit cautious"; real discussion occurred at the cocktail parties afterward.[83]

Despite its unsatisfactory commercial position, Power Jets did not hold back from collaborating with other firms. The firm's board agreed that the firm should "collaborate to the fullest extent," but it insisted that all the parties involved in any exchange of information should acknowledge Power Jets' unique "legal and contractual position."[84] In January 1942, Whittle wrote a memo to the secretary of the Gas Turbine Collaboration Committee, Wing Commander Watt, requesting the "recognition of certain essential differences in the status of the various firms collaborating."[85] In the memo, Whittle included a list of the many beneficiaries of Power Jets' work. He gave a detailed account of those firms that had access to and those that had contributed to Power Jets' work—although he was more generous in the first than the second.[86] Under the broad interpretation of invention voiced by Johnson, who became the firm's patent authority, Power Jets claimed not only ownership of turbojet designs based on its work but in effect also—quite in contradiction to patent practice—on all improvements made to Whittle's ideas and on solutions found to problems first described by Power Jets.[87] Such broad claims for ownership became more frequent as the firm's future commercial position was increasingly undermined during the war, and more and more firms began making turbojets.

Power Jets' work with Rolls-Royce took a slightly different track than its earlier relations with Rover. Collaboration on the WR.1 engine in

1941–42 was cordial, in no small part because Rolls-Royce consented to take subcontracts from Power Jets—the situation that Power Jets had demanded but had not been granted with respect to Rover. The congenial relations between the two firms led Power Jets in late 1942 to approach Rolls-Royce to produce its newest engine design, the Power Jets W.2/500. When it was a short time later decided that Rolls-Royce should take over W.2B engine production from Rover, it at first seemed that Power Jets' relationship with Rolls-Royce would continue to be friendly. Yet the combination of Rolls-Royce's growing confidence in turbojet design, supported by direct contracts from MAP, with Power Jets' prickly commercial concerns, including its insistence on the primacy of its inventive work, meant that relations between the two companies soon followed the same pattern as previously between Power Jets and Rover. Only this time, Power Jets found itself outmatched.

Power Jets' relations with Rolls-Royce may have started out warmly, but its claims to patent rights were no less extensive than earlier. In April 1943, when Rolls-Royce had taken charge of the government's turbojet factories, Power Jets argued that patent arrangements would be the single "major difficulty" preventing close collaboration between the two firms.[88] In a letter to Ernest Hives, Rolls-Royce's general manager, on May 10, 1943, Tinling, one of Power Jets' managers, sought to explain Power Jets' position, appealing to Hives as a fellow businessman:

> In regard to Patents which you may take out, we have not suggested anything like so severe a position as we contend applied to Rover in the past. Our purely tentative suggestion was that you might join us in Patent Applications which you may make arising out of development of our designs . . . we are largely dependent on Patent rights in the monopoly sense as a main asset, and therefore necessarily take a . . . view . . . that patents relating to designs of our parentage should not be freely available to parties not having a title in them.[89]

Tinling claimed that Power Jets was the "parent designer" of the entire W.2 series of turbojet engines, including not only the W.2/B, the W.2/500, and W.2/700 engines (three Power Jets developments of the W.2 engine), but also the W.2/B26 engine (as Tinling referred to the Rover STX, a heavily revised version of the W.2B that had been designed by Rover to ease W.2B engine manufacture and which Power Jets strongly opposed)[90] and the W.2/B37 engine (as Tinling referred to Rolls-Royce's B/37, which was based on the Rover's STX and included some features from Power

Jets' W.2/500 engine). He insisted that Rolls-Royce make its views on the patent situation known, for "it is more likely that future difference of opinion may turn on the accuracy of this view than on any other single question we can envisage."[91] Tinling's appeal made clear that Power Jets' interests lay in strengthening their patent position as much as possible.

Rolls-Royce was unfazed by Power Jets' claims, for despite Tinling's veiled threats that Power Jets could legally withhold information from Rolls-Royce, MAP sided with its trusted contractor. In February 1943, Archibald Rowlands, the Ministry's permanent secretary, assured Arthur Sidgreaves, Rolls-Royce's business manager, that "there can be no question of Power Jets being able to restrain other Companies from performing [Government contracts] . . . by the exercise of patent rights."[92] The questions of ownership, control, credit, survival, and patents were connected for Power Jets, yet the firm had little success in establishing or defending its patent rights. It was seriously hampered in doing so by the extraordinary level of government funding that it depended on.[93] Patents, it became increasingly clear, were not enough to sustain a company as an independent inventive firm in a wartime environment of increasing state expenditure on research and development.

Rapid Growth

Power Jets' unique nature as a company meant that it fitted uneasily into the Air Ministry's existing structure for the development and production of new aero-engines. Firms that designed but did not produce aero-engines were, as an official historian later remarked, "something of an anomaly in the field of Air Ministry R&D."[94] The Ministry's standard arrangement for funding aero-engine research was to make progress payments based on hours of test-running or prototypes built, but as the prewar delays in funding Power Jets had demonstrated, such a model was inappropriate for a small company that needed working capital.[95] In 1941, the Air Ministry entirely abandoned its attempt to fund Power Jets through development contracts and began funding the company's work by the "rather unorthodox procedure" of simply paying all of the firm's operating expenses.[96] Expenditure at Power Jets more than doubled each year between 1939 and 1942, when it exceeded £500,000. The rapid rate of expansion of the firm led to internal disagreement between Whyte, who advocated a more cautious approach to firm building, and Whittle, whose desire for quick expansion to facilitate faster development was undoubtedly fueled by the well-publicized service of his fellow RAF officers during the Battle of Britain in 1940. Overruled, Whyte resigned as managing director of Power Jets on July 4, 1941.[97]

Through paying the firm's expenses, the state directly funded Power Jets' expansion. During the latter half of 1940, the firm had made some modest additions to its facilities after it was given a government authorization of £27,000 for equipment and building requirements (particularly test houses).[98] This allowed it to build a small workshop capable of making engine components by September 1940. When the size of the factory at Lutterworth began to constrain the rapidly expanding company, the firm appealed to the government for assistance. On April 11, 1941, Power Jets' management submitted a request to MAP for an additional 49,000 square feet of factory and office space.[99] The firm had already rented rooms for its design team in Brownsover Hall near Rugby to relieve the competition for space at Lutterworth, but otherwise, temporary extensions to the Ladywood works were the norm. The firm's desire to expand occasioned a MAP meeting on April 23, 1941, to determine what should be done, and a range of possibilities was discussed. Roxbee Cox argued that a resolution had to be met quickly because the lack of space would soon interfere with Power Jets' "vital research and development work." He judged that the firm "desperately" needed 13,000 square feet of additional space—double the factory space that the firm had at Ladywood.[100]

By July 1941, the proposal for Power Jets' factory expansion had grown from 49,000 to 80,000 square feet. Of this, 27,000 feet were estimated for an office staff of about 180. Just under 47,000 square feet were allocated for key facilities: a sheet metal department, welding departments, machine shop, jig fixture and tool room, fitting shop, electrical department, various stores, auxiliary test rigs, inspection and viewing department, model and pattern shop, lofting room, carpenter's shop, five engine erection bays, heat treatment and metallurgical section, burner department, five combustion bays, and combustion laboratories. The new factory would also require additional outbuildings for four test houses and two observation rooms, a building to house the Dartford compressor (later abandoned in favor of two smaller Browett-Lindley compressors),[101] a canteen, a garage, a boiler house for heating apparatus, and a gate house in addition to service roads, fencing, and a parking lot. An additional 25,000 square feet, bringing the total to more than 100,000, were allowed for the future installation of the machines needed for limited engine production.[102]

At a further meeting on July 31, 1941, MAP decided that Power Jets would get a new factory. The ministry had failed to negotiate a satisfactory agreement to acquire a share in the nominally private company, so the company's relationship with the state was to carry on as before.[103] Power Jets would be maintained as an independent organization in order to enable Whittle to keep working as a chief engineer for MAP.[104] Whittle's appeal to be allowed to expand "his organisation" to include the production of engines was refused, but the Ministry allowed that the firm's "research work" could be extended to "include the construction of prototype experimental engines." As long as Power Jets continued to collaborate with "any persons nominated by the department," it was promised it would be given the facilities that it wanted to produce twelve experimental engines per year.[105] Moving Power Jets to Farnborough or Bedford (both future sites of the government's Royal Aircraft Establishment) was suggested, but the firm requested that a site be found near its existing factory in Lutterworth. Because a suitable existing factory building could not be found in the area, and Power Jets threatened that the existing conditions at Lutterworth would "delay production," MAP agreed in October 1941, despite the severe restrictions on building during the war, to erect an entirely new factory for the firm less than ten miles from Lutterworth at Whetstone, Leicestershire.[106] In the end, Power Jets' new "development factory" cost £325,000 to build and equip.[107] It remained—in keeping with an existing model of MAP company relations, government property, with Power Jets as the factory's shadow management.[108] By October 1942, the Whetstone factory was established and capable of making about four engines per year,[109] but it wasn't yet quite clear what the factory would be used for.

The new facilities at Whetstone were accompanied by an influx of workers assigned by the Ministry of Labour that drastically changed Power Jets' identity. The company's labor force expanded from 330 employees at the end of 1941 to almost 1,000 by the end of 1943 (see figure 3.1). Despite assurances that workers in Leicestershire would be well suited to the new factory because it was experienced in the "batch production of hosiery machines . . . and the men therefore had the right mentality,"[110] Whittle complained about the quality of the new workforce. Meanwhile, Power Jets' management struggled to instill a feeling of belonging in the larger organization, which contrasted with the close-knit team that had been employed at Lutterworth. The management tried illustrating the importance of Power Jets' work to the war effort through demonstrations of a turbojet engine—which was still secret—to the staff.[111] Whether or not it contained communist sympathizers who staged strikes, as Whittle later alleged,[112] the workforce was certainly less loyal to Power Jets than

FIGURE 3.2 Power Jets' factory at Whetstone under construction in 1942. The factory's basic construction followed wartime building standards. Hives wrote on November 17, 1942: "some decision should be taken as regards the Power Jets factory at Whetstone. I was astonished at the size of it, and the emptiness of it, when I visited it a short time ago." (RR HC, November 17, 1942, Letter from Hives to MAP) *Source*: Smithsonian National Air and Space Museum (NASM 9A08601-4)

the company's core engineering team. In the latter half of 1943, the new workers denounced the competence of the company's leadership to Stafford Cripps, the minister of aircraft production, during a factory visit. This complaint occasioned an investigation by MAP that confirmed the leadership's inability to manage a large firm and questioned the firm's freedom from Ministry oversight given the large government investment in the company.[113]

Whittle, meanwhile, had become more and more distant from the day-to-day affairs of the larger company, in part because of problems with his health. In December 1941, Whittle suffered a nervous breakdown that kept him from the company for one month. In mid-1942, he spent about three months abroad working with companies in the United States. After attending a three-month course at the Royal Air Force Staff College in May 1943, he returned to Power Jets with reduced duties, concerning himself primarily with long-range projects in the firm's new special projects section at Brownsover Hall headed by L. J. Cheshire. It was by no means clear that Whittle would even return to Power Jets in mid-1943, but MAP wanted to keep him involved in the program and Whittle refused to take a position in the ministry.[114] After the construction of the factory at Whetstone, Power Jets' work was still dedicated to the design and development of turbojets, but its agenda was no longer driven by Whittle. It was now the engineering team that he had recruited that was responsible for perpetuating Whittle's work.

Whittle's trip to the United States in June to August 1942 reinforced his view that pilot production, or the development of manufacturing technique through the production of small batches of engines, was an essential part of development. He returned to Britain determined to redefine the divisions between "design," "research," "development," and "production" in order to extend Power Jets' activities.[115] In early 1943, he proposed the extension of Power Jets' activities to include pilot production. Hives responded that Power Jets could "best contribute at present" (Rolls-Royce had just started taking over the Barnoldswick factory from Rover) not by developing production technique but by "producing improved designs."[116] So although Power Jets received two MAP contracts for W.2/500 engines in January and February 1943, Whittle didn't get his way, and the company retained a circumscribed role in the nation's turbojet program.[117]

The production facilities ultimately provided at Whetstone meant, however, that Power Jets was no longer reliant on other companies for engine parts.[118] It was in the new factory that Power Jets built the first W.2/500 engines—the first turbojet engines that the firm had ever manufactured in their entirety. The W.2/500 engine was based on the W.2B but incorporated improvements, some of which came from organizations outside Power Jets. The engine ran a mere six months from when Power Jets started the design drawings, and it performed precisely as predicted, reaching its design thrust of 1,750 pounds on the first day of testing, September 13, 1942.[119] Although this seemed to be a demonstration of Power Jets' superior expertise, Rolls-Royce declined to put the Power Jets engine into production in favor of its own turbojet design, the B/37.

During 1943, Whetstone continued development of the W.2/500 engine and started on the design of the firm's next engine, the W.2/700, which again used Power Jets' typical reverse-flow architecture. Neither engine went into large-scale production, however. Despite Power Jets' arguments, MAP decided to produce and deploy the Rolls-Royce B/37 engine instead of the Power Jets W.2/700 engine.[120] The much more extensive Power Jets organization at Whetstone thus failed where the smaller unit had succeeded, for the W.2B engine did become the first British turbojet engine to reach production.

A National Resource

The large and growing investment of public funds in Power Jets was not accompanied by any corresponding government control, and this became of increasing concern to MAP over the course of the war as the firm's expenses continued to increase.[121] The government's expenditure on Power Jets, including running costs, machinery, and buildings, came to almost £1,300,000 between the second half of 1939 and the end of 1943.[122] In late 1943, Stafford Cripps, the minister of aircraft production, started pushing outright for the purchase of Power Jets by the government. He saw this as the best way to gain some benefit for the nation from the investment of public funds in the company, while simultaneously securing an advanced technical base for the British aviation industry in the postwar period.[123] Cripps, who was a qualified patent lawyer, also argued that purchasing Power Jets would straighten out the convoluted patent situation "very much to the country's advantage" (likely ensuring the government's total rights).[124] MAP had previously declared that "the patents position could not ... be cleared up without an intensive enquiry lasting for about a year," thus the validity of Power Jets' patents and their complicated relation to the work of other firms were by no means assured from MAP's point of view.[125]

Cripps proposed that the nominally private Power Jets firm be replaced by a government-owned company. This plan departed from all precedent. The first part was clear. Like a government establishment, the nationalized Power Jets was to do work for the Ministry, concern itself with the advance of knowledge, carry out research, and share information. It would be "regarded as the recognised National authority in this field," investigating the application of gas turbine not only to aero-engines but also to marine engines, power stations, and locomotives.[126] The company's planned role as a national center would make it necessary to incorporate into the new company the section of the RAE that was devoted to gas turbines. In fact, it was Roxbee Cox, who was in charge of turbojet development at MAP, who made the combination of Power Jets and the turbine section of the Royal Aircraft Establishment a condition for his accepting the directorship of the new company, Power Jets (R&D).[127]

The peculiar thing about Cripps's proposal was the second part, that the company was to "take all reasonable business measures to establish itself on as nearly a self-supporting a basis as possible by arranging profitable outlets for its intellectual or physical products."[128] Because commercial disinterest was an essential attribute of government establishments, this seemed directly to contradict the organization's primary role as a national research center. Cripps nevertheless insisted on making the institution commercially competitive. This, he thought, would facilitate negotiation with other firms, allow the management a greater liberty of action, and most importantly, make possible the retention of the best technical personnel through higher salaries.[129] The new company's convoluted legal status notwithstanding, the arrangement required no immediate changes at Power Jets.

Although it was agreed that Power Jets should be nationalized, Cripps's proposal met with objections from all sides. The aero-engine industry objected vehemently to the firm's proposed commercial engine design activities as unacceptable state-supported design competition, and many civil servants agreed with the industry position that the best innovations emerged fastest from private enterprise. This was not surprising, as already in the interwar period, both civil servants and industry had viewed government design capacity with suspicion.[130] Viscount William Weir questioned the implications of the nationalization for the government's role in development, which he thought should not be allowed to become a function of the state.[131] Power Jets' other shareholders were distrustful of the nationalization because they suspected that it was an opportunistic use of war powers to cheat them of profits on a risky investment (the government ultimately paid £3.125 for each £1 share, which might be considered a generous return on what turned out to be a short-term investment that could not have succeeded without government support). Tinling managed to engineer a unanimous vote among the company's shareholders to wind up the company in 1944 by convincing them of the real possibility that the company would be shut down completely if they refused to sell.[132] Weir and his brother used the profits from their Power Jets stock to create a private fund for research.[133]

In the end, Cripps's proposal went ahead, and the firm was nationalized. The only way to overcome the strong Treasury opposition to the establishment of a long-term government-owned company, however, was a compromise that would bring Power Jets (R&D)'s existence back to the top of the government agenda in the future: the new company was agreed to be a temporary expedient, arranged to avoid disruption in gas turbine research during the war, but up for review when it ended.[134] Because of this compromise, the government company ended up being more of an experiment in government ownership than a change in policy. The immediate resistance raised to the formation of a government company goes against David Edgerton's interpretation of the nationalization as representing a widespread new conception of the state's role in aero-engine design that was quickly reversed. Instead, the resistance to state design in aviation that Edgerton charts from the 1930s continued to be strong during the war, and war conditions led only to an apparent divergence of policy.[135] In this light, Cripps emerges as Power Jets' defender rather than its undoing, for he gave the firm a new purpose at a time when it could no longer continue working in the way in which it had earlier in the war. Indeed, Cripps's advocacy was crucial in overcoming resistance; despite disagreeing with the scheme (interestingly for fear of upsetting work at the Royal Aircraft Establishment), Cherwell later wrote to Churchill, "I think it would be difficult to overrule [Cripps], as he has set his heart on this new government company."[136] Without Cripps's scheme, Power Jets likely would have been converted to an establishment even sooner than it was, for Power Jets' relationship to MAP had already been moving toward that between a research establishment and its ministry.[137]

On April 28, 1944, Power Jets (Research & Development) Limited was founded as a government-owned company, created from the combination of the nationalized Power Jets Ltd., which had about 1,000 employees in early 1944,[138] and the Royal Aircraft Establishment's turbine section, numbering about fifty people.[139] The new company's capital was increased from Power Jets' £20,000 to £200,000. Its board of directors was nominated by MAP and included a number of powerful industrialists in addition to some of Power Jets' previous directors.[140] Over the first year of the government firm's existence, its staff and annual expenditure continued to increase, reaching 1,327 employees and about £600,000 pounds.[141] Whittle was disappointed by the nationalization, but he went along with it. He held out hope that the formation of the government company would still allow Power Jets to become a successful engine designing and developing firm as he had always wanted.[142] Although his ill health and frequent lecturing kept him away, Whittle was made chief technical advisor to the board of the new company and a director in January 1945.[143]

The Royal Aircraft Establishment

The Royal Aircraft Establishment's (RAE) turbine section, which was unceremoniously combined with Power Jets in the new nationalized company, advocated and facilitated the development of Britain's second line of turbojet development, that of axial gas turbines. The first serious research on gas turbine aero-engines done in Britain was done at the Royal Aircraft Establishment. It was based on the work of the establishment's senior scientist, Arnold A. Griffith. In an important 1926 report, he disagreed with the infamous conclusions of W. J. Stern, arguing that an axial compressor could reach high enough efficiencies to be suitable for an aero-engine (and therefore a gas turbine could be suitable as an aero-engine) if it were designed according to aerodynamics. The paper's theoretical prediction was later proven by experiments done at the RAE.[144] The establishment's early work on axial gas turbines was, however, abandoned in 1930 and only taken up again when Hayne Constant returned to the establishment as Griffith's assistant in July 1936.[145] Constant had spent the previous two years working as a lecturer at Imperial College in London, where he had carried out experiments that convinced him of the utility of Griffith's ideas. At Imperial, he had been encouraged to pursue the work particularly by the college's rector and chairman of the Aeronautical Research Committee, Sir Henry Tizard.[146] When the government's advisors on the Aeronautical Research Committee lent their full support to gas turbine development in Britain in March 1937, concluding that "the time was ripe for departures in power plant design of this type and that a special effort should be made to foster them,"[147] their decision was based in no small part on the testimony and experimental experience of Griffith and Constant.[148]

The Aeronautical Research Committee's recommendation resulted in increased government support for work on gas turbine aero-engines in parallel at Power Jets and the RAE.[149] In a stroke, Britain thus had two officially sanctioned turbojet projects: one led by an RAF officer and the other by a civil servant, both Cambridge trained engineers. Although different in emphasis and focus, both institutions relied on outside contractors for manufacture and testing.[150] MAP continued to support both programs throughout the war in order to maintain work on multiple, competitive approaches to gas turbine aero-engines—the two progressed roughly in parallel.[151] At the same time that MAP authorized Power Jets' new factory in the summer of 1941, the Ministry also paid for new specialized turbojet research facilities to be built at its establishment. The new facilities of the two jet teams were significantly different in both scale and purpose. The Royal Aircraft Establishment's new turbine section at Pyestock included an engine test house, a workshop, an office block, and a laboratory. These facilities, well suited to a research establishment, were very much in contrast to the extensive Power Jets facility at Whetstone. The expenditure on buildings and research equipment at Pyestock was £150,000—less than half of the £325,000 spent at Whetstone.[152]

The formation of the RAE's independent turbine section (separate from its engine department) in August 1941 was one of a group of sweeping changes ushered in during 1941 by the establishment's new director, William Farren. As the first director of a newly unified, larger RAE, he worked to expand and strengthen the establishment even further.[153] Farren appointed Hayne Constant to head the establishment's engine department in 1941, thereby placing an ambitious man with a particular interest in gas turbines in a position to revivify the establishment's insufficient and deteriorating engine research facilities.[154] Farren's ambitions for the establishment resonated with Constant's. Constant wanted the establishment's turbine section to become the nation's leading research center on the subject and MAP's main advisor on gas turbine aero-engines, carrying out experimental research, performance calculations, project studies, and flight trials.[155] In this way, Constant sought to use the establishment's advantages in gas turbines to expand the RAE's role in aero-engine development, which had been strongly restricted since the end of World War I.[156]

Farren's earlier career had brought him into contact with the first work on turbojet aero-engines and aircraft in Britain and must have made him alive to the turbojet's potential for increasing his establishment's influence.[157] In January 1942, Farren argued to MAP that the evaluative role of the Royal Aircraft Establishment should be extended from jet airframes to include jet engines. The new engines, unlike reciprocating engines, were particularly well suited to calculation, he argued, and thus to the establishment's particular expertise.[158] David Pye, MAP's director of scientific research, agreed that the Royal Aircraft Establishment's research capability gave it a unique and legitimate role in British gas turbine aero-engine design. He argued that "to design jet propulsion engines successfully, the designer must have at call certain fundamental aerodynamic knowledge. It is an art in which technical experience alone is not enough."[159] Despite some opposition, it quickly became MAP practice to submit all new gas turbine aero-engine plans to the Royal Aircraft Establishment for evaluation.[160] In November 1942, the future of the British jet engine program was left to rest on a "complete survey" by the RAE of the virtues of all of the several turbojet engines then under consideration for production.[161] Constant's particular interest in the turbine section that he helped to create ensured that it continued to grow as an increasing number of personnel were diverted from piston engines to gas turbines. By 1944, it represented half of the RAE's engine department.[162]

The creation of Power Jets (R&D) as a national gas turbine center from the merger of the RAE's turbine section and Power Jets was perhaps more disruptive for members of the RAE's turbine section than for those working at Power Jets. After years spent building up the RAE's expertise and authority in gas turbines, the establishment's director, the head of its engine department and the head of its turbine section—Farren, Constant, and Hawthorne, were not best pleased at the prospect of one of the RAE's sections being merged with an institution with very different research priorities and a very different working ethos.[163] What Cherwell described as "the Minister's new sociological experiment"[164] left the Power Jets organization largely the same but required serious adjustments from the staff from the Royal Aircraft Establishment, who were changed in a stroke from civil servants to private employees. Some members of the Royal Aircraft Establishment decided to stay on in public service at MAP rather than join the new company.[165] Constant, however, chose to leave his many responsibilities at the RAE and remained with the turbine section as it became part of Power Jets (R&D).[166] He became head of the new company's engineering department, where he was influential in defining the firm's technical policy; Constant was well positioned to perpetuate the agenda of the RAE.[167]

Regardless of their new unified leadership in a government-owned company, the former Power Jets and the RAE turbine section remained physically separate, with the director of the company, Roxbee Cox, spending half of his time in Pyestock and half in Whetstone.[168] Despite having worked closely since the late 1930s, Constant and Whittle now found themselves leading competing factions at meetings to establish the new company's technical policy. Constant thought the new company should emulate a national establishment by primarily supplying industry with research data. Whittle, in contrast, wanted the company to remain the leading gas turbine design agency in Britain, which he argued required designing and developing prototype engines. Industry sided with Constant; what aero-engine manufacturing firms wanted from the government, according to Rolls-Royce, was not new turbojet engine designs, but "to be supplied with more basic data on the many research and aerodynamics problems involved [in jet engines] ... presented in a manner directed for specific application after confirmation by intensive testing."[169] When it became clear that he would not be able to influence the company's engineering policy as he and Roxbee Cox had hoped, Whittle resigned from the firm's Technical Policy Committee (but did not leave the company), sometime between April 18, 1945, and May 8, 1945.[170]

The End: Government Company and Back Again

Power Jets (R&D) was active despite the outstanding questions about its duties. Whittle had deeply instilled in his firm his view that only "one thing could save Power Jets from complete extinction—superior technical competence."[171] In its new guise, Power Jets' members continued their research work but increasingly without reference to the work being done by other firms. As part of Power Jets (R&D), the Whetstone factory worked further on thrust augmentation; created new turbojet designs including a supersonic engine and an axial engine, the LR.1, which Whittle contributed to; tested German engines and confirmed their "marked inferiority" to British engines after the war; set up a training school at Lutterworth; and instructed numerous visitors from Britain and abroad.[172] The government-firm's terms of reference gave the company a chance to begin limited aero-engine production, and the factory at Whetstone achieved a new success in 1945: pilot-producing engines for the first time. It began to manufacture W.2/700 engines at a rate of thirty engines per year, with spares equivalent to another fifteen engines per year, although these engines were used for development and never deployed.[173]

In order to reassure the aero-engine industry that the new government-owned company would not unfairly compete with it, Cripps established a "Gas Turbine Technical Advisory and Co-ordinating Committee," at the same time as Power Jets (R&D) was formed.[174] The committee included members from MAP, the Admiralty, Power Jets (R&D), and the aero-engine industry.[175] It was through this advisory committee that the aero-engine industry agitated to restrict the new firm's activities. Because of the "imperative need to restore the confidence of the jet engine firms" and because the government felt that it needed to reassert control over the state-owned company, officials agreed in May 1945 to redraft Power Jets (R&D)'s terms of reference and eliminate the firm's manufacturing activities. That such a move would make an important difference to the scale of the company (making it smaller) weighed in the proposal's favor. Power Jets (R&D) was 1,400 strong in May 1945; a national establishment, in contrast, would employ only 300–400 people, releasing many technically competent individuals into the wider postwar economy.[176] On December 21, 1945, it was finally decided that Power Jets (R&D) should be turned into a national research establishment. Cripps, then president of the Board of Trade, was present at the negotiations that ended his government-owned company.[177]

On July 1, 1946, the National Gas Turbine Establishment was opened. Its responsibilities were more akin to those of the former Royal Aircraft Establishment turbine section than the former Power Jets. Power Jets' original engineering team saw the transformation into a government research establishment as a great injustice. Despite the similarity in some of the work that had been done by Power Jets and the Royal Aircraft Establishment, Whittle and his closest engineers felt very strongly that doing research without also applying it to engine design was worse than useless. Whittle argued that his engineering team (his memoir makes no reference to the rest of the Whetstone organization) was "temperamentally unsuited" to "pure research."[178] Whittle resigned from Power Jets (R&D) in January 1946 before it was turned into a research establishment, complaining that "I cannot arouse much interest in a myriad of small-scale and relatively unrelated experiments which end merely in a mass of reports"—a revealing description of his understanding of the work of research organizations.[179] His objection went beyond the distasteful nature of "pure research," however. In his memoir, Whittle later explained that he and his engineers "would not tolerate a situation in which those *who had founded an industry were deprived of the right to design and make experimental engines*."[180] But what was being denied to the engineers, whether or not they had founded an industry, was not "the right to design and make experimental engines"—any of his engineers would have been taken on by the aero-engine industry at once, but rather the opportunity for the group to continue to design and produce experimental engines on particular terms that they themselves chose (a freedom that had been allowed by war conditions), rather than on the terms that Power Jets' wartime expansion at government expense dictated. Whittle felt, according to his memoir, that "there was nothing unreasonable in [trying to restore Power Jets' former lead in all branches of gas turbine technology] ... Power Jets had clearly established a moral right to include [the development of manufacturing methods and limited engine production] as one of its functions because ... not only had Power Jets pioneered engine development itself, but had also contributed more to the evolution of special production methods than all the rest of the firms put together."[181] Whittle

thus invoked the record of work at Power Jets to defend his team's right
to a privileged position. The entitlements due to technical pioneers, he implied, trumped the dues of public funding. Despite its reliance on government support, Whittle thought that the firm's pioneering work and engineering excellence gave it a "moral right" to pursue its work as it saw fit.
Later, he argued that although Rolls-Royce and de Havilland had both
"put more jet engines into the air than Power Jets . . . no other firm had yet
equalled Power Jets' record for the number of *types* of engine which had
powered experimental aircraft (W.1, W.1A, W.2B, W.2/500, W.2/700)."[182]
Still, as had been demonstrated in Power Jets' nationalization in 1944, the
moral rights that had increasingly replaced patents in the firm's strategy
didn't obviously translate directly into commercial power.

Not all of the employees at Power Jets (R&D) felt as negative about
being part of a national establishment as Whittle and his close engineering team. It is well known that Whittle and sixteen of his original engineers left Power Jets (R&D) in April 1946, but it is less advertised that
some one hundred "scientific and experimental officers" from the company stayed on. Most of the Power Jets engineers who left did not return
to aero-engine work, choosing instead to work on industrial and marine
gas turbines.[183] Although there was some public worry about the seriousness for British competitiveness of the loss of the firm's most visible engineers, Roxbee Cox and the Ministry of Supply were confident that they
would be able to fill the vacancies at the new establishment from a large
pool of applicants.[184] In fact, the decision to turn the company into a research establishment was made in full expectation that Whittle and Roxbee Cox would resign because both had earlier insisted on the importance
for the company of being allowed to produce aero-engines.[185] Whittle did
resign, but although he hesitated at first, Roxbee Cox stayed on as the director of the National Gas Turbine Establishment. Hayne Constant became his deputy. (Roxbee Cox recalled that he had needed the job, and
the terms of his five-year contract were very favorable.)[186] In 1948, Constant ascended to the directorship of the establishment, where he remained until 1958. As director, Constant oversaw the end of Power Jets'
research program on centrifugal gas turbines and the triumph of the axial
compressor in British aero-engines.

After the formation of the National Gas Turbine Establishment, Power
Jets (R&D) remained in existence to administer the British jet patents
now owned outright by the Air Ministry. Between November 1949 and
October 1951, the United States government agreed to pay one million

pounds (four million dollars) for the use of all of the British patents during the war and "over the next 20 years." No one ever undertook to unravel the many patent claims.[187] The British Treasury decided against using the patents in its Lend-Lease settlement and classified them instead as a legitimate contribution to the country's debt payments. The total conceded by the United States for use of the British jet patents was much less than the amount that the British government had spent on the turbojet aero-engine during the war, and it was quickly exceeded by the sale of British production jet engines and airframes overseas.

The Ernst Heinkel Flugzeug Werke

The Ernst Heinkel Aircraft Company (Ernst Heinkel Flugzeug Werke, EHFW) was unique in Germany as the only company outside the established aero-engine industry that developed a turbojet engine during the war. Because of the firm's record-making turbojet flights, it is as familiar in the story of the German turbojet as Power Jets is in the British. Before von Ohain joined it, the firm was already known for its many innovations, its importance as the second-largest German aircraft company in the interwar years,[188] and its flamboyant proprietor, Ernst Heinkel— indeed in the literature, the company and the man are almost synonymous.[189] Heinkel's public visibility was reinforced by the publication of his ghost-written memoir in 1953 (released in English in 1956), which laid claim to having invented the turbojet, among other things. His account has, however, been proven markedly unreliable.[190]

Accounts of the turbojet in Germany have taken two different approaches to Heinkel's jet record. Accounts that focus on the first turbojet flight like those of M. M. Postan, Edward Constant, and Margaret Conner acclaim the company's success in turbojet development. The British official history, authored by M. M. Postan, argues that von Ohain progressed faster than Whittle "due very largely to the fact that he had the great good fortune to be able to develop his jet propulsion unit in surroundings which, if not exactly perfect, were considerably more propitious than those in which Air Commodore Whittle's unit was developed." The official history's inadequate account of Heinkel uses a very brief and flawed rendition of von Ohain's early story mostly as a foil to emphasize the difficulty of Whittle's path.[191] Constant similarly asserts that von Ohain's work advanced more quickly because of Heinkel's "tremendous (compared to

Power Jets) financial resources" and "usually, the unrestrained support of Heinkel himself."[192] In her biography of von Ohain, Conner too characterizes the relationship between Ernst Heinkel and von Ohain as a constructive one.[193] These authors treat von Ohain as if he were an independent inventor, free to do as he chose after securing funding for his work. They depict the relationship between Heinkel and its turbojet designer as positive and enabling and argue that von Ohain was lucky to have found in Ernst Heinkel a willing sponsor for his work—especially as compared with Whittle.

The second view of Heinkel's turbojet work in the literature is a negative one. Accounts that portray a mutually satisfactory turbojet program at the company, in which both von Ohain and Ernst Heinkel wanted to achieve a world first, do not capture the complex nature of work at the company. Authors who consider the firm's entire turbojet program, rather than just von Ohain's earliest work, such as Robert Schlaifer, and aviation writers Dieter Köhler and Volker Koos, judge the company's jet work a failure. Although Ernst Heinkel was an important early supporter of jets, no Heinkel jet engine or airframe was ever produced in quantity or deployed by the Luftwaffe. All three authors argue that the firm's resources were made less effective by being divided between too many projects. Köhler, however, maintains that Heinkel had a positive influence on the program at his firm, interpreting the division of resources as a by-product of the uncertainty of pioneering.[194] Schlaifer, in contrast, describes the Heinkel Company's turbojet program as a mistake given the airframe firm's lack of resources for and expertise in aero-engine manufacture.[195] Koos's recent account of turbojet work at Heinkel goes even further, attributing the failure of the Heinkel Company's turbojet program primarily to Heinkel's behavior.[196]

Rather than limiting its focus to Heinkel or von Ohain, as previous authors have done, this section looks in detail at the growth of the Heinkel Company's turbojet capability over time and how this influenced von Ohain's work. Instead of just studying the time before the legendary flight of the world's first turbojet fighter, the He280, it considers the entirety of von Ohain's time at the company, which stretched for many years after that flight. Its conclusions support the negative impressions of Schlaifer and Koos—allegations later made publicly by von Ohain himself.[197] This section argues that the reason that von Ohain's designs had no major legacy is directly attributable to the nature of the company where he worked and the fact that he was an employee there.

Both von Ohain and Ernst Heinkel were undoubtedly dedicated to the development of the turbojet, but their goals within that field were often opposed. As an employee, von Ohain was in the weaker position, and his work was in fact hugely disadvantaged through alliance with Heinkel, with whom he nevertheless remained on good terms. The flight of the He280 jet fighter was in this view not an important accomplishment for von Ohain, but marked instead the culmination of Heinkel's vision of invention as a quick route to success. It was only with developments at the firm in 1942–43, when Heinkel was removed from control of the turbojet program of the Heinkel Company, that von Ohain finally got the favorable conditions and autonomy for development that he had desired since he joined the company in 1936.

The Beginning of Heinkel's Turbojet Work

In 1936, the Ernst Heinkel Flugzeug Werke (Ernst Heinkel Aircraft Company) became the second German company to fund turbojet development after the Junkers Aircraft Company. Junkers had begun work in 1935 under the leadership of Herbert Wagner, but the airframe firm gave up its turbojet work in 1938 after the German Air Ministry made clear its policy of only giving aero-engine contracts to firms with aero-engine experience, forcing its team to go elsewhere (it would later go to the Heinkel Company).[198] Although the Heinkel Company had never produced an aero-engine before 1936, the high-speed turbojet fitted well into the company's image as a pioneering, speed-record-breaking company, with design rather than production at the heart of its business. Furthermore, Ernst Heinkel was not averse to joining the aero-engine industry.[199] When he received a letter introducing Dr. Hans von Ohain and his design for a turbojet aero-engine in March 1936, he jumped at the chance to develop an aero-engine, with the turbojet an opportunity to secure another world record for his company and gain advantage over established aero-engine firms, none of which were at that time working on a turbojet engine.[200] Thus began what would become the most extensive (in terms of number of turbojet acro-engine and airframe projects) but also one of the least successful (in terms of production) turbojet programs in Germany.

Von Ohain had approached Ernst Heinkel because he feared a negative response from established (and he presumed conservative) aero-engine firms and because he knew that Heinkel had a reputation for being interested in high-speed flight and unorthodox ideas, both of which well described the turbojet.[201] Indeed, despite some skepticism from the company's engineers, von Ohain received a friendly reception at the Heinkel Company. He hoped to be able to develop his turbojet ideas at the company, but in the end, he was hindered by the very things that had made the firm seem a good place for an inventor to work: Ernst Heinkel's rabid interest in new projects, his expansionist tendencies, his independence, and the company's financial self-sufficiency. Because von Ohain's work became central to Heinkel's ambition to join the aero-engine industry, von Ohain was never entirely free to develop a turbojet engine as he chose so long as Heinkel was in charge, which was for the first seven (of nine) years of von Ohain's employment. By becoming part of a successful firm, von Ohain did gain access to ready financial support, but his work was seriously obstructed by the inappropriate facilities and expertise available at the company as well as its owner's ethos and demands.[202]

The terms of von Ohain's first contract with the Heinkel Company, signed on April 3, 1936, made it clear that further support for his work was contingent on quick successes. If von Ohain successfully demonstrated the principle of jet propulsion, he was to be hired to continue work on the project for at least one year at a salary of 400 Reichmarks per month, in which case von Ohain also agreed to transfer his turbojet patents to the Heinkel Company.[203] (His later patents were indeed owned by Heinkel with von Ohain listed as inventor.) Thus the newcomer immediately felt the compulsion to quickly achieve results, the so-called Heinkel tempo that was de rigueur at the company. Heinkel's inexperience with aero-engines contributed to the pressure because, as von Ohain remembered, Heinkel was "enormously impatient" and expected that a complex new engine could be finished in as short a time span as a new airframe—an entirely unreasonable expectation.[204]

With his employment and the opportunity to further develop his turbojet ideas hanging in the balance, von Ohain sacrificed his systematic development plans (particularly for combustion) in favor of a design that could be built and run as soon as possible. Because centrifugal elements were better known and promised to be easy to build, von Ohain's first demonstrator unit at the Heinkel Company, the HeS1, had a centrifugal compressor and turbine; von Ohain had employed the same elements in the turbojet model that he had built before moving to the aircraft firm.[205] The fact that the engine had both a centrifugal compressor and centrifugal or radial-flow turbine was unique among all early turbojets and remained a feature of all of the engines that von Ohain designed up to 1942—Whittle's first engines also had a centrifugal compressor but not turbine.[206] Von Ohain's decision to use hydrogen gas to power the engine was also made in the interest of speed and was another unique feature of his engine. The choice was not suitable for a service engine (as Whittle was making), but it meant that the inventor could run his first demonstrator, the HeS1, without the delay that would be incurred by the additional development that a liquid fuel combustion system would require. This judgment proved to be correct; problems with combustion delayed all of the early turbojets.[207] The engine was built in the Heinkel Company's existing facilities, as there was no money to get additional machines. The lack of machines suitable to aero-engine production dictated that von Ohain's demonstrator was made mostly from sheet metal. Systematic component testing, which was necessary to develop high performance aero-engine parts but not needed in building airframes, was made impossible by the lack of equipment as well as Heinkel's constant pressure for speed.[208] Some parts of the demonstrator were even subcontracted out to a nearby shipyard for lack of capability at the firm.

In March 1937, the HeS1 engine ran, demonstrating the practicality of a self-sustaining jet engine. In so doing, it simultaneously secured both a world record and von Ohain's job. Von Ohain's group's second demonstrator engine, the HeS2, was built soon after. It produced more thrust than the HeS1 and included the firm's first liquid fuel burning combustion system, which was designed with important contributions from Max Hahn, the man who had helped von Ohain to build his model in Göttingen and who had joined the Heinkel Company with von Ohain. Hahn remained one of von Ohain's chief assistants throughout the war.[209] The salary stipulated in von Ohain's next contract was not very large for an engineer with his duties, but he was promised a bonus after each new engine ran. This was typical of the way that Heinkel used monetary incentives to get quick results; von Ohain was similarly promised an immediate raise after the flight of the firm's first jet plane.[210]

After the success of the first two demonstrator units, Heinkel pushed von Ohain to develop a flight engine as soon as possible, "regardless of the quality of the engineering." For Heinkel's purposes, the engine just had to run; it didn't need to be (in von Ohain's words) a "real mass production engine."[211] At the same time, Heinkel ordered that an experimental fighter airframe be built for von Ohain's first flight engine.[212] He hoped to deploy the same method that he had already successfully used in his airframe work to enter the aero-engine field: gaining commercial and political concessions by using stopgap means to achieve a surprise, first-time performance success (rather than a production success). The proprietor's conception about the difference between demonstration and production prototypes (and the speed with which they could be developed) was drawn from his experience in airframe practice and proved to be misguided in the aero-engine field.[213] Regardless of von Ohain's intentions, his turbojet assumed a key place in Heinkel's business strategy. Heinkel promoted a first jet flight as the surest way to get the resources that von Ohain needed for further turbojet development, so von Ohain went along with his employer's strategy.

Thus in 1938, von Ohain found himself again working without tools or experienced aero-engine personnel and sacrificing his systematic development plans for a quick result—albeit now in pursuit of resources for aero-engine development from which he expected to benefit directly. His first flight engine, the HeS3 engine and a second version, the HeS3b engine, were, like the two demonstrator units before them, quickly built from materials the firm had at hand. The new engine used the liquid combustion system of the HeS2 demonstrator, which was, however, only partly developed. In addition to dealing with ill-suited equipment, von Ohain struggled to recruit competent engineers to his small engine team.[214] Meanwhile, Heinkel assigned his best aerodynamicists, Siegfried Günter and Heinrich Helmbold, to design the single-engined experimental He178 jet airframe meant to carry von Ohain's engine. Although Helmbold designed aircraft, he had earlier developed a two-stage contra-rotating compressor for a high-speed wind tunnel at the Heinkel Company, so he could give von Ohain some useful advice although Heinkel had assigned him elsewhere.[215] The He178 airframe was designed by the same team as the Heinkel He176 rocket-powered plane and was similarly made largely out of wood.[216]

Despite the HeS3b engine's rushed design and construction, the engine propelled the He178 airframe on the world's first turbojet flight on August 27, 1939. Von Ohain later recalled of this achievement, for which he would become famous: "we flew the first time with many prayers ... that was *not* really a sincere engine business. It was simply a demonstration of the principle."[217] The HeS3b was only just sufficient for a first flight[218] and included numerous trade-offs that went against von Ohain's notions of how to go about developing a turbojet engine. (That Heinkel valued firsts can be deduced from the fact that on February 13, 1939, Heinkel de creed that a museum should be established at the firm to house all of von Ohain's engines—and thus the firm's successes.)[219]

By presenting jet flight as a fait accompli in late 1939, Heinkel hoped to force the German Air Ministry, which had initiated its own turbojet program in 1937–38, to support the Heinkel Company's aero-engine development activity, against the ministry's avowed policy of only giving contracts to aero-engine manufacturers because they had the facilities and expertise for aero-engine development.[220] Despite the performance of the He178, which surely boosted the RLM's jet plans, the Ministry nevertheless refused to grant the airframe firm an aero-engine contract.[221] Thus although the engine secured a record for the Heinkel Company, the practical results of the feat turned out to be less than Heinkel had hoped. Although the German Air Ministry acknowledged the Heinkel Company's work in the aero-engine field and granted the firm a contract for a twin-engined fighter prototype (which became the He280) and two engines,[222] these contracts didn't give the Heinkel Company access to aero-engine manufacturing facilities or engineers, both of which were in extremely short supply in wartime Germany.[223] So after the flight of the He178 powered by the HeS3b engine, von Ohain was little better off in terms of resources for developing his engine than he had been before. In fact, because Heinkel's acquisitive actions dramatically expanded the firm's turbojet program in late 1939, von Ohain soon found himself in a distinctly worse position.

The Inexorable Expansion of Heinkel's Turbojet Program

In late 1939, Heinkel made von Ohain's life dramatically more difficult by hiring a second turbojet development team, which competed with von Ohain's group for the firm's already insufficient aero-engine development resources.[224] The same spirit that had made Heinkel welcome von Ohain and his centrifugal turbojet now made him eager to recruit the team led by Max Adolf Müller, who had worked with Wagner at the Junkers Aircraft Company, on an axial turbojet design. Until enough information was available to choose definitively between them, Heinkel wanted his firm to pursue as many variants of the new engine type as possible. In fact, in autumn 1938, Heinkel and von Ohain had initiated work on a turbojet engine (the HeS9) based on an axial compressor designed by Walter Encke at the Aerodynamisches Versuchsanstalt. Von Ohain had attended Walter Encke's lectures while he was a student in Göttingen, and the physicist recommended Encke to Heinkel as a consultant on axial compressors. Encke remained a consultant to the Heinkel Company until the end of the war, although work on the HeS9 engine was abandoned in favor of the axial engine design that Müller brought with him.[225] Müller's team started work at the Heinkel Company on September 1, 1939, five days after the first flight of the He178.[226] With the addition of the Müller team, the Heinkel Company became the only company in Germany to have two turbojet teams working independently to develop centrifugal and axial engines.

In addition to its advanced axial turbojet design, the so-called Wagner-Müller engine, Müller (a very enthusiastic supporter of the Nazi Party)[227] brought with him something especially valuable to a firm lacking aero-engine resources: a dedicated group of about fourteen engineers, who had been working together for several years.[228] The arrival of Müller's team more than doubled the size of the turbojet program at the Heinkel Company, which had grown from a start of two men in 1936 to about ten in 1939. By 1940 it had reached about thirty (including Müller's team)—an improvement, but still far less than Power Jets' staff at the time.[229]

Further exacerbating the shortage of resources for von Ohain, Müller indulged Heinkel's tendency to support new projects by beginning a range of new jet engine developments. His group's most successful engine was the "extremely advanced" and impressively efficient axial engine known internally as the HeS30 engine (German Air Ministry number 109-006), which was based on the group's previous work at Junkers.[230] In addition to the two HeS30 units that were eventually completed at the Heinkel Company,[231] Müller and his engineers also devoted resources to work on a bypass axial turbojet engine, the HeS30a; a constant volume combustion (as opposed to constant pressure) turbojet engine built from HeS30 parts, the HeS40; and a motorjet project begun at Junkers, which resulted in the HeS50 and HeS60 engines.[232] None of these engines was, however, successfully completed, and all were devised seemingly without regard for the limits of the firm's existing capacity.[233] Most of them were given up during 1942. Müller left the company in June 1942 (after Heinkel was no longer in control, see below).[234]

In 1940, an opportunity for the Heinkel Company to gain both more experienced aero-engine personnel and aero-engine facilities presented itself, when the German Air Ministry put the small Hirth Motor Company of Stuttgart-Zuffenhausen up for sale. (Hellmuth Hirth, its founder, had died in 1938.) Several firms competed for the purchase.[235] Heinkel

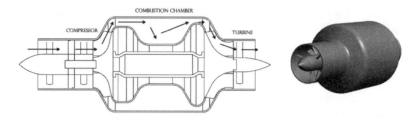

FIGURE 3.3 The Heinkel 109-001 or HeS8 jet engine designed by Hans von Ohain. Approximate length 2,400 cm, approximate diameter 780 cm. *Source*: Models by Denie Technology

argued that Hirth Motors would be especially useful to the Heinkel Company's turbojet work because Hirth, a piston engine maker, built small in-line, air-cooled engines, which could be used in Müller's motorjets, and because under government ownership, Hirth had begun making turbo superchargers, which shared many elements with turbojets.[236] The development department at Zuffenhausen had a staff of about eighty-three, which easily exceeded the staff devoted to aero-engines (turbojets) at the Heinkel Company.[237] Once again, the importance of piston aero-engine expertise and facilities to the turbojet is evident.

Ernst Heinkel hoped to secure the sale of Hirth Motors for the Heinkel Company through the demonstration of the company's turbojet fighter prototype, the He280. Once again, he was relying on a spectacular first to convince the German Air Ministry to favor him. This ambition increased pressure on turbojet development at the firm. The original choice for the He280 airframe's engines was the promising, smaller-diameter HeS30 axial engine. When this engine proved incapable of running self-sustaining under its own power, however, von Ohain was diverted from the compressor research that his group had begun in order to design and build a centrifugal engine that would enable the airframe to fly in the spring of 1941. This engine was the HeS8.[238]

Still struggling with a lack of resources and personnel, von Ohain now found himself building an engine in order to acquire an engine company. He was coming to see that aero-engine development at the Heinkel Company was not necessarily about making production turbojet engines. As he later told it: with the HeS8 "again there was a string attached."[239] The design and construction of the engines for the twin-engined He280 was hurried, but they successfully powered the world's first turbojet fighter on March 30, 1941. With the flight, von Ohain achieved yet another turbojet record; however, it was again in the service of Heinkel's agenda not his.

The HeS8 engine was yet another distraction from von Ohain's more systematic development program aimed at resolving the basic problems of the new engine type.

The demonstration flight of the jet powered He280 for Ernst Udet, head of the German Air Ministry's Technical Office, in April 1941, seemed indeed to have made more of a political difference than the earlier flight of the experimental He178 aircraft. The Heinkel Company was subsequently allowed to purchase Hirth Motors. The German Air Ministry also gave the firm a production contract for one hundred preproduction HeS8 engines (Air Ministry number 109-001) and for the preparation of facilities for HeS8 manufacture. The contract was given on Ernst Udet's orders. Udet was a friend of Heinkel and a supporter of the Heinkel Company's turbojet work. He is better known, however, for his unsuccessful guidance of the German Air Ministry's engine development program during the war and his suicide in November 1941 (his position was taken over by Erhard Milch).[240] Heinkel, confident about the future of his turbojet program and his achievements to that point, wrote to a friend on August 8, 1941, that "the start from the airframe field into the aero-engine field has so far been very promising."[241]

In the process of trying to turn the HeS8 engine into a production design, the faults of the engine's rushed design became apparent. Despite an additional year of development, the engine never reached suitability for production. Like the HeS3b engine, the new engine had incorporated temporary solutions that were sufficient for a first flight but not for routine service, and von Ohain was unable to resolve them with his existing resources.[242] The first order for the HeS8 engine was reduced to twenty-four engines for experimental work, and any notion of preparing the engine for production was canceled in March 1943, after the Jumo 004 was ordered into production.[243] Von Ohain attributed the HeS8 engine's failure to the Heinkel Company's lack of the "fundamental prerequisites" for engine development: test stands, manufacturing equipment, and quality control specialists.[244]

Heinkel boasted that the He280 had been decisive in clearing the way for the Heinkel Company's acquisition of Hirth Motors, as he hoped it would be,[245] but the reasons for the sale were less dramatic. The decision to sell Hirth Motors to the Heinkel Company was the product of a campaign by the firm's management during 1940 (based on the firm's inventive record), but even more importantly, the sale relied on internal Air Ministry support for the turbojet program of the Heinkel Company. Already in mid-1940, most of the German Air Ministry supported the Hein-kel Company's bid to continue developing turbojet engines.[246] The last re-maining contender for purchasing Hirth Motors withdrew his bid in early 1941,[247] and by March 6, 1941, some three weeks before the He280's first turbojet flight, Heinkel had already been given approval for the acquisi-tion.[248]

A crucial factor that swung the sale of Hirth Motors for the Heinkel Company was the advocacy of Helmut Schelp, the leader of the Ger-man Air Ministry's turbojet development program, who recognized that the lack of appropriate facilities was hindering the work of the skilled turbojet designers the firm employed. The acquisition of an aero-engine manufacturing firm would finally connect what he saw as the Heinkel Company's important development teams with facilities and personnel experienced in aero-engine production, an essential prerequisite for ef-fective turbojet development.[249] The sale was certainly a gain for Schelp's program; Hirth Motors represented a "fairly large addition of manpower" to turbojet development at a time when the German Air Ministry's de-partments were competing for limited resources.[250] Indeed, after the sale, Schelp undertook to steer the use of the facilities at Hirth Motors accord-ing to his own plans, rather than Heinkel's. Heinkel's impetuous manage-ment style was blamed for spreading the resources of the Heinkel Com-pany too thinly among multiple aero-engine development projects,[251] and a condition of the Hirth Motors sale was that the Heinkel Company's turbojet work would be transferred to the smaller firm.[252] Schelp forced Heinkel to concentrate his resources on a single engine by instructing the firm to cancel, during 1942, all of its existing projects, including the HeS30 engine, which ran for the first time in May 1942, duplicating the perfor-mance of the axial Jumo 004 and BMW 003 engines, both of which were closer to production at the time (if less efficient).[253]

In place of all its previous projects, Schelp initiated work at the Hein-kel Company on the development of the HeS11 engine, which was to be a more powerful successor of the BMW 003 and Jumo 004 turbojets.[254] Von Ohain was put in charge of the engine design and with Schelp de-signed its unique diagonal compressor, a mixed centrifugal and axial ele-ment, which Schelp argued would not be vulnerable to icing up or stone hits during takeoff and landing and would also beneficially preheat the in-coming air before it reached the engine's axial compressor.[255] The design of the HeS11 engine was completed by autumn 1942, but much remained to be ironed out during manufacture and testing.

New Prospects for von Ohain

The benefit of the purchase of Hirth Motors in 1941 was not uniformly felt by the Heinkel Company's different turbojet projects. Müller's group moved to Zuffenhausen and its aero-engine facilities, while von Ohain was left in Rostock.[256] This state of affairs was altered only through the actions of Diplom-Ingenieur Harald Wolff, who took charge of turbojet engine development at Heinkel-Hirth on August 1, 1941.[257] Wolff, whom Schelp had suggested to Heinkel for the post, was an erstwhile director of Bramo and BMW and was, unlike Heinkel, an "experienced and acknowledged specialist" of aero-engine development.[258] In June 1942, without consulting Heinkel, Wolff fired Müller, in part because his team refused to work with the men at Hirth Motors.[259] Müller left and was followed by most of his engineers.[260] Von Ohain was made the leader of the firm's remaining turbojet work, which focused on the HeS11 engine. Heinkel's irresolution continued to prevent the transfer of von Ohain's group to Hirth Motors, however, even after the German Air Ministry insisted in July 1942 that Heinkel finish transferring all of his turbojet work to Zuffenhausen.[261] Meanwhile, Wolff was getting increasingly frustrated by Heinkel's intrusive second-guessing of his decisions and tendered his own resignation on February 24, 1943.[262] Through the intervention of the Air Ministry, however, Heinkel was instead removed from oversight over the program. Wolff was convinced to return to Zuffenhausen, where he was made Kommissar of Heinkel-Hirth Motors on March 27, 1943. This appointment was made in the hope of speeding up work on the HeS11, which Milch declared to be "decisive for the entirety of air armaments."[263]

Now in charge, Wolff quickly moved von Ohain's work to Zuffenhausen, where von Ohain at last found himself at the head of a "first class engineering staff," with access to testing and manufacturing equipment suited to aero-engine development. With Müller's group gone, about 150 engineers, designers, and draftsmen were made available at Heinkel-Hirth for work on the HeS11 engine.[264] Key among the Hirth staff was its talented chief engineer, Max Bentele, who was put in charge of engine components for the new engine type.[265] Heinkel-Hirth's production facilities were kept partially occupied through the continued manufacture of turbo superchargers—von Ohain didn't yet have an engine ready to be manufactured.[266]

From mid-1943, the conditions for von Ohain's work were better than at any time before. Heinkel continued to pressure development of the

HeS11 engine, but Wolff now shielded von Ohain from Heinkel's demands. In contrast to the HeS1, HeS3 and HeS8 engines, which made history, the HeS11 engine was von Ohain's first real engine project. His team collaborated productively with the personnel at Heinkel-Hirth, and the development of the engine moved ahead rapidly. Although some of the problems of the Heinkel Company's earlier centrifugal engines could now be overcome or avoided, the new engine was not without shortcomings.[267] Bentele recalled that the sheet-metal-heavy annular combustor of the HeS11 engine, which owed much to earlier turbojet development at the Heinkel Company, "barely fulfilled its purpose."[268]

Although mostly free of Ernst Heinkel's influence, work on the HeS11 was put under pressure by the war situation and by the Air Ministry's decision in 1943 to manufacture and deploy jet engines as soon as possible. By the end of 1943, the company was under political suspicion for unnecessary delays with the engine.[269] Pushed from all sides to once again sacrifice thorough development for speed, von Ohain's team worked to move the engine toward production. In November 1944 (after Milch had decreed that Daimler-Benz should collaborate with Heinkel-Hirth as described in the previous chapter), Schelp threatened Heinkel's directors that the German Air Ministry would confiscate Hirth Motors and give it to Daimler-Benz, where the 011 engine design would be completed "with greater élan and tenacity."[270] Schelp's lack of faith in development at the firm was demonstrated by his decision late in the war to develop an engine for the Ar 234 jet bomber by scaling up an existing engine rather than developing the HeS11 engine to higher thrust.[271] It became clear that even with appropriate facilities and experienced personnel, one-and-a-half years were not enough to get the novel engine into quantity production.[272] The Jumo 004 and BMW 003 engines, which Heinkel hoped that the HeS11 engine would catch up with, had been in development at least two years longer. Heinkel later denounced the engine as a mistake ("Irrweg") because it was not the quick route to a production aero-engine that he had expected it to be, whereas the more systematic von Ohain viewed the HeS11 as the first step toward a new generation of high-powered German turbojet engines.[273] The contrast between these views reveals Heinkel's and von Ohain's fundamental disagreement about the purpose of the firm's turbojet program.

That the HeS11 did not reach quantity production during the war was in no small part due to the nature of development at the Heinkel Company during the first seven years of von Ohain's turbojet work there. At the end of the war, Hirth Motors and the Heinkel Company—then the Ernst Heinkel Aktien-Gesellschaft [public limited company], because on April 1, 1943, financial emergency had forced Heinkel to reorganize his company—stopped producing turbojets. Apart from some interest from American investigators after the war in some of the HeS11 engine's elements, including especially its air-cooled turbine,[274] and the construction of some six of the engines for the US Navy (the project was shut down in response to Soviet complaints),[275] von Ohain's turbojet work at the Heinkel Company had no direct successors. After the war, von Ohain immigrated to the United States, where he remained an active and successful aero-propulsion engineer. Hirth Motors became independent once more. When the remainder of Heinkel's firm was allowed to take up aviation production again in the mid-1950s, Heinkel's efforts to get the HeS11 engine into license production in Spain failed. Between 1951 and 1956, when the work ended, only three engines were built under the Spanish designation INI-11.[276] The Heinkel Company never managed to become an aero-engine producer as Ernst Heinkel had wished, and von Ohain's fame rests to this day on the records established with stunt turbojet engines rather than on the outcome of careful turbojet development, as he would have preferred.

Conclusion

Despite their very different natures, Power Jets and the Heinkel Company had several things in common. Turbojet work at both firms began early. In order to retain their independence of action, both of the inventors, Whittle and von Ohain, chose not to ally themselves with the established aero-engine industry, and both men began by designing centrifugal engines, which were simpler and faster to develop than axial engines (and thus quicker to reach the point of demonstration). Both men were the first in their countries to get their engine designs built, although as production engines their designs were quickly abandoned, and neither managed to get further designs into production nor to get their designs taken up by other firms. The lack of success of both men was less a comment on their inventive achievements, however, than on the nature of the institutions where they worked. Whittle's early work was shaped by the necessity of impressing the British government with a viable service engine. Von Ohain's first engines ran before Whittle's, but they were designed to meet a very different set of requirements, because Heinkel's first interest was in producing inventions in order to gain resources, rather than in producing engines in order to gain revenue.

Power Jets' later goals were shaped by its institutional insecurity and by its strong belief in its uniqueness and centrality to the British jet effort. The firm was dependent on government funding and weakened by the (required) dissemination of its work to commercial firms in the British and American aero-engine industry. The effect of the peculiar mix that faced the firm was that it began to pursue excellence in turbojet design with little reference to the requirements of the market or the military. Power Jets had long been at a disadvantage because war had intervened before the firm had a chance to establish itself as a commercial enterprise. War accelerated and shaped the firm's evolution into a research and development company and largely negated the management's ability to establish a viable commercial position for the company after the war.

In contrast to Whittle's challenges in finding funding, von Ohain found ample funding at the Heinkel Company. Yet von Ohain still struggled throughout the war to find appropriate facilities at an airframe company whose owner implausibly insisted that engine development should run like airframe development. Even in his improved position at Heinkel-Hirth, von Ohain still had fewer resources than were made available to Whittle and Power Jets. Power Jets already had 140 employees at the end of 1940 and subcontracted its construction work to a steam turbine and engineering company, whereas the Heinkel Company kept its engine production activities mostly in house, where they were bedeviled by the lack of experienced staff and appropriate facilities. Unlike Power Jets, the Heinkel Company was an established airframe manufacturer, for which aero-engine manufacture was at best a secondary pursuit. Von Ohain was much less central at the Heinkel Company than Whittle at Power Jets, and the German became part of an existing culture that the firm's owner, Heinkel, defined. Because of Heinkel's strategy to attract resources to his firm through inventive success, the Heinkel Company did not pursue a production engine in the first five years of its turbojet work. Under Heinkel's impetuous leadership, von Ohain's systematic work was repeatedly sacrificed for stunt engines. By joining an existing and fully fledged firm, von Ohain was able to go to work on his idea immediately. Yet the Heinkel Company proved ill-suited to turbojet development, whereas within the constraints of time and money, Whittle was able to shape Power Jets according to his own needs for more than five years.

The fact that Power Jets received virtually a blank check from the Air Ministry to design and develop turbojets during the war did not make it more successful than the Heinkel Company in the long run; both had to fit into a broader industry full of strong players. Instead, dependence on the British state made the company defensive and militated against the formation of deeper alliances with established companies, a remoteness and suspicion that was mutually reinforced by manufacturers' increasing self-reliance in designing turbojet engines. The opening of Power Jets' large Whetstone factory in 1943 led the company to reject the possibility of becoming a consultancy like Ricardo's successful firm, insisting instead that it be allowed to produce engines itself.[277] In the end, this was not a successful business strategy. Other models were more successful. In contrast to Power Jets, which ceased to exist in 1946, Ricardo still designs and develops high-powered piston engines. Yet another model was provided by Frank Halford, an important British engine consultant who decided to ally himself with an existing company rather than continue to work independently. Before the end of the war, that company spun off the de Havilland Engine Company and Halford was put in charge of it.

After the war, when Power Jets was turned into a national establishment, the company had few supporters like Cripps left. The firm's aspiration for preeminence in commercial aero-engine design was lost to the competition along with the centrifugal turbojet designs it had favored throughout the war. Indeed, the opening of the National Gas Turbine Establishment on July 1, 1946, with a research program dominated by staff from the Royal Aircraft Establishment, was the ultimate expression of the establishment's influence in the gas turbine field. With the new national establishment, the Royal Aircraft Establishment's more flexible understanding of its role as an advisory government research institution ultimately prevailed over Power Jets' commercial ambitions.

The Heinkel Company followed a different trajectory than Power Jets, but its turbojet work reached a similar, unproductive end after the war. It was Heinkel rather than von Ohain who felt a paternalistic attachment to his firm and sought to extend its activities through turbojet invention. The conviction of the centrality of inventors was shared as much by Heinkel as by Whittle's supporters and was ultimately also a disservice to the Heinkel Company. In contrast to Power Jets, which shifted from designing a production engine to doing research work without reference to production engines, the Heinkel Company tried first to survive on turbojet design, and later shifted to focusing its efforts on production. Despite being forced to change its focus after buying Hirth Motors, the Heinkel Company remained much more research oriented than production ori-entated to the end of the war. When von Ohain was finally able to turn his attention to seriously developing an engine, too little time remained for the HeS11 engine to reach quantity production and von Ohain's early turbojet legacy ended with it.

The Construction of a Hero

The foregoing chapters sketched a picture of the world of activity that Whittle's and von Ohain's inventive work was a part of, a world peopled by a multitude of innovative engineers and institutions in many countries. Beside this, the heroic story of Whittle that was later reproduced for von Ohain looks distinctly peculiar. Yet the fondness for telling invention stories in the model of individual genius continues to ensure that although public interests have shifted since the postwar period, the center of the popular jet engine story—the inventors—has never seriously been challenged. The unchanging nature of the story suggests that our understanding has become static, disconnected from more recent concerns that should produce new forms of understanding that reflect current values.[1] Examining how one particular story came to dominate the public story of the jet engine reveals the way in which cultures of memory, variable access to (or interest in) the historical record, and cultural predilection influence historical narrative. This chapter examines the popular reception of the jet engine story over time, which looked very different from the perspective of different countries.

The first story of the heroic invention of the jet engine appeared in Britain, where the inventor Frank Whittle was honored. The strains of the earliest jet engine propaganda in Britain were strongly nationalist. After the war, the jet engine was a leading example of the nation's technological prowess that could be readily sold abroad at a time when the realities of heavy indebtedness put pressure on increasing foreign exports. Wartime collaboration with the United States, Britain's key commercial competitor in jet engines, was portrayed as unambiguously one way, with the United States having benefited from Britain's work.

Whittle's elevation fit Britain's tradition of publicly celebrated inventor-heroes (he has since been the subject of several public memorials), and his role as a heroic inventor was quickly established for both domestic and international audiences after 1944. The notion of a "heroic inventor," borrowed from Christine MacLeod, is used here to refer to Whittle's public profile and does not reflect a judgment of his actions. The popular story of the jet engine first appeared in a wartime setting in which the spread of information about technical development was carefully controlled. Official government restrictions on the publication of official information about the jet engine in Britain and the United States left an opening for the story's presentation to the public by the press through a straightforward narrative of a lone, visionary inventor, which Whittle came to embody. The heroic conception of invention that made this story so plausible was rooted in nineteenth-century debates in Britain that linked patents, individuals, and invention in the public mind—debates that may well have informed Whittle's supporters.[2]

Examining the national dimensions of the jet engine story in the first postwar decades helps to explain many of the peculiarities of existing accounts of the jet engine. The German story, for example, was first revealed through postwar intelligence efforts by the Allies, who tried to locate every person who had been involved with Germany's jet engines. The result was a remarkably inclusive record of activity in Germany. This contrasts with the story in Britain, where detailed information about wartime work remained dammed up behind corporate walls. The story that emerged there was shaped by political and popular demands for a heroic invention story. In the United States, where firms were rapidly developing and selling jet engines after the war, the two foreign stories were combined, with the German story reworked in the same heroic model as the British. Hans von Ohain, the German inventor, became a counterpart to Frank Whittle. Many of Germany's leading jet engine engineers moved to the United States after the war.[3]

The Allied intelligence story of the German jet engine remained decisive because there was little attempt in Germany in the immediate postwar period to piece together the history of German jet aviation under National Socialism. The first real attempt to write a German history of German aviation came in the early 1980s. By that time, Hans von Ohain had already been accepted as Germany's jet engine inventor in the United States. This had little effect in strongly techno-nationalist Britain early on, but today the "dual inventor" story first popularized in the United States is internationally dominant.[4] The dual inventor story served a specific rhetorical purpose in the United States, which was a party to Britain's development during the war and learned about Germany's work after it. The dual inventor narrative explained that the country's industry rested on a dual inheritance that was improved upon by American expertise. Importantly, this allowed the story of the jet engine to become a story of the postwar success of American engineering.[5]

The Jet Story Enters the Public Domain

The existence of the jet engine in Britain and America and its connection to the work of Frank Whittle were first made public on January 6, 1944. The release was controlled by MAP's jet publicity policy, which insisted on a ban on jet publicity apart from approved releases. In addition to concerns about the danger of revealing information to the enemy, a key focus of MAP's jet publicity policy from the start was ensuring that Britain get credit for the invention. Indeed, the reason that MAP originally requested from the United States a ban on jet publicity in November 1943 was largely because the Ministry was afraid that credit for developing the jet would be contested, with the American press claiming the invention for America. As Air Marshal Wilfrid Freeman, chief executive at MAP, reminded his colleagues, the same had earlier happened with radar, another British machine shared with United States for production reasons. In that case, he argued, American claims to have invented radar had created "complete confusion in the minds of the American public" and had irreversibly damaged British credit for the invention.[6] Defending the need for a stringent ban in the case of the jet engine, Freeman argued that "unilateral publicity from USA would undoubtedly give the impression that since the Americans were first to broadcast the facts they had been first in the field of development. This, of course, is absolutely untrue, and it would be most unfair to our industry to allow the idea to gain currency."[7] Freeman's interest in invention claims rested on their perceived economic consequences but had broad effects.

The proposed ban went through in November 1943. It was agreed between the chief of the air staff, Air Marshal Sir Charles Portal, and the commanding general of the United States Army Air Forces, General Henry "Hap" Arnold, that any Allied jet publicity in the future would take the form of jointly agreed releases that were published simultaneously. The first press release about the jet, released on January 6, 1944, was issued, like all other wartime releases about the machine, by the British Air Ministry and American War Department on behalf of the Royal Air Force and the United States Army Air Forces.

The first release was not a result of British grandstanding, however. Instead, it came about through American pressure when, in November 1943, General Arnold urged that an official press release be made as soon as possible in order to prevent unauthorized press speculation about test flights that were taking place over an ever-larger land area in California. The British Air Ministry would have preferred not to publicly reveal any information about the jet, but it was afraid of losing control of the story through voracious, unilateral publicity in the United States. Citing security concerns, the Air Ministry instructed the Royal Air Force delegation in Washington, DC, responsible for upholding the ban, to try to get American agreement not to make any disclosures to the press. MAP wanted (as Freeman had argued) to prevent "premature publicity [being] given to American developments" as it "might lead to world-wide misunderstanding" of British and American contributions to the jet.[8] So while acquiescing to Arnold's request, MAP sought nevertheless to use the unwelcome release for its own purposes, hoping that a joint announcement could be used to ensure its central objective: American recognition of Britain's pioneering role in the new technology.[9]

Toward the end of December 1943, a joint statement for release in January began to be drafted via cypher between Washington and London. The text was traded back and forth, with each side proposing amendments, until it was finally approved by Arnold and Portal. The Royal Air Force delegation objected to the first text sent from the American War Department because it did not give enough credit to Britain. An amendment suggested by the British delegation addressed the problem by adding that the first work on jet engines had been begun in Britain by Group Captain Frank Whittle, that the first engine had run in Britain in 1937, and that a Whittle engine had been flown successfully in Britain in 1941. The Air Ministry in Britain passed the new text to Freeman, to ensure that it was acceptable to give Whittle full credit for the invention. The Air Ministry was cautious in this because it had earlier run into trouble over the attribution of the so-called Leigh Light to Wing Commander Leigh alone. Freeman saw no problem with the attribution to Whittle, however, arguing that it was justified because Whittle's jet patents had been granted at a time when the Air Ministry was uninterested in Whittle's work.[10]

The first press release of January 6, 1944, was issued after programs to develop jets had been underway in both countries for several years, but it was in that release that a first, abbreviated account of the secret project's history was made public. The press release reported that jet aircraft would shortly begin being produced in both countries, but gave full credit to Britain for its pioneering role. It noted that the British Air Ministry had ordered a jet airframe from the Gloster Company in 1939—a sign of foresight and an act of confidence in the new technology. Appended to the press release were notes on the careers of Frank Whittle, "the inventor of the first successful jet propulsion engine for fighter aircraft," and P. E. Gerry Sayer, the test pilot who flew the Gloster E.28/39 experimental airframe on its maiden flight.[11] Even American achievements were cast in light of British work; although the release reported that the first American experimental jet aircraft had flown successfully on October 1, 1942, for example, it stated unequivocally that Britain's engine design had been passed to the United States (the full text is given in Appendix B).[12] The press lost no time in reproducing the official release and publishing the corresponding story of heroic invention.[13] Thus the public revelation of the jet propelled Whittle into popular celebrity and claimed the invention for Britain.

Whittle was taken aback by his dramatic rise to public prominence after the release. Less than a week later, on January 11, 1944, he made his objections to the contents of the official press release clear in his yearly address to the employees of Power Jets. He assured the works that his remarks on the BBC on January 7, in a radio interview in which he endorsed the government line, had been strictly circumscribed by the broadcasting company. He angrily complained that by mentioning only himself and Power Jets and neglecting the teamwork that had been essential to the development of the first British jet engines, the release gave a "lopsided" view of jet development. Whittle found it inexplicable that the press had not published the statement given to them by Power Jets, which acknowledged, for example, Power Jets' continued collaboration with the British Thomson-Houston Company.[14] Indeed, despite being praised by the first release, Whittle was not a party to the government agreement on jet publicity and had little direct influence over it.

The Air Ministry chose not to officially rectify the exclusion of the British Thomson-Houston Company from its release. The Ministry chose not to act despite the fact that the firm had already written to MAP expressing its indignation about the omission, which seemed especially egregious since the release had named the American company General Electric.[15] The British government's decision was based on the fact that public interest in the jet was dying down, and MAP was afraid that extending recognition to the British Thomson-Houston Company would provoke additional, undesired public attention. In addition, any further publicity in Britain would risk endangering the Ministry's ability to pressure the United States Army Air Forces to keep its side of the publicity bargain. In February 1944, the British Thomson-Houston Company was given public acknowledgment in a letter to *The Times* from Sir Maurice Bonham Carter (a director of Power Jets) and an article in the British trade magazine *Flight* ("B.T.H. and the Jet: British Manufacturer's Share in Development") and the matter was left at that.[16]

Publicity after the First Press Release

Giving Whittle credit and through him Britain for the invention of the jet engine was a weapon in the technological rivalry between Britain and the United States.[17] During the war, the battle over jet publicity was a spirited contest between the Allies because of the substantial postwar stakes, when inventive genius presumably would be transfigured into commercial success. The story of Whittle was taken up with more alacrity in Britain than in the United States, but it still had an effect across the Atlantic. Although the American press didn't foreground Whittle as the British press did, contriving to mention the work of the two American firms Bell and General Electric as frequently as the British inventor, it did acknowledge the British contribution to the American jet engine as having been critical. On January 10, 1944, for example, the *New York Times* ran a Bell Aircraft advertisement that asserted that "Bell Aircraft has built the first American fighter planes powered by jet propulsion engines. The engines, constructed by the General Electric Company, were developed from British designs.... Like the Wright Brothers' flight at Kitty Hawk, we believe that the hundreds of successful flights made by Bell jet propelled ships open a new chapter in American aviation history." Arnold was soon informed of the unilateral advertisement's infringement of the jet publicity agreement, and to keep up the British end of the bargain, Portal instructed the Air Ministry that everything possible must be done to prevent similar lapses in Britain. It was essential to avoid the charge that Britain wasn't playing by its own rules for fear of losing power over the American press.[18] On March 31, 1944, all British manufacturers were instructed that their advertisements had to be submitted to censorship.[19]

Despite complaints from Britain, the American press continued to give publicity to Bell and General Electric's jet work in the United States. In April 1944, the minister of aircraft production, Sir Stafford Cripps, wrote to Portal explicitly on the matter, quoting several articles in the American press that infringed the joint jet publicity policy. Cripps argued, in support of Britain's jet firms, that the effect of American publicity was to give the impression that "the origin of this revolutionary development is not British but American. . . . Already there is considerable feeling in this country . . . that the United States are being allowed to walk away with a British Invention."[20] Cripps also objected to American publicity for the Italian Caproni-Campini motorjet (in which the compressor was powered by a piston engine rather than a turbine), which had first flown in August 1940. The motorjet that powered the airframe provided an alternate invention story to Britain's, and Cripps suggested countering American publication of the story by publishing an "obsolete" photograph of the Gloster E.28/39, the first (true) jet to fly. Cripps's request was turned down by the Air Ministry, however, which was unwilling to risk jeopardizing the joint ban on publicity through a unilateral release. It felt that the publicity ban, which it continued to defend, was the only thing—however partially effective—protecting British interests from the mighty and nationalistic American press.[21] It did not see the Caproni-Campini story as a threat.

During the first half of 1944, MAP brought at least seven American articles on the jet to the Air Ministry's attention.[22] In July 1944, Freeman wrote to Portal about an article that had appeared in the American *Evening Standard* that featured General Electric's jet engine production plans. He argued that the article proved that the publicity ban was not being enforced in the United States. Simultaneously, he suggested that the impending operational deployment of the Meteor offered an "excuse" for some "limited publicity" for British work. In addition to reasserting Britain's invention claims in the United States, such a release of information, he argued, would be useful for Britain because it would motivate British workers in airframe and engine production as well as reassure the British public that the Germans weren't technologically ahead of the Allies.[23] The Royal Air Force delegation also felt that the Meteor's success deserved to be publicized. At the same time, the delegation also wanted to douse American clamoring for permission to advertise American turbojet research and development.[24]

When on August 14, 1944, General Arnold suggested relaxing the joint Anglo-American publicity ban in order to respond to reports of the deployment of the first German jets, the British government agreed that releasing some information on Allied jets was desirable.[25] With some mis-givings, the War Cabinet approved the release of a second joint statement on September 11, 1944.[26] Published on September 27, the release was much shorter than its predecessor. It reassured the public that the arrival of German jets had been expected, that the Allies' jets had made progress since the first release in January, and that British jet fighters had already been "employed with success against the flying bombs [V1s]." No manu-facturers were mentioned (the full text is given in Appendix B).[27] The re-lease was accompanied by photos of the first Allied jets, neither of which was to go into service: the single-engined British Gloster E.28/39 and the double-engined American Bell P-59. A caption explained that the E.28 was the first Allied jet to have flown.[28] Afterward, the Royal Air Force delegation in Washington, DC, reported approvingly that a full American media frenzy about the Bell P-59 aircraft had only been prevented by the "wonderful co-operation and magnaminity [*sic*] of the War Department," which refused to allow the radio or film news to cover the story, "espe-cially as their aircraft is a better looking job than the Gloster" (the P-59 was a fighter prototype, whereas the E.28 was an experimental aircraft). Such cooperation was not to be taken for granted. The cancellation of an impending trip by Whittle to New York was a relief to the delegation, as the War Department "would have gone up in smoke if he [Whittle] had come out with a fresh burst of publicity for British J.P. [jet propulsion]." Like Portal, the delegation wanted to ensure that both nations contin-ued to respect the ban because it too believed that the ban's existence was advantageous to Britain, in its continued work in "keeping the Brit-ish achievement before the American public."[29] So although it agreed to the release, the British government insisted that the ban should remain in place.[30] Despite appeals from the American War Department for ad-ditional publications, the Air Ministry refused to relax the Arnold-Portal agreement without a new publicity policy being agreed on in its place.[31]

The continued enforcement of the publicity ban made American jet manufacturers restive, and they began to put a conspiratorial interpreta-tion on it. *Aviation News*, an American weekly, published confident specu-lation on November 13, 1944, that the British authorities were forcing the jet publicity ban on the United States Army Air Forces in order to repress reports of American progress that might "possibly overshadow" Britain's work.[32] The British responded to this provocation with lightning speed. By the next day, the publisher, McGraw Hill, had been asked to print a cor-rective statement in its next issue giving the true details of the ban, "mak-ing clear that were it not for security considerations Britain would be the first to desire full publicity showing the scope of her whole achievement and current production in this field."[33] The Air Ministry did not feel that it was unjustly looking out for Britain's industry, but that it had real accomplishments to publicize.

It was only in February 1945, that the British government finally conceded that it was time to revise the Arnold-Portal agreement.[34] Its hand was forced by the United States Army Air Forces' determination to release a statement on the most promising American jet aircraft to date, the Lockheed P-80, in order to improve morale and attract workers to America's jet factories. After some debate, the British decided to release a detailed statement about the Meteor simultaneously with the announcement of the United States Army Air Forces about the P-80 on February 28, 1945. The Air Ministry insisted that the American announcement mention that the first engine for the American P-80 aircraft was built by the British firm, de Havilland—a fact also included in the British release.[35] Releasing information about the Meteor, the Royal Air Force delegation, judged, would certainly "overshadow" the American release as the Meteor was already in active service.[36] Indeed, the Air Ministry published details of the Meteor's operational role in the "Battle of the Flying Bomb" (against German V1 flying bombs) as well as giving specific details about British jet airframe and engine manufacturers. Indeed, the statement ultimately read like an extended advertisement for British industry. Again Whittle was foregrounded. It reported that:

> The first and so far the only jet propelled aircraft of the United Nations to go into action against the enemy is the Gloster "Meteor." ... The R.A.F. [Royal Air Force] "Meteor" proved to possess a greatly superior speed to the pilotless German flying-bomb and many tactical lessons were learned from these early combats.
>
> Like the Gloster E.28/39, the first turbine jet aircraft in the world to fly (in May 1941), the "Meteor" is also a product of the Gloster Aircraft Company (Hawker-Siddeley Group). The "Meteor" is powered with Rolls Royce engines manufactured to the basic design of Air Commodore Frank Whittle, R.A.F. [Royal Air Force], in collaboration with Power Jets Ltd., and the British Thomson-Houston Co., Ltd. The first engine supplied to the U.S. Army Air Force [*sic*] in Oct., 1941 was built by Power Jets Ltd. Air Commodore Whittle visited the U.S.A. in order to assist our Allies to initiate their development programme.
>
> ... It is known that the Rolls-Royce engine of this type is more efficient and of longer life than the Jumo engine of the German Messerschmitt Me 262 (the full text is given in Appendix B).[37]

FIGURE 4.1 Rolls-Royce jet engine advertisements from 1945. *Source*: *The Times*, April 19, 1945; *Flight*, October 25, 1945

Indeed, Rolls-Royce published an advertisement in *The Times* in April 1945, which quoted the official press release exactly.[38] A few months later, in October, the firm ran an ad in the periodical *Flight* that made no mention of Whittle but referred to jet propulsion as "another British invention." Yet in contrast to the advertisement in *The Times*, the second ad did not mention invention and focused instead on Rolls-Royce's achievement in designing and producing "the first Jet-propulsion engines to be put into production in this country [Britain]."[39] Rolls-Royce's new focus was made possible by the replacement in June 1945 of the original Arnold-Portal agreement by a new policy that set out a list of items that could be publicly discussed without mutual consultation. It explicitly prohibited, however, the release of detailed information or performance data on engines.[40] The agreement remained in force even after the war and was set to extend until the future of all British jet patents had been secured. It was canceled only on January 18, 1946.[41]

After the end of the publicity ban, the British aero-engine industry continued for some time to invoke its connection to Whittle, both demonstrating and reinforcing the continued familiarity of the story in Britain. In 1950, de Havilland placed an ad for its jet engines in *The Times*, which boasted that when the authorities "felt that we could afford to divert some serious engineering capacity from piston-engine design to concentrate upon Sir Frank Whittle's invention of the gas-turbine jet engine for aircraft," "the de Havilland Company through Major Frank Halford was the first of the established engine builders to enter the new field with a jet intended for quantity production." The advertisement closed on a nationalistic note coupled to a (perhaps self-serving) warning: "Britain holds the lead in jet-engine development, but to retain it will call for strong and sustained effort."[42] British companies could promote engine sales and other support by connecting their work to national interests. One popular way to do this was by invoking the Whittle story that the British government had so carefully defended during the war.

Relations between the British Government and Its Inventor-Hero

From the day of the first press release about the jet, Whittle began receiving awards and honors from the British government as well as academic and industrial bodies. Each public award was a testament to and an enhancement of Whittle's public profile. In January 1944, Whittle was granted a CBE (Commander of the Most Excellent Order of the British Empire) by the government and a master's degree from Cambridge University.[43] In 1947, he was made a Fellow of the Royal Society,[44] and in 1948, he was knighted (like several other wartime inventors).[45] His recognition spread beyond Britain. In 1946, he was made a commander of the United States Legion of Merit and received the American Guggenheim Gold Medal for achievement in aeronautics.[46] In 1951, he was granted the gold medal of the Fédération Aéronautique.[47] The distinction between recognizing the importance of the jet engine and of its inventor was fluid; some awards were granted explicitly for creating the jet engine while others were for achievement in aeronautics more generally.

Contrary to Whittle's expectations, however, the rise of his public stature was accompanied by his decreasing importance to the British government as a technical expert. Indeed, as the previous chapter illustrated, at the time of the government acquisition of Power Jets in 1944, Whittle was already an increasingly marginal figure at the company. Nevertheless, he expected to be involved in "the counsels of the Ministry" on turbojets. He complained to MAP in 1944 that "in my opinion, I have not carried anything like the weight I should have done in matters of policy."[48] In late

1945, when it decided to turn Power Jets (R&D) into a national establishment, the Ministry of Supply (MAP's successor) no longer saw the loss of Whittle and his engineering team (the possibility that had prevented any suggestion of turning the firm into a national establishment in 1944—see previous chapter) as an insurmountable problem.

Despite not asking his advice, however, the government was not indifferent to Whittle's welfare. In June 1946, a few months after the acting air commodore resigned from Power Jets (R&D), the Ministry created a new position for Whittle: technical advisor on engine design and production to the controller of supplies (air).[49] Retaining Whittle in an advisory position meant that the government could maintain its connection with the inventor-hero. It could thus ensure that the engineer was financially secure, while continuing to use the connection to Whittle's continually growing renown to promote the British-ness of the jet engine. Whittle spent much of 1946 and 1947 on lecture tours around the world, including in Holland, Belgium, France, the United States, and Canada.[50]

The Ministry of Supply's generous treatment of Whittle after the war testified to the obligation the government felt toward him. It was not unusual for an inventor to be rewarded for his service to the Crown, and there was long-standing precedent for the secretary of state for air to pay remuneration to a company if its designs were used in government contracts.[51] For its part, MAP had decided that Whittle should be given an ex gratia award after the war. Thus its successor, the Ministry of Supply, made an interim payment to Whittle of £10,000 on April 4, 1945, while deferring the determination of the amount of the final award to be paid to Whittle to the anticipated Royal Commission on Awards to Inventors.[52] The initial award was "assessed on a conservative judgment" of the Royal Commission's future ruling, and its significant size reflected the unusual importance of the jet engine case.[53] Britain's second Royal Commission on Awards to Inventors (the first was established after World War I in 1919) was set up by royal warrant on May 15, 1946, according to plans made during the war, as the mechanism for making awards to inventors for the use of their patents during the war. The commission was composed of government officials and senior engineering experts, and it heard 440 cases during its existence from May 15, 1946, until March 17, 1955.[54] By mid-January 1954, it had awarded about £600,000 (interestingly, significantly less than the first Royal Commission on Awards to Inventors).[55]

When the time came, Whittle made it clear to the Ministry of Supply that it would be "extremely distasteful" to him to submit a claim to the

Royal Commission. Rather, he believed that "the State should look after its own."[56] (This distinction was important: Whittle later made repeated attempts to clarify to the public that he had not made a claim to the Royal Commission on Awards to Inventors himself and had instead been recognized by the Ministry of Supply for his earlier contribution.)[57] The Ministry of Supply applied to the Royal Commission on Awards to Inventors on Whittle's behalf, and the commission consented to hear the unique case, although an application by a government ministry lay, strictly speaking, outside of its terms of reference.[58]

On May 26, 1948, Whittle's case was presented to the commission formally by E. L. Pickles, a member of the Ministry of Supply, who had been a party to the early turbojet program at MAP.[59] It was heard by Lord Justice Cohen (chairman), Mr. K. Swan, King's Council (deputy chairman), Sir William Stanier and Sir George Lee (both well-known engineers), Sir John Rae (representative of the Treasury), and Dr. G. M. Bennett (government chemist).[60] The day before the hearing, the commission visited the National Gas Turbine Establishment at Whetstone (Power Jets' erstwhile factory), where they saw "various types of these devices" and heard positive testimony from the establishment's director and deputy director, Harold Roxbee Cox and Hayne Constant, about the importance and extent of Whittle's contribution to jet propulsion and gas turbine development in Britain.

Pickles had two aims: firstly, he sought to establish the degree to which Whittle, a serving officer, had been helped by government support in his invention, and secondly, he tried to demonstrate the contemporary importance of the aero-engines that Whittle's work had given rise to. Thus he began his case by presenting a brief history of the turbojet aero-engine. This emphasized Whittle's service record and the cost to the officer physically and mentally of his work on the turbojet during the war. Pickles then outlined the Ministry's involvement in Whittle's early work. He defended the importance of the turbojet engine by highlighting its economic value to Britain, quoting a value of approximately £7 million for the 1,410 "Whittle-derived" engines that had been exported from Britain to date. Whittle's importance, Pickles argued, lay in the fact that his work had hastened and "spectacularly advanced" the lucrative gas turbine industry in Great Britain (notably not in making jet engines).[61]

The Royal Commission on Awards to Inventors ruled on Whittle's case on May 28, 1948. Both the Ministry of Supply and the Treasury agreed to the suggested award of £100,000 in recognition of Whittle's "achievement in the science of aeronautics by devising and developing practical means of applying the principle of jet propulsion and for his improvements in the design of gas turbines and air compressors for use in conjunction with jet propulsion."[62] In 1948, Whittle retired as an air commodore from the Royal Air Force on health grounds, but the handsome award from the state alleviated any financial worries that he might have had.[63]

The award granted to Whittle caused a stir because it was more than eight times as much as the highest award made by the Royal Commission up to that point.[64] One rumor was that the award's value was calculated to reimburse Whittle for his shares in Power Jets, which Whittle had given up shortly after the first public revelation of Power Jets' work because he was convinced that a serving officer should not gain commercially from work that was part of his official duties.[65] The award did take into account the value of the Power Jets shares that Whittle had gifted to the Air Ministry, which was almost £48,000, but the shares were understood as evidence of the value of his work rather than gifts needing reimbursement, and Whittle made great efforts to make this fact publicly known.[66]

Whatever the reasoning behind it, the award was important for the public story of the jet engine because there was no more public affirmation of an inventor in Britain than an endorsement from the Royal Commission on Awards to Inventors. The first Royal Commission had already set a precedent for making authoritative rulings on important inventions like the tank.[67] Regardless of the commission's process of evaluation or its grasp of the elaborate nature of invention,[68] the values of the awards granted by the commission were understood by the public to directly reflect the relative value of different inventions and inventors to Britain. By this measure, the award placed Whittle incontrovertibly among the small coterie of Britain's most exceptional engineers. Although the conditions of both invention and award varied widely, in the press, Whittle's award was compared to those of other great wartime inventors, including Sir Donald Bailey (£12,000 for the Bailey bridge) and in January 1952 Sir Robert Watson-Watt (£50,000 for his work on radar)—the only single award that came close to Sir Frank's.[69] The largest amount that was awarded by the Royal Commission for a single machine was £120,000 for inventions to do with the undercarriage of the Halifax aircraft, but this was divided between several people.[70] (The armed forces also made some payments independently of the commission with respect to various weapons.)[71] Whittle's was the only award given for jet propulsion. In response to a letter questioning the amount of the award, the Treasury summed up the government's rationale: "we were satisfied that an award was fully justified particularly in view of the fact that his [Whittle's] invention has led up to a considerable volume of new export trade.... This is, admittedly, a handsome sum but we do not regard it as too much to pay to a member of the community who has rendered such a service to the state as Whittle has done."[72] Whittle's award was unusual in that the Royal Commission was explicitly interested in the postwar more than the wartime value of the invention.

New Medium, Old Story

After the publicity surrounding Whittle's award, the next big public celebration of his work came in an official government film about the creation of the jet engine, *The Wonder Jet*. The film was commissioned by the British government in the late 1940s in response to demands made in the national press, and it exemplified Whittle's centrality to the British story.[73] Completed in December 1949, the film was envisaged as a contribution to a proposed series of eleven films on economic subjects that was to be made by the Central Office of Information, the 1946 successor to the wartime Ministry of Information.[74] The British government hoped that the films would gain wide exposure and be distributed around the world, where they would promote British exports. Almost twenty minutes in length, *The Wonder Jet* was longer than all of the other films in the series. Its length, which added to distribution costs because it ran onto two film reels, underscored the importance attached to its subject.[75]

The film was produced for the Central Office of Information by the Crown Film Unit. It was an exemplar of the British "story documentary" tradition of documentary filmmaking, which had been pioneered in the interwar years by the Crown Film Unit's predecessor, the General Post Office Film Unit. In films made in the "story documentary" style, nonactors portrayed themselves in realistic narratives based on their own experiences.[76] Thus Whittle not only contributed content to *The Wonder Jet* but also appeared in it. In the film's segment that dealt with the "inventor's long road of discouragement and danger," for example, Whittle played himself in a reenactment of one of his favorite stories about an early engine that ran alarmingly out of control when fuel that had collected in the engine caught fire. He included the same story in his memoir (although there is no other confirmation that it took place).[77]

FIGURE 4.2 Stills from the British government film, *The Wonder Jet* (1949). Frank Whittle and W. E. P. Johnson portrayed themselves in the film, which re-created several of Whittle's anecdotes. *Source*: *The Wonder Jet*, 1949

Although an official narrative of Britain's development of the jet engine had already been completed for the nation's official historians by 1949, it was not among the film's sources. The Cabinet Office refused to release the existing narrative to the filmmakers for reasons of confidentiality.[78] An official commented that the "very full and frank account of relations between Government departments and private firms, also of quarrels between individually named private firms" in the official history might be used to make a "highly-coloured film in honour of Air Commodore Whittle," which might portray the government in a poor light. The Air Ministry, he suggested, might prefer to "have some say in how much of the pre-war story of official encouragement of jet propulsion development (or lack of it) is made public at this stage."[79]

The official narrative that the Cabinet Office refused to release to the film company was almost exactly reproduced in the official British history of the war. It was written by Cynthia Keppel based on official files and wartime interviews.[80] Keppel was untrained in history or aero-engines and had worked as a secretary at MI5 before becoming a research assistant to the official historian M. M. Postan in 1942 (she married Postan shortly after the war).[81] Keppel's narrative was naturally focused on the British case, but also included an account of the German developments based on British government reports and interrogations. Although a unique account of the British story, by the time that Keppel's narrative was finally published in *Design and Development of Weapons* in 1964 (the volume was allegedly held up for security reasons), it added relatively little to the published accounts that proceeded it: those written by Schlaifer (1950) and Whittle (1953).[82]

In the event, the film's invention story did honor Whittle's wartime work, but the story presented was a generally positive one. Resistance to the innovation was represented by conservatives (shown playing golf) rather than members of the government. Indeed, the government's involvement in the engine's development was not widely known and was hardly mentioned in the film. Because of the Cabinet Office's fear of the consequences of giving a filmmaker a view into government-industry relations, the film presented a traditional story of invention, in which the Whittle story remained front and center, and the actions of industry and the government remained in the background.

The Wonder Jet presented the jet engine not primarily as a secret, military technology (a role given to other typically British wartime inventions like Barnes Wallis's bouncing bomb), but as a postwar economic boon. Throughout, the film emphasized that patents were the most valid type of claims to invention. Thus the film functioned to promote (the sale of) British jet patents abroad by portraying the story of invention behind them. Power Jets (R&D), which remained in existence after 1946 as a patent holding firm for the government-owned British jet patents, ordered a special edition of the film that included an additional sequence depicting the shipment of the Power Jets W.1X to the United States in 1941. The company wanted to use the film to refresh the connection between American engines and British jet patents in the American mind.[83]

The same objective motivated the gift by Power Jets (R&D) of the W.1X turbojet engine to the Smithsonian Institution in Washington, DC, in November 1949.[84] (The first flight engine, the W.1, found a secure home in the Science Museum in London at the same time.)[85] W. E. P. Johnson, patent agent and director of Power Jets (R&D), hoped that the presentation of the historic engine to a leading American museum would, like the film, "authoritatively put on record" the fact that the American jet industry owed the technology on which it was based to Britain.[86] Arriving in Washington, DC, not long after the original Wright flyer, which had since 1928 been housed in the Science Museum in London, the induction of the W.1X into the Smithsonian would simultaneously be an induction into the annals of American history. Official support was indicated by the fact that the engine, like the pioneering aircraft, was presented to the American museum by the British ambassador, Sir Oliver Franks.[87] Johnson counted on the fact that the obsolete technical artifact would, in the Smithsonian, continue to contribute historical authority to Whittle's preeminence and bolster Power Jets' (R&D's) patent sales.

The First Academic Histories

The first full account of the jet engine to be written by a professional historian was published by the Harvard Business School in 1950 under the title *Development of Aircraft Engines: Two Studies of Relations between Government and Business.* The study formed part of the business school's research program on military and civil aviation, which it had inaugurated in 1942. Robert Schlaifer, the author, was then a young assistant professor of business administration at Harvard. After receiving a PhD in ancient history from Harvard in 1940, he had worked during the war writing technical reports for the American Underwater Sound Laboratory.[88] In his study of aero-engines, both piston and jet, Schlaifer had the technical help of S. D. Heron, a veteran of engine development in both Britain and the United States. Schlaifer's study contrasted the government's role

in developing piston aero-engines—an area in which American companies were ahead of European companies—and jet engines—an area in
which it judged Europe (Britain and Germany) had achieved more. The
Harvard program's interest in government and industry relations dictated
the government program as Schlaifer's unit of analysis,[89] while the book's
foreword made the connection of the investigation to early Cold War concerns about weapons development explicit.[90]

Completed in late 1949, just a few years after the events that it chron-
icled, Schlaifer's study was based (unlike Keppel's) only on interviews and
published documents. A significant part of Schlaifer's account of the British turbojet program was made up of a detailed financial history of Power
Jets. Schlaifer chose not to reference his sources in the published study,
but they certainly included many of the same men Keppel interviewed,
including Whittle. Among them were leading ex-members of Power Jets,
who were no doubt eager to get their story on record after their firm
was nationalized in 1944 (Keppel wondered "why Power Jets gave him
[Schlaifer] so much dope without asking the Ministry").[91] They also included the civil servant Harold Roxbee Cox, who left MAP and joined
Power Jets (R&D) in 1944. For the German part of his study of turbojets,
Schlaifer relied on interviews with German expatriate engineers who had
moved to the United States after the war.[92] The pages of *Development of
Aircraft Engines* juxtaposed the stories of Griffith, Constant, Whittle, von
Ohain, Heinkel, Wagner, Müller, Oestrich, Franz, and Schelp.

Detailed archival research has since challenged some of his arguments,
but Schlaifer's conclusions (many of which are still valid) are all the more
exemplary because his source material was so limited.[93] His account has
been used as a key empirical source by virtually every account of the jet
engine both popular and professional written on the British, German, or
American programs, although his key arguments about the importance of
the aero-engine industry to technical development have not been taken
up by later authors.[94] Schlaifer's interest in design and development kept
him from giving much attention to the question of invention, although his
analysis acknowledged the difference between the creation of novel ideas
and development.

Historiographical Success

One reason for the continued dominance of the story of Whittle told in
The Wonder Jet was the lack of alternatives available to the public. Schlaifer's study was never widely publicly available, even in the United States.

It was only with the publication in 1953 of Whittle's memoir, *Jet: The Story of a Pioneer*, that a fuller version of the British jet engine story appeared publicly in Britain. It was published by the popular press of Frederick Muller. Whittle's account of what he referred to as his "association with the development of the turbo-jet engine in Great Britain" (never in his book did Whittle refer to the jet as an invention nor to himself as an inventor) was embraced by the public—coming as it did after enormous press interest, a generous award, and a long government film—as the inventor's authoritative story of his invention.[95] In a later edition the book was even marketed in this way.[96] The book brought Whittle's story to many outside of Britain too. In 1954 it was republished in New York by Philosophical Library and in 1957 as a trade paperback by Pan Books.

Jet transformed the public understanding of the development of the jet engine in Britain. In the book, Whittle publicized for the first time the intimate and messy relationships between the British Air Ministry, industry, and Power Jets—topics that the government had studiously avoided discussing in public. His narrative expanded on a familiar and negative trope of the heroic inventor story; it was not just nature and technical caprice the inventor had to combat, but also more modern problems like lack of funds, government short-sightedness, industry incompetence, and distrust from financiers at every minor technical setback. Given the atmosphere of disaster newly surrounding the de Havilland Comet and the apparent failure of the second generation of British jet fighters, Whittle's account was widely understood in Britain as an objective and justified indictment of government mismanagement of a seminal invention.

In the decades following its publication, this interpretation of the Whittle story was seized on and furthered by declinist historians in Britain, like Glyn Jones, who emphasized Whittle's heroic status in their attempts to demonstrate the British government's constant squandering of the nation's inventive resources.[97] Indeed, the story of the jet engine is still invoked today by critics of the British government bemoaning the lack of support for science and innovation.[98] That historians continue to use *Jet* as their primary empirical source on the British jet story attests to its continuing influence, its impressive amount of detail, and the lack of alternatives. The first edition of Whittle's authorized biography, written by Golley in collaboration with Whittle, borrowed whole passages from *Jet*; it was published in a fifth revised edition in Britain in 2009, with a new supplement by Whittle's son.[99] The biography shares the same point of view as *Jet*.

Whittle's book gives a one-sided view of events, but the few alternatives never challenged Whittle's central place in the public imagination. Immediately after the war, many of the engineers involved in turbojet development gave public lectures and published papers in technical journals and the proceedings of professional societies. The technical details of work done by aero-engine firms appeared for another expert audiences in aviation journals like the British *Flight*.[100] These presentations served to inform the wider British engineering community of important developments that had taken place in secret during the war and focused therefore primarily on technical details. Because they required a relatively high level of technical fluency to be understood, however, they had little impact on the public story. Over time, other actors' accounts and biographies contributed other personal views of the jet engine's early years, but these too had little power to shift the fixed hegemony of Whittle's account.[101]

Alternative stories about the wartime work of engineers who led jet engine development in British industry such as de Havilland's Frank Halford, the Royal Aircraft Establishment's Hayne Constant, and Metropolitan Vickers' David Smith are known to the British jet engine community but not beyond it. In contrast to Whittle, who quickly became a public figure and soon after left the jet engine field, Halford, Constant, and Smith remained little-known but active aero-engine engineers. A biography of Halford published in 1999 by the Rolls-Royce Heritage Trust, for example, devoted only a single chapter to the designer's work on the turbojet—much more space was devoted to the longer period in which he worked on piston engines.[102] Many other men had claims to being important turbojet pioneers, but their contributions remain known only to small groups of fellow engineers and enthusiasts.

Arnold A. Griffith has perhaps the strongest claim to challenge Whittle's dominance in the British invention story. As a government scientist, he wrote an important paper on gas turbines in 1926 that suggested the design principles for constructing an axial gas turbine aero-engine. The work that his initial research gave rise to at the Royal Aircraft Establishment ultimately became, in the words of Roxbee Cox in a lecture on December 17, 1945, to the American Institute of the Aeronautical Sciences, the second "distinct stream" of gas turbine work in Britain.[103] Yet even this second stream, which had come to dominate gas turbine development in Britain by 1946, was eclipsed by the Whittle story. So well known was the Whittle story that in 1951, an article in *Flight* on British turbojet and turboprop "designs of a decade" could refer simply to

Whittle's "struggle" as an adequate account of the turbojet's early years in Britain.[104] In 1999, an ex-employee of the Royal Aircraft Establishment, Richard Dennis, tried to match the popularization that *Jet* had achieved for Power Jets' work by presenting the Royal Aircraft Establishment's jet work, following Griffith's early paper, for a popular audience for the first time.[105] Despite such promotion by their supporters, however, the "backroom boys" of the Royal Aircraft Establishment never aroused much public interest.[106] Griffith is commonly cast as a jealous scientist, while Whittle is known as one of the world's great inventors, alongside Watt, Stephenson, Edison, Marconi, and the Wright Brothers.[107]

No accounts of the development of the jet engine appeared from the perspective of the Air Ministry. The only account of events to be written by a government official was part of an autobiography that was written in 1960 but not published until 2002. It was written by Major George Purvis Bulman, head of aero-engine development in the British Air Ministry from 1928 to 1944. Despite contradicting Whittle's account, the framing of the book's section on the jet engine reflects both the dominance of Whittle's view of wartime events and his demonization of the government in *Jet*. Bulman responded explicitly to Whittle's portrayal of MAP's actions, reacting to the canonical status of *Jet* by structuring his own story around a series of passages taken from Whittle's memoir. He followed each passage with his own interpretation. Nevertheless, Bulman did not expect his account (even if it had been published) to change what was by then deeply entrenched public opinion. Instead, he appealed to the "future judgment of historians."[108]

The German Challenge

An obvious challenge to the British story was offered by the story of the German jet engine. Yet although the German story became quickly known to the Allies through intelligence reports (Germany's jet program was identified as an area in which to extract "intellectual reparations"), it had little impact on the public discourse. Engineers from Britain and the United States toured Germany, collecting experimental equipment, technical drawings, documents, and prototypes and interviewing any engineers they found. New research was done for the Allied occupiers using German equipment run by German engineers, and German jet engines and engineers were taken by all of the Allies.[109] The collected information from Allied intelligence operations was later widely published in numerous reports intended primarily to assist industry. Compiled by military and industrial personnel, the reports drew on information extracted from interrogations, tours of German industry, captured artifacts, and documents. Because of these reports, which included details of all of the major German jet engine projects, the German wartime jet engine program was uniquely well technically documented. Supplemented by documents found after the war, these reports still inform accounts of the German jet engine.[110]

Early British reports began to build a narrative of the German program in Britain, but this was overshadowed by publicity given to the British story.[111] The treatment of the jet story in Britain is a prime example of what David Edgerton has labeled "techno-nationalism." Non-British achievements in jet propulsion were simply not publicized.[112] No romantic invention story was popularly associated with the German weapon—for that reason too it was not well known. Whether or not they knew that the Heinkel He 178 had flown before the Gloster E.28/39, for example, MAP ignored the fact that a German turbojet had been the first in the world to fly, and this remained the case even after the German official Helmut Schelp brought it to their attention during his interrogation in London.[113] In their release on the Meteor in February 1945, the Air Ministry had claimed the honor of the world's first jet flight for the Gloster E.28, and this was never publicly modified.[114] *The Wonder Jet*, although it appeared after the war, told an exclusively British story, making no detailed mention of American or German jet work. The lack of information about the German story in Britain is suggested not only by the briefness of the official history's account of German work but also by the fact that it misspelled the German inventor's name as "von Chain." The fact that the official history reproduced Keppel's mistake when it was published in 1964 suggests that the German inventor's name was still not well known to the British public almost twenty years after the war.

Thus although knowledge of German work (often quite technical) circulated in Britain, it was not much advertised to the general public. Whittle's memoir, which did reach the public, mentioned only briefly that German work on jet engines had proceeded in parallel to his own. Whittle's only reference to von Ohain's work was a single sentence that noted: "while the B. T.-H. [British Thomson-Houston Company] were building my engine, the aircraft firm of Heinkel were building a turbo-jet engine to the design of a young engineer named von Ohain."[115] Later in the book, Whittle devoted a brief chapter to German jet engines in which his main argument was that they were much inferior to British engines. Further neutralizing any possibility that the German story might offer a challenge to his claims, he strongly implied—although he admitted that could not make a "definite claim"—that the Germans had copied his 1930 patent, which was made available in Germany.[116] (He only relinquished this conviction after meeting von Ohain in person.) The German story was simply not formulated in Britain in a way that might challenge the British story.

Birth of the Dual-Inventor Narrative in the United States

In the United States, which did not claim its "own" national jet inventor but had a successful jet industry, the story of the jet engine played a very different rhetorical role than in Britain or Germany. The American press continued to praise the development and production prowess of American companies. After having collaborated enthusiastically with Britain during the war, American authorities were eager to make full use of German engineering expertise after the war (bolstered no doubt in part by officials who needed a compelling case for importing German engineers). Hans von Ohain moved to the United States in 1947, where he began working for the United States Air Force at Wright Field in Dayton, Ohio, along with other eminent German turbojet engineers, including Helmut Schelp (responsible for jet engines at the German Air Ministry) and Anselm Franz (the head of Jumo's turbojet project).[117] Promoting multiple origin narratives was not only supported by the high standing of German technical expertise in the American view but also by those who objected to a narrative of debt to Britain for jet engine technology. In its film *Jet Power* (1952), General Electric stated that its first jet engine bore only a "basic resemblance" to the Whittle design, rather than being a copy of it. (The film's technical supervision was provided by the company's postwar Gas Turbine Department.)[118] Thus more narratives about the jet engine were available to the American public than the British, and the public could embrace them all without changing the role of American companies in the story.

Immediately after the war, a dual narrative of German and British invention always prefaced remarks on American jet progress, which was (and still is) understood as building on and surpassing both German and British jet work. This can be seen in a 1948 book on jet propulsion. Meant to inform engineers, students, and "a rapidly increasing number of executives, businessmen, and professional men who are vitally interested in the problems of the air age that is now upon us," the book's chapter headings give a good indication of how the American authors approached jet development in each country.[119] They revealingly included: "How the Nazis Beat Us to It," "The British Were Early, Too," and "The AAF [Army Air Forces] and American Industry Pull a Miracle." The book's account of the German story relied entirely on the published record of Schelp's interrogation in London. British work was reviewed based on lectures by the British figures Whittle, Roxbee Cox, and Watt (Roxbee Cox's deputy), and began with a discussion of the axial gas turbine (Griffith, Constant, Metrovick, and Armstrong Siddeley) before moving on to Whittle's story. A similar treatment of the transnational story can be found in a popular history published a few years later called *Jet Pioneers* by Grover Heiman. Heiman's first chapter was devoted to von Ohain and Heinkel, his second to Whittle, and the remainder of the book to American pioneers.[120] Although the same information was readily available, no equivalent popular, transnational account was published in Britain or Germany.

With American accounts juxtaposing secret German and British work, it was only a matter of time before the German story was refashioned to parallel the already popular British story. The early date of the first German jet flight gave the Germans a claim to an aviation "first," which was politically and socially powerful. Von Ohain was a plausible counterpart to Whittle, a similarly young and unaligned inventor and resident in the United States. Indeed, the German physicist was important to American jet work; he became director of the Air Force Aeronautical Research Laboratory in Dayton in 1956 and by 1975 he was the chief scientist of the Aero Propulsion Laboratory. His colleagues in particular pushed for the recognition of his wartime work. There was no equivalent to the Royal Commission on Awards to Inventors for von Ohain, but his hero-dom was confirmed and promoted by his friends and acquaintances in the United States. An acquaintance from Dayton wrote the only biography of von Ohain to date, which was published in 2001, after his death, by the American Institute of Aeronautics and Astronautics. Von Ohain was made a fellow of the institute in 1970.[121]

It was in the United States that Whittle and von Ohain were first feted as coinventors of the jet engine. The two men first met in America in January 1966, when von Ohain was awarded the American Institute of Aeronautics and Astronautics' Goddard Medal for the "first successful application of turbojet propulsion to aircraft" jointly with two American jet

FIGURE 4.3 Frank Whittle (1907–1996) and Hans von Ohain (1911–1998) at the Smithsonian Institution on October 24, 1974. *Source*: Smithsonian National Air and Space Museum (NASM 74-11297-07)

pioneers.[122] (Whittle had received it in 1965.)[123] They met again at the Smithsonian Institution's National Air and Space Museum in Washington, DC, on October 24, 1974.[124] This meeting preceded a string of joint public appearances that took place after Whittle moved to the United States in 1976, in which the two inventors' dual act became increasingly well rehearsed, with von Ohain arguing that Britain could have prevented World War II if only the British government had supported Whittle earlier.

On May 3, 1978, the commander of Wright Field's Aero Propulsion Laboratory arranged for Whittle and von Ohain to publicly recount the early history of the turbojet at an event titled "An Encounter between the Jet Engine Inventors."[125] A symposium organized by the Smithsonian Institution's National Air and Space Museum on October 26, 1979, to commemorate the fortieth anniversary of the He 178's flight, included lectures by Whittle, von Ohain, and Franz beside important American figures.[126] The two inventors appeared once again very publicly together to celebrate the "Golden Anniversary of Jet Powered Flight, 1939–1989" (again

the German date was used to calculate the anniversary) at an event organized by the American Institute of Aeronautics and Astronautics on August 23, 1989.[127] In 1991, von Ohain was given the Guggenheim Medal
(jointly sponsored by three American engineering associations: American
Society of Mechanical Engineers, the Society of Automotive Engineers,
and the American Institution of Aeronautics and Astronautics) "for pioneering development of turbojet propulsion resulting in the first flight in
1939 and lifetime achievements in aeronautical propulsion dynamics."
(Whittle had received the same award in 1946.) The men's shared status
was soon after again publicly recognized with the joint award in 1992 of
the Charles Stark Draper Prize to both "for engineering innovation and
individual tenacity in developing and reduction to practice of the turbojet
engine."[128] Like the jet engine, America made its inventors her own.

The inclusion of the dual-inventor narrative in the Smithsonian Institution's National Air and Space Museum's new exhibit on jet aviation presented the contributions of the two men to an even wider public. The museum's Jet Aviation Gallery opened in its building on the National Mall in
July 1981 (the museum's building on the mall opened in 1976).[129] Its display case on the "Pioneers of Jet Propulsion" declared that Whittle and
von Ohain are "rightly regarded as the inventors of the first practical jet
engines," although the jet engine "reflected the efforts of many individuals from many nations" (these individuals were illustrated in the following
display case, "Milestones in Jet Aviation History"). The new gallery drew
on the museum's particular resources. These included not only the inventors, who were frequent visitors, but also the museum's engine collection,
which provided key objects for display.

The authoritative gallery neatly juxtaposed American jet history with
its foreign origin stories. Its claim for the equivalence of the two inventors' work was made through the display of a full-scale reconstruction
of von Ohain's first flight engine, the HeS3b, beside the W.1X, the much
earlier donated prototype of Whittle's first flight engine. The two foreign
engines were surrounded by American jet engines of later manufacture.
Opposite the engines in the gallery was a confiscated Me 262 and the
Lockheed XP-80, the prototype of America's first operational jet fighter.
The Bell P-59, the first American jet airframe, which was never deployed,
was at the time hanging in the museum's flight testing gallery and was
thus not considered for the jet aviation gallery, despite its claim to be a jet
aviation first. (Ironically, losing the war increased the historical footprint
of German artifacts in America's leading museum.)[130]

FIGURE 4.4 The Jet Aviation Gallery in the Smithsonian Institution's National Air and Space Museum, Washington, DC. The line of jet engines on display includes the Power Jets W.1X (second engine from the left) and the re-creation of the Heinkel HeS 3b (third from the left). The mural behind the engines, created by Keith Ferris, depicts the first jet aircraft in the world. The Heinkel He 178, Gloster E.28/39, and Bell XP-59A are depicted in the lower left of the mural. The noses of the Lockheed XP-80 (left) and Me 262 (right) are visible in the foreground. *Source:* Smithsonian National Air and Space Museum © 1980 Keith Ferris

It was no accident that the museum collection included the world's two first jet flight engines. The W.1X had been donated to the museum by Power Jets in 1949 with the express intention of shaping the American narrative.[131] It was well suited to do so because it was a copy of a historic engine that had been sent to General Electric during the war. The W.1X symbolized Britain's crucial contribution to the birth of jet propulsion and was a powerful symbol of transatlantic technical cooperation—a useful political message in the postwar years.[132] Simultaneously, the donation was also an economic stratagem by a war-weakened country eager to promote the sales of its jet engines and patents.[133] Although the acknowledgment of General Electric's debt to the engine has since disappeared from the display, the W.1X is still today centrally located in the gallery, where its materiality continues to make a strong statement.[134]

The re-creation of von Ohain's first functioning turbojet engine was also donated to the museum, albeit in 1980. When the Smithsonian's jet aviation gallery was first conceived, the museum did not have any engines related to von Ohain's invention; plans to recreate von Ohain's first flight engine were begun in the early 1970s but were eventually scrapped, leaving the German side of the story to be represented with only photographs and drawings.[135] In 1980, however, in time for the new gallery, the Deutsches Museum in Munich offered the American museum a copy of the re-creation that it was having made for its own collection.[136] The donation of the re-creation, built by latter-day German aero-engine firms, was welcomed in Washington. Alongside the museum's confiscated Me 262 (powered by Jumo 109-004 engines), the reconstruction successfully promoted German jet engine history. At the Smithsonian, even the re-creation of an engine had a special power to put von Ohain's inventive achievement on record, and the donation ensured a central position for the German story in the museum's gallery. The donation, like that of the W.1X before it, thus allowed external groups to shape the narrative of a powerful American museum. Both donors sought to use the authority of the Smithsonian to buttress their own historical claims about the jet engine because they had important political and commercial implications.[137]

The Smithsonian Institution's Jet Aviation Gallery offered the first permanent, public testimony to the dual-inventor-dom of the two men. Despite their very different histories, today the two early jet engines are presented on the National Mall as equivalent in their respective national contexts as well as in the American aviation context. Von Ohain and Whittle were both involved in the donations of their engines to the museum as well as in the opening of the Jet Aviation Gallery and thus gave their implicit support to the exhibition and its message. Not long after the exhibit opened, in 1984–85, von Ohain served as the Charles Lindbergh Professor at the National Air and Space Museum, where he was encouraged to record his memories of inventing the jet engine.[138] Through their continued presence at the museum, the two inventors actively shaped the public story of their work and ensured their central place in jet engine history.

The Jet and the Rebirth of the History of German Aviation

The surge of public interest in the jet engine in the United States in the 1970s and 1980s paralleled an increased German interest in the history of German aviation. Immediately after the war, even as reports of Germany's wartime jet activity made the rounds in Britain and the United States, a veil of silence descended over the recent past in Germany, as the population's focus turned to rebuilding daily life. The German aviation industry was closed down along with the rest of the German weapons industries. The first accounts by and of participants in the German jet engine program appeared slowly after 1950 and without much fanfare.[139] These narratives were aimed at professionals and jet engineers, and the story remained with them. It was only in the 1960s, when the first concentration camp inmates began to give public testimony of their experiences during the war, that the public indifference to or agreed silence about the past in Germany began to dissipate.[140] By the late 1970s it had begun to recede, as narratives of German victimhood began to challenge the dominance of the interpretative framework that had connected the resurgent West German economy with Weimar Germany, by entirely leaving out World War II.[141] It was in the late 1970s that those same men who had rebuilt the German aviation industry in the 1950s and 1960s brought new attention to the history of German aviation.[142] German authors used the intelligence reports that had pieced the German jet story together after the war in Europe as well as the accounts of the original actors, many of whom were still alive and active, working in the aircraft industry around the world.

The Deutsches Museum, Germany's leading technical museum, contributed to the renewal of awareness of German aviation accomplishments when it opened a new Air and Space Hall in 1984. Not only did the museum organize the reconstruction of von Ohain's first flight engine, but it was also busy at the time trying to collect examples of all significant German aero-engines, many of which it had to borrow from foreign collections.[143] In addition, the museum began releasing a series on the history of German aviation in 1980 in collaboration with the German Aviation, Space and Supply Industries Association and the German Space Agency. Many of the series' authors had worked in the German aviation industry. The second volume, published in 1981, gave the first German account of the turbojet engine in Germany. Edited by Kyrill von Gersdorff and Kurt Grasmann, a civil servant and an engineer respectively, it gave a technically rich and impressively comprehensive history of German aero-engines that traced an unbroken line from the beginning of German aviation to contemporary aero-engine development. Much of the book's information on jet engines was drawn from Allied intelligence and German company reports, and the book was written with the help of technicians who had worked with the engines being described.[144] Whereas a comprehensive format is common in retelling the German story, the British story has rarely been approached in the same way. Such a format was recently copied for the first time for the British jet engine story.[145]

In the 1980s, important jet anniversaries coincided with a sea change in the historical scene in Germany that further strengthened German interest in the jet story. The appearance of the reconstruction of the HeS 3b in Germany in 1981 and the fiftieth anniversary of the engine's first flight in 1989 resulted in von Ohain being publicly recognized in Germany for the first time.[146] Von Ohain himself was a key figure in the Deutsches Museum's new display. Not only had he re-created the drawings for the HeS3b reconstruction from memory, but he also supplied biographical information for the museum's posted text and was the featured speaker on the occasion of the reconstruction's arrival at the Deutsches Museum on May 15, 1981.[147] Indeed, von Ohain became an important node in writing the history of German aviation, as other engineers contacted him.[148] Although never receiving the same level of attention in Germany as Whittle in Britain, von Ohain remained a central figure among German jet engineers, in the United States and in Germany, who began seriously trying to reconstruct their wartime work in the 1980s.[149]

Professionals Borrow the Tale

The second transnational history of the jet engine to be written by a professional historian appeared in the United States in 1980, perhaps part of the same wave of interest in the jet engine that was sweeping through the United States and Germany at the time. After at least ten years of work, Edward Constant's *Origins of the Turbojet Revolution* was published in 1980.[150] It presented the idea of a technological revolution carried out by revolutionaries (rather than inventors). Reacting to the status of individual inventors in the United States and perhaps the dual-inventor narrative of the jet engine, Constant presented his work as radically changing the German story. With respect to the German jet engine, he highlighted the work of three "revolutionaries": Schelp, Wagner, and von Ohain. (He corresponded directly with von Ohain and Wagner.) Yet the work of all three was mentioned earlier in Schlaifer's book, so Constant did not with this selection add new empirical research to the invention story.

In the British case, Constant did not revise the dominant story of Whittle (whom he corresponded with) but in fact contributed to its strength by giving an explicit rebuttal of Griffith's potential claims to invention. Despite what he described as Griffith's "intimate connection" with the turbojet revolution in Britain, Constant dismissed his claim to be a revolutionary because, although Griffith had understood that a gas

turbine could be used in an aero-engine, he had not grasped the importance of high-speed flight, which made jet propulsion (according to Constant) practicable.[151] Unlike Whittle, whom Constant was able to ask in
1970 about the detailed steps in his earlier thought process, Griffith died
in 1963 and was thus unable to contribute to Constant's study. In making
his judgment about the British turbojet revolutionaries, Constant relied
on the existing British literature on the jet engine, which either supported
the Whittle story explicitly or failed to challenge it. It is thus unsurpris-
ing that, revisionist claims notwithstanding, Constant too concluded that
Whittle was the key figure in the turbojet revolution in Britain.

The first German academic historian to tackle the story of the jet was
Ralf Schabel. Writing in the mid-1990s, he could still point to legends that
grew up in the postwar period as the historiographical frame that he was
breaking out of. The technical histories of the 1980s had added some detail to postwar stories but had steered clear of politics. Schabel's study was
interested in politics, but not technology.[152] Importantly, he challenged
and extended the German historiography of the jet fighter by observing
that the jet was not sought after by the leadership of the German Air
Ministry until 1943, and that production problems rather than political
problems ultimately delayed the last-ditch deployment of the country's
jet fighter, the Me 262. Against much German popular literature, Schabel
argued that rather than too late, the jets were deployed too early—before
they were matured technically. For Schabel, the real tragedy of the German jet was not the failure to win the air war, as nostalgic Luftwaffe officers had argued early on, but rather the last-ditch deployment of the jet,
which served only to increase the war's cost.[153] Schabel's work is important for its interest in the politics that accompany technical change and for
adding new empirical evidence to the existing literature.

The first rigorous British academic research on the early history of the
jet engine appeared some years after Schabel's work. That narrative, by
Andrew Nahum, employed new empirical research to challenge the accepted story of the British Air Ministry's poor treatment of Whittle. It was
published in a slightly extended version as a popular book, *Frank Whittle:
Invention of the Jet*, in 2007 (showing the popular interest in Britain).[154]
The literature on the jet engine has progressed so little since the postwar
period that Nahum too, writing the first critical history of the British jet
engine program, found it necessary to challenge tropes that arose in the
immediate postwar period. At the same time, the centrality of Whittle in
the story of the British turbojet has become so well established that even

Nahum's revisionist history was structured around the story of Whittle's engine. There is much to criticize in Whittle's account, yet Nahum's critique was less of Whittle's place in the historiography than of Whittle's interpretation of events.

Conclusion

That Whittle made crucial contributions to the creation of the jet engine is uncontested. Nevertheless, it is revealing that he became famous after he ceased to be a significant force in British jet engine development. By 1944, when the first jet publicity appeared, Whittle was already a shrinking figure at Power Jets. Despite this, the public story of the jet's early years came in Britain to be dominated by the story of Whittle's wartime work. This contrasted to the development of the German story of the invention of the jet engine, which was from its start more broadly focused.

The historiography of the jet engine has undoubtedly been shaped by the politics of memory. Within a decade of World War II, a particular story of Frank Whittle had become the dominant popular history of the jet in Britain. This was not true in Germany. Until recently, the average schoolboy in Britain knew of Frank Whittle, whereas Germany's national pioneers were first introduced to the German public in the 1980s.[155] In contrast to his obscurity in Germany, von Ohain's story first became publicly known in the United States, where he settled after the war.[156]

The continued dominance of Whittle's story shows the public power of a particular narrative of heroic invention. At first an important means of buoying the postwar British economy, the story found a deep resonance in Britain. The same style of invention story was also influential in countries like the United States, where von Ohain's story was fashioned into a parallel German version—albeit deployed for a different rhetorical purpose. Together these rhetorically powerful stories have obscured the stories of other actors and reduced the story of the jet engine to the work of two inventors. Their stories are so deeply rooted that they continue to hide much of the story of the jet engine, even from revisionist historians.

To note that the public visibility of Whittle and von Ohain, a product of forces far removed from the jet engine, took a particular form, is not an attempt to detract from their very real achievements, which are in fact only caricatured in the well-known standard narratives. Tracing the historiography of the jet engine demonstrates that the relationship between an inventor's work and his public visibility is by no means as obvious or as direct as the familiar heroic inventor genre implies. It also suggests the existence of an uncritical affinity between popular histories of invention and professional histories of technology. Indeed, it is the similarities between the popular treatment of invention and the narrow outlook of professional historians writing about technical change that make the fact that professional historians frequently borrow conclusions from popular accounts of invention—whether unknowingly or not—possible and even common.

The Jet Engine and Innovation

This book has presented a new way to write about technical change. It has shown how setting aside the constraints of the transition narrative, which argues that innovation comes from individuals or corporate and academic research, opens up the analysis of technical change—many important sources of creativity lie in other places, including particularly in industry. Looking at new places makes us aware not only that creativity emerges from these places and the people who work there, but also that creativity builds on and is shaped by expertise. Arguing that cases of technical discontinuity are underpinned by important continuities of engineering culture, machinery, method, and personnel, this book has shown how, far from holding back novelty through promoting conservatism, the old facilitated and shaped the development of the new. Indeed, existing expertise not only influences how new machines are made but even their very design. Important types of expertise include familiarity with elements that are recombined in new things as well as expertise in the very process of (re-)combination. In other words, industry is important in stories of new machines because of continuities of design and manufacture and also continuities of personnel and methods—particularly methods of dealing with uncertainty. Industry is a crucial repository of the skills, confidence, and resources necessary to successfully and continually take on major creative challenges. Revealing these continuities often requires a deep level of technical detail. What exactly Whittle or von Ohain invented may be difficult to say, but the question is crucial. We have to be as careful about what is invented as about where, when, how, and by whom. Similar questions might be asked about the seemingly most academic-centric innovations, like the atomic bomb.[1] The transition narrative of invention dominates our thinking about invention. Yet its pervasiveness is a product of its conventionality and its influence over the choice of new research topics rather than any positive analytical power. It has too often led the analysis of technical change to retreat into the unproductive dichotomy between individuals and corporate labs, blind to where and how technical novelty really emerges. This has left professional research, in turn, too open to the influence of popular discourse, which is buffeted by the contingencies of politics and economics. As historians, we cannot continue to adopt without critical thought the evaluations of unreflective nonspecialist narratives.

Assuming that technical change takes place because more advanced technology is necessarily desirable prejudices our accounts. We need to recognize that many considerations lie behind the adoption of every new technology—many that may counterintuitively be conservative, especially in the case of wartime, for example, when innovation carries significant risks for the adopter—and recover them. Understanding how firms deal with technical change relies on an intimate understanding of the full range of uncertainties facing every firm, not only technical but also personal, political, and economic.

This book told an invention story in a new and unfamiliar way by telling the story of production first. By exchanging our focus on invention and on "firsts" for a focus on production, this account centered on industry and highlighted the concurrent and intimately connected stories of each element used in a new machine. The story of each is best told separately, but their connections too must be explored. Although each chapter of this book focused on a different topic, each chapter covered the same years and were thus able to show how different decisions and elements came together, each exerting its influence on others in a nonstraightforward way.

*   *   *

This book's different approach transforms the history of the jet engine, which has been told as the story of radical, disruptive innovation by individual outsiders in which success is measured by production. The standard story of the jet engine insists that Germany was first and most successful with jets. Britain is generally dismissed as less successful, although its early start is often recognised. This lack of success is attributed to the British government's reluctance to support, or even negligence in failing to support, jet engine *development*. The United States is seen as the least

successful—a failure even—because it did not *invent* the jet engine, forcing its engineers, supposedly, to copy a British one. This book has challenged all of these interpretations both in general and in detail. As we
saw, it was only Germany that ended up pursuing invention, development,
and production successively—others took a different route. The production story revealed a new story in which Britain, not Germany, was first
in deciding to produce jet engines. Germany was surprisingly the last of
the three countries to decide to produce jet engines. It was in trade-offs
between development and production and in their consequences for production and design, not in the notion of a jet engine, in which one can see
the different regimes that created each specific jet engine.

German jet engines were easier and cheaper to make than existing
piston engines; they were produced in such large numbers because they
could be and were successfully adapted to the extraordinary conditions
of scarcity (of both labor and material) that characterized German weapons production in the last years of World War II. Jet engines fit well into
the regime's late-war strategy, which relied on brutally producing as many
weapons as possible, as cheaply as possible—even technically compro-
mised weapons if necessary. In the end, the German jet engine was not
so much a cutting-edge, war-changing weapon as an ersatz piston engine,
an engine of desperation, and a technological embodiment of the state's
National Socialist values.

Britain, by contrast, abandoned the aim of early production and focused instead on development for the postwar period. British jet engine
production was stimulated late in the war, not by military necessity, but by
the need to justify early investments, reassure the British (and German)
public of the country's military might, and influence the postwar world
market in aero-engines. And indeed, Britain was the world's leader in jet
engines at the end of the war; many of its engines were produced under
license by foreign firms, including in Sweden, France, Italy, Switzerland,
Australia, Canada, and the United States.[2]

In the United States, a real stimulus to intensive production at the cost
of development never really appeared during World War II. The United
States Army Air Forces seized the opportunity to use the British W.2B
engine design because the engine appeared to offer a fast route to production. Yet low production numbers and copying a foreign design were
not indications of incompetence. Indeed, the jet engine offers a story of
intimate Anglo-American collaboration, similar to, if not going beyond,
that seen in the development of other shared weapons, like radar. And

the decision to borrow was not without consequence. It was because of this that General Electric ended up extensively developing a centrifugal jet engine—a design not otherwise proposed by any American firm. The resources brought to bear on turbojet development in the United States enabled a different approach to the new engine type from that taken in other countries.

This book has shown how much more can be understood about the jet engine and about technical change more generally by putting successful and unsuccessful designs back into the context out of which they emerged—both the national programs and the expertise of the firms that made them. The decisions made in each country to produce jet engines in large numbers were based on prosaic things: how long before the new engines could be produced and put in aircraft, what was the cost of production, how was the war proceeding? The answers to these questions in the case of the jet engine agree with what other historians have written about the place of new technology in Britain and Germany's war efforts.[3] The fact that production decisions largely did not depend on, although they influenced, development and invention was revealed by exploring production before anything else. By starting the story of the jet engine in a new way the narrative thus revealed the complex relations between production, development, and invention.

The new story of development given pride of place here emphasized industry and continuities. Although one might expect existing expertise to be irrelevant in the realization of novel ideas, in fact existing expertise was indispensable for making jet engines a reality. Relevant expertise was tacit, embodied by designers and workers, and formal; subject-related as well as methodological; to do with manufacture and development as much as with design. Studying each firm that produced or developed jet engines revealed which types of expertise were most important to the emergence of the new technology—not, as generally expected, expertise with steam turbines, but expertise with aircraft engines.[4] In the end, the aero-engine industry's familiarity with the development and production of aero-engines was difficult to reproduce. Aero-engine firms benefited from important continuities, but this was no guarantee of success. For some firms, the continuities that were advantageous for other firms proved to be toxic. While firms outside the aero-engine industry were unable to exploit the positive synergies of aero-engine makers, they conversely did not suffer from problematic continuities and were often successful in development and invention if not production.

Looking at firms and considering their interests revealed that the most important single reason that firms were reluctant to take up jet engine development was the press of work on piston engine production and development. It was thus not primarily the success of a group of revolutionaries dedicated to jet engines that converted the aero-engine industry to making jet engines, but rather the lessening pressures of piston engine production that freed firms to work on the new engine type. No firm abandoned a successful piston engine program for a jet engine program during the war, but some identified (and invested in) the jet engine as the future of the field.

A key tool that this account made use of to trace both continuities and discontinuities was the expository power of engine layouts, distinguishing in detail between different machines that standard stories lump together under a single name. During World War II, "the jet engine" was many things: different types of compressor and turbine; a variety of combustion systems; different metals and fuels; and the product of a range of industries with distinct engineering standards, goals, and material circumstances. It was only later, after winners were picked, that the field became more homogenous. This suggests the dangerous degree of simplification in popular accounts in which what Whittle and von Ohain invented was "the jet engine," what Jumo and Rolls-Royce produced was the same power plant.

These new stories of production and development suggested the need for a new approach to invention, a new approach that would better elucidate the connections between the three activities. Putting the stories of the well-known inventors into a larger frame made it possible to understand the roles the well-known inventors played in a more detailed way for the first time. By telling their stories in terms of the institutions where they worked, this account revealed how Whittle and von Ohain and the realization of their ideas were constrained by factors other than physical nature or government support: Whittle by the commercial needs of the company that he helped to found and von Ohain by Heinkel's mercurial style of management and extensive experience producing airframes. Both of these things were far more difficult to overcome than any resistance to innovation these men encountered. The broader view taken in this book showed the importance of institutional context to the fate of new ideas. No matter how supportive, neither Power Jets nor Heinkel was part of the aero-engine industry. Like the process of development, the process of invention can usefully be understood in terms of the nature and capability of organizations. In the end, it was precisely those characteristics that were so welcoming to new inventors that led ultimately to the inventors' failures and the ascendancy of other jet engine designers. It is only by looking at the stories of Whittle and von Ohain from an institutional perspective that one can begin to understand why the work of these two men was not more widely adopted by others.

The contradiction between the central position of these two men in accounts of the jet engine and the meager effects of their work on subsequent designs can best be understood by looking at how the mechanisms of memory fashioned these two men as heroic inventors, generating the stories with which we are familiar—but stories that bear little resemblance to the world of work and political realities revealed in this book's first sections. The standard story's insistence on the independentness of the jet engine's inventors reflects the cultural potency of stories of heroic inventors more than their importance to the narrative.[5] Indeed, it is not unusual for scientific and technical communities to become attached to certain individuals through the establishment of origin stories.[6] Regardless of their connection to the work these men actually did, the popular stories that emerged enabled them to continue their key contribution to the jet engine: inspiring future work. It is not surprising to see that popular memory privileged the stories of individuals. What is startling, however, is that the popular story of the jet engine has been allowed to shape the revisionist accounts of professional historians.

*   *   *

By not letting the transition narrative define the most important aspects of this story and abandoning and revising key assumptions, this book has shown how much traditional narratives miss—and thus how the transition narrative blinds historians to important sources of technical novelty. This account has done many things differently from standard accounts: it has looked at failures as well as successes, it has considered continuities as well as discontinuities, it has looked at institutions rather than individuals, and it was not distracted by looking for science or laboratories that do not feature in the story.

It is revealing that this book made no references to academic science. Much has been written about the importance of science for causing technical change, not least Constant's arguments about the role of science in the creation of the jet engine. Yet science was not available as a source of inspiration for the jet engine as a machine. As Scranton put it in the case of jet engines during the Cold War: "rather than basic science providing the foundations for technological applications, technological creativity . . . led both to propulsive innovation and to richer scientific knowledge."[7] That some elements of jet engines have still never successfully been analyzed by scientists has not kept firms from building and developing jet engines to ever-higher powers. The development of jet engines emerged as a nexus for using and producing science, as historians like Walter Vincenti have argued by drawing attention to the systematic research and testing done by industry as a serious source of knowledge.[8] The jet engine relied more on scientific inputs than piston engines, yet the most fundamental formal expertise for making the new aero-engine type was never the lab-based science so prominent in the historiography of the early twentieth century, never chemistry or physics, but rather mathematics, projective geometry, mechanics, and material science.

In its genesis, the jet engine is not unique. During the twentieth century, streams of new products issued continuously from engineering industries without research laboratories in the academic sense.[9] Despite being creative, innovative outputs, jet engines were not produced by corporate research or by academic science. Instead, creativity was exercised in design offices and workshops, production lines and boardrooms, government laboratories and test stands. Many different people made many different decisions. Technical novelty came from government experts, engineers, mathematicians, entrepreneurs, bankers, and salaried inventors. An array of industries that were active and successful in the mid-twentieth century—trains, shipbuilding, automobiles, airframes, machine tools—similarly testify to the fact that we need to radically change our understanding of innovation in heavy engineering in the twentieth century.

Far from the products of gifted individuals working alone, the jet engine was created using all of the trappings of twentieth-century business: from banking houses to government contracts, from general and managerial staffs to huge production organizations. Indeed, the jet engine could only have been produced by a large engineering organization. The new, high-powered aero-engines of World War II were not something that could be made by a small group of people in a mechanic's shop. Jet engines were complex machines, produced by extensive systems that relied on a range of skills, thousands of workers, extensive supply chains, and tangles of interconnected subcontracting companies. Novelty appeared during all stages of the jet engine's emergence and in a range of interconnected locations, many of which were in industry. Production, development, and invention were closely connected.

The great distance between the story told in this book and the standard story of the jet engine, which ignores industry and focuses on individuals, suggests that the popular judgment of which stories should be told is not the same as historians'. A crucial first step in changing our historical accounts of technical change is to question traditional and popular measures of technological success. We can no longer countenance the great loss of understanding that our treatment of invention has brought with it. Making changes to our stories of invention and the assumptions behind them is not only helpful, it is crucial to our understanding of technical change.

Power Jets Ltd., Schedule of Shareholders, November 1, 1943

Shareholder	Shares (£1)	Additional Notes
Mrs. Shelia Emmet **Anscombe**	10	
Edward Robson **Atkinson**	50	BTH draftsman
Harry Edward **Burgess**	300	
Miss Ethel Georgina **Cox**	25	Possible relation of Roxbee Cox
V. E. **Crompton**	20	Employee
I. D. **Davidson**	50	
Bonamy **Dobree**	100	
Messrs O.T. **Falk Nominees Ltd.**	5,170	
G. B. R. **Fielden**	50	Employee
Messrs. **Gunn and Company, NY USA**	550	Includes 300 transferred from O. T. Falk and 250 from L. L. Whyte
Miss L. M. **Herringshaw**	20	Probably secretary
His and Her Highness Yeshwant Ras and Marguerite **Holkar, Maharaja and Maharani of Indore**	600	Client of Whyte*
Denys Quinton **Holland**	4,500	For the British Thomson-Houston Company
James Dayton **Imrie**	5,000	For the Weir brothers
A. B. S. **Laidlaw**	100	For services rendered
Messrs **Laidlaw, Drew & Co Ltd**	370	
Mrs. S. **Lees**	150	Possible relation of Wing Commander Lees
H. B. **Leney**	450	
Messrs. **Lindsay, Long and Partners**	1200	Williams, Tinling Directors; appointment of Lindsay, Long and Partners for management services terminated on December 31, 1941
Messrs. **Lloyds Bank City Office Nominees Ltd.**	900	
Mrs. L. **Lubbock**	100	Possible relation of Roy Lubbock
Mercantile Bank of India (Nominees)	600	Transferred from L. L. Whyte (Director), originally for service
R. D. **van Milligen**	40	Employee
Ernest Kew **Mills**	200	Employee
W. M. **Ogston**	20	Employee
William Albert **Randles**	150	Draftsman at the British Thomson-Houston Company

Shareholder	Shares (£1)	Additional Notes
John D. **Richardson**	50	
John Ryan **Strugiss**	200	
T. A. **Taylor**	50	Chief metallurgist
Mrs. I. F. **Taylor**	20	Reissued from I. F. Lucas (maiden name)
J. C. B. **Tinling**	21	Director, A-Shares**
R. G. **Voysey**	50	Employee
Reginald Newnham **Waite**	200	Transferred from Mrs. F. M. Chapman (possible relation of chief draftsman C. J. Chapman)
W. S. **Wasserman**	125	Of New York, Whyte client*
Mrs. I. V. **White**	200	Custodian of enemy property 1943
Frank **Whittle**	56***	Honorary Chief Engineer, A-Shares**
Rolf Dudley **Williams**	21	Director, A-Shares**
Guy Richard Charles **Wyndham**	100	
Total	**£21,818**	

Shares held by O. T. Falk as a Nominee Company on behalf of the following beneficial owners:

A. J. Beamish	50	
Bonham Carter Pool	400	
Lady Charles Bonham Carter	300	
General Sir Charles Bonham Carter	100	
Gerard Bonham Carter	100	
Sir Maurice Bonham Carter	700	Includes 500 for services; Director
A. J. Bryant	20	
Durani Syndicate	250	
Oswald T. Falk	700	Includes 500 for services
H. M. Farrer (dead)	250	
Mrs. E. J. Leney	150	Also invested in Napier
Sir Ronald C. Lindsay	800	Ambassador to USA
Mrs. H. M. Lubbock	400	Possible relation of Roy Lubbock
James Collingwood Burdett Tinling	250	Director
Mrs. Vickery	200	
Rolf Dudley Williams	500	Director
Total	**£5,170**	

* Whyte, "Whittle and the Jet Adventure," 115.
** The three "A"-shareholders (Whittle, Tinling, and Williams) held with ninety-eight shares 49 percent of the voting rights in the company as well as the right to appoint the chairman of the board. The remainder of the shares was issued as "B" shares.
*** Initially, Whittle held forty-two shares for himself and fourteen for the Crown (Air Ministry). After 1940, he held thirty-two for himself and twenty-four for the Crown.
Source: BL Add 61931; NASMT, Shareholder Lists in Box 1 and Box 2; Schlaifer and Heron, *Development of Aircraft Engines*, 340–42.

Air Ministry Jet Publicity (1944–45)

First press release, January 6, 1944

NOTE.

THIS STORY IS BEING ISSUED SIMULTANEOUSLY IN U.S.A.

Not for publication or broadcast or use in club tapes before 2300 hours B.S.T., Thursday, 6th January.

Not to be transmitted before that time, except to the countries and in accordance with the conditions set out hereunder:

May be transmitted forthwith to the Dominions, India, U.S.A. and Egypt by cable only and endorsed: "Unpublishable, unbroadcastable and untransmissable (ex-country addressed) before 2200 hours G.M.T., Thursday, 6th January".

Air Ministry Bulletin No. 12544.

JET PROPELLED FIGHTER SUCCESSFULLY DEVELOPED.

FIRST FLIGHT IN BRITAIN IN 1941.

The following joint statement, which has been agreed with U.S. War Department in Washington, is placed at the disposal of the press by the Air Ministry and the Ministry of Aircraft Production:

According to a Joint U.S. Army Air Force–British Royal Air Force statement issued to-day by the War Department, Washington, Jet propelled fighter aircraft have successfully passed experimental tests and will soon be in production.

Originally to British design, the improved jet propulsion engines eliminate propellers on the new aircraft. Work was started on the engines in

Great Britain in 1933 by Group Captain Frank Whittle. His first engine ran successfully in April 1937. The Air Ministry placed their first order in 1939 for an aircraft using jet propulsion engines with the ~~Gloucester~~ Gloster Aircraft Company Limited, Gloucester, England; the engines were to be built by Power Jets Limited, special factory in England, to whom Group Captain Whittle was loaned. The first successful flight of this aircraft was in May 1941. The pilot was the late Flight Lieutenant P.G. [Philip Gerry] Sayers [*sic*], chief test pilot of the Gloucester [*sic*] Aircraft Company. The greatest credit should be given to Group Captain Whittle for this fine performance, for it was his genius and energy that made this possible. [-2-]

Full information about this jet propulsion engine was disclosed in July 1941 to General Arnold who, like the British R.A.F. [Royal Air Force] and M.A.P., had the foresight to appreciate the tremendous possibilities of this new form of aircraft power unit. He at once asked for an engine to be sent over to U.S.A. The engine which had made the first flight was sent to General Electric Company in September 1941.

As the result of close cooperation between the U.S.A.A.F. [United States Army Air Forces], R.A.F. [Royal Air Force], A.A.F. [Auxiliary Air Force], Material Command and the Ministry of Aircraft Production and General Electric Company, a number of these engines were built. The first was ready for test in less than 6 months. At the same time the Bell Aircraft Company were given an order to build an aircraft suitable to take two of these engines and the first flight in the US was made in under twelve months. Several hundred successful flights have been carried out since then by British pilots in the U.S. and by British pilots with the British aircraft in England, many of them at high altitudes and extreme speed, all without a single mishap.

In view of this successful record and the obvious advantage of this new type of aircraft, General Arnold, commanding general of A.A.F. [Army Air Forces] and the British Air Ministry and Ministry of Aircraft Production have directed that plans be made for the production of a sufficient quantity for training purposes both in the US and Great Britain.

The U.S.A.A.F. are allotting a number of these to the U.S. Navy for additional trials and experimentation. The maiden flight of the first experimental aircraft in U.S.A. took place on October 1st, 1942. This was the first successful operation of a combat plane using the jet propulsion principle. Robert M. Stanley, chief test pilot of the Bell Aircraft Company, was at the controls on the initial flight. The next day Brigadier-General (then

Colonel) Lawrence C.D. Craigie, flew the aircraft, thus becoming the first army officer to fly a jet propelled
[-3-]
military aircraft in the United States. Among others who have tested the new aircraft are Brigadier B.W. Chidlaw, Chief of the Material Division, Office of Movement Control; Chief of Air Staff Material Maintenance and Distribution and Major-General William E. Kepner, an Air Corps fighter pilot formerly commanding 4th A.F.

In the United Kingdom the J.P. aircraft has been flown by a number of test pilots of aircraft firms and R.A.F. officers.

NOTES ON CAREERS.

GROUP CAPTAIN FRANK WHITTLE.   Group Captain Frank Whittle is the inventor of the first successful jet propulsion engine for fighter aircraft. His first engine ran successfully in 1937, when he was placed on the Special Duty List for research work. Group Captain Whittle had been engaged in perfecting his invention through all its stages—except for a period at the R.A.F. Staff College in 1943—whilst on loan to Power Jets Ltd. in a special factory in England.

He was born at Coventry on 1st June, 1907, and was educated at Leamington College, leaving there in 1923 to become a Royal Air Force apprentice. While an apprentice he made a number of remarkable model aircraft, and at the end of his apprenticeship was awarded a cadetship.

From 1926 to 1928 he was a flight cadet at Cranwell, and at the end of his course he gained the Abdy-Gerrard-Fellowes Memorial Prize for aeronautical sciences.

After leaving Cranwell in 1928 he became a Pilot Officer in No.111 Fighter Squadron, and at the end of 1929 he took the instructors' course at the Central Flying School. During this period he began to work on his jet propulsion ideas.

Promotion to Flying Officer took place in January 1930, and in the same year he became a flying instructor at No.2 Flying Training School, Digby. It was during this period, that
[-4-]
with the late Flying Officer G.E. Campbell, he gave one of the most thrilling exhibitions of crazy flying at the Royal Air Force Display at Hendon.

He next became an experimental test pilot on float seaplanes at Felixstowe during 1931–2, specializing in catapult work. Following this he went tot Henlow on the Officers Engineering Course, 1932–3, being promoted to Flight Lieutenant in Jan, 1934.

He then spent three years at Cambridge University, obtaining a First Class Honours Mechanical Science Tripos in 1936, followed by a year's post graduate work at Cambridge, during which he was associated with Professor Sir Melville Jones. He became a Squadron Leader in December 1937, Wing Commander in June, 1940, and Group Captain on 1st July, 1937.

Group Captain Whittle married in 1930, and has two sons, aged 12 and 9 years.

FLIGHT LIEUTENANT P.E.G. SAYER. "One of the finest test pilots in the country", was the description applied to Phillip Edward Gerald Sayer, at the time of his death on October 22nd, 1942.

He was born on February 2nd, 1905 and was educated at Colchester Grammar School, obtaining a short service commission in the Royal Air Force on June 1924. After learning to fly in an Avro 504K, he flew Snipes and Gladiators, his outstanding qualities as a pilot resulting in an appointment as test pilot at Martlesham Heath.

Flight Lieutenant Sayer left the R.A.F. on the completion of his five-year commission to join the Hawker Company, and was appointed as assistant to Group Captain P.W.S. Bulman. Here he was engaged in testing Harts, Furies, and other aircraft, and on the acquisition of the Gloster Aircraft Company by the Hawker group, was appointed in 1935 chief test pilot of the Gloster Company. In this capacity he tested Gladiators, Hurricanes and Typhoons.

[-5-]

Flight Lieutenant Sayer was awarded the O.B.E. in the New Year's Honours List in 1942, for his work with the Gloster Company.

He has been described as a typical British test pilot, modest to a degree and disinclined to talk about his work. A tribute paid to him by Mr. Michael Daunt, one of his assistant test pilots stated:

"I have worked with 'Jerry' for five-and-a-half years and although he was virtually my boss he was the type of boss who was so much more a personal friend that one did every job of work as a piece of co-operation. As a test pilot he was one of the foremost in the country. He had a terrific ability of being able to fly extremely well and smoothly but he had also the type of orderly brain that never misses a detail. When he landed after an important test, the information he gave would always be of value to the future of that aeroplane and of help to its designer".

F/Lt. Sayer was a son of Wing Commander E.J. Sayer, Retired.

Directorate of Public Relations,

Air Ministry,

King Charles Street,

Whitehall, S.W.1.

6th January 1944.

Photographs of Group Captain Whittle are available at B.I.P.P.A. Reference Nos. CH 11867, 11868 and 11869.

Source: NA AIR 2/7070, 27B

Second press release, September 27, 1944

27.9.44 — No. 44

Air Ministry News Service Air Ministry Bulletin No. 15768

THIS STORY IS BEING ISSUED SIMULTANEOUSLY IN U.S.A. NOT FOR PUBLICATION OR BROADCAST OR USE ON CLUB TAPES BEFORE 2000 HOURS B.S.T. WEDNESDAY, 27TH SEPTEMBER

===

PROGRESS OF JET-PROPELLED AIRCRAFT   The following joint statement, which has been agreed with U.S. War Department for issue in Washington, is placed at the disposal of the press by the Air Ministry and the Ministry of Aircraft Production:

The Press has already recorded that Allied aircraft in the European Theatre have been in action on several occasions with German jet-propelled fighters.

The appearance of these aircraft had been expected, and their design and operational characteristics appear to follow closely the estimates which had been formed of them.

In spite of their high speed and rate of climb, they have shown themselves to possess poor manoeuvrability, and our aircraft have had satisfactory exchanges with them in the engagements that have occurred.

It must however be expected that increased numbers of jet-propelled aircraft will appear in service and that they will become more effective as greater esperience [*sic*] is obtained.

Meanwhile, we are able to report that the development of British and American jet-propelled aircraft previously described in the announcement issued on January 6, 1944, has made progress.

Details of these aircraft and their engines must still remain secret but research scientists, aircraft technicians and workers in both Britain and America may take pride in their work.

British aircraft of this type have already been employed with success against the flying bombs.

Most valuable experience has thereby been gained.

+++

Source: NA AIR 2/7070, 108A

Third press release, February 28, 1945

Air Ministry News Service Air Ministry Bulletin No. 17692.

Not for publication, broadcast or release on club tapes before 2000 hours B.S.T. on Wednesday, 28th February 1945.

A statement on jet-propelled aircraft is also being released in America.

R.A.F. "JET" FIGHTERS IN ACTION. The first and so far the only jet propelled aircraft of the United Nations to go into action against the enemy is the Gloster "Meteor". These "Meteor" jet propelled fighters were first employed by a squadron of R.A.F. Fighter Command against flying-bombs launched by the Luftwaffe last summer from France so the first combats of British jets were not against conventional aircraft. The R.A.F. "Meteor" proved to possess a greatly superior speed to the pilotless German flying-bomb and many tactical lessons were learned from these early combats

Like the Gloster E.28/39, the first turbine jet aircraft in the world to fly (in May 1941), the "Meteor" is also a product of the Gloster Aircraft Company (Hawker-Siddeley Group). The "Meteor" is powered with Rolls Royce engines manufactured to the basic design of Air Commodore Frank Whittle, R.A.F., in collaboration with Power Jets Ltd., and the British Thomson-Houston Co., Ltd. The first engine supplied to the U.S. Army Air Force in Oct., 1941 was built by Power Jets Ltd. Air Commodore Whittle visited the U.S.A. in order to assist our Allies to initiate their development programme.

In addition to the "Meteor", Great Britain has another jet-propelled fighter in an advanced stage of development. This has been designed and engined by the De Havilland Aircraft Co. Ltd. These engines are also manufactured on the basic principles used by Air Commodore Whittle.

One built by this company was supplied to the U.S.A.A.F. in July 1943 and was used by the Lockheed Company as the power unit of a proto-type aircraft which was built by that firm. This prototype later engined by a Unit of American construction [GEC Type I-40] was developed into the Lockheed P-80A.

Other research and development work is in active progress with a view to progressive increase of the performance of British built aircraft using jet propulsion units. These are of a highly secret nature.

Just as full information was provided to the United States Air Forces of the original Whittle design, so in conformity with the unswerving policy of the British Government to make all technical information immediately available to our American Ally, full details of the progress made by British Aircraft firms in developing jet aircraft and engines had been freely communicated to the United States. Information on American development progress is similarly being made available to us.

The original British turbine jet aircraft was a single-engined aeroplane but the "Meteor" is a twin-engined monoplane of very clean design. It first flew experimentally in March 1943, and since then the production types have been considerably improved.

The engines of the "Meteor" take in enormous quantities of air (hundredweights a minute) which are sucked in, compressed, heated with burning paraffin and ejected through the turbine and then through a large roar nozzle. There are several immediate advantages to be found in the jet engine or gas turbine. First and foremost it is simpler in almost every respect than the piston engine; it is lighter; it is far more easily serviced and it possesses a rotary as distinct from a reciprocating movement. It is known that the Rolls-Royce engine of this type is more efficient and of longer life than the Jumo engine of the German Messerschmitt Me 262. [-2-]

The turbines emit no flame, as did the jet-propulsion units of flying-bombs, and only under certain rare conditions do they leave any smoke trails. The passage of a jet plane on the ground leaves in its wake the typical smell given off by a hot paraffin oil stove or a hurricane lamp.

The Gloster designers, chief of whom is Mr. W.G. Carter, had to take account of a new crop of aero-dynamic problems in order to achieve with safety the high speeds at which the "Meteor" flies, but in spite of this the aircraft is highly praised by the R.A.F. as being very manoeuvrable, easy to fly and with no such penalty as high landing speed. Its extreme smoothness of running, absence of the usual vibration, and simplicity of the engine controls are very much welcomed by pilots.

The first R.A.F. Squadron to be equipped with "squirts", as the pilots call their jet planes, was a squadron which had previously been flying Spitfires.

The squadron's pilots, not specially selected in any way and representing the average cross-section of any fighter unit in the Royal Air Force, began their jet training by converting from single-engine aircraft to "twins", learning the multi-engined technique in Oxfords' standard RAF [Royal Air Force] twin-engined training aircraft.

Spurred on by the new job in prospect, they went solo on "twins" in record time, and after an average of six hours' solo multi-engine flying time they were judged ready to pass on to the next stage—the jet aircraft.

Travelling secretly, and in small batches, pilots and key ground staff went to an R.A.F. [Royal Air Force] Experimental Establishment, where jet fighters fresh from the factory assembly line were awaiting collection.

While the pilots, fortified with a few hours' ground instruction from test pilots and experimental personnel, flew their first solo jet flights and did a few hours' practice flying on the new type, the ground crews learned the care and maintenance of the prototypes.

Then, after a few days, the pilots flew the jet aircraft back to their base, where the 'planes were guarded every minute of their earth-bound time by special security police.

The RAF's first operational jet patrol was flow during the Battle of the Flying Bomb and had its first success on August 4, 1944. Subsequently the Meteor shot down a substantial number of flying bombs.

"They are really beautiful aircraft, and I should hate to return to normal flying", said a pilot. "When they start up and taxi, our 'squirts' make a noise rather like an oversize vacuum cleaner, but when they take off, or fly at full throttle, they sound almost like a normal aircraft.

"The cockpit layout differs very little from the conventional type, and it is very comfortable with good visibility all round. There is plenty of armour to give one a sense of security, and a remarkable contrast to a normal type of aircraft is the almost complete absence of noise in the cockpit when one is flying, it is just like driving about on a cloud.

"The 'squirts' have plenty of power, and if you open the throttle suddenly you get a kick in the back from your seat. They go up like a lift, the faster the higher. They're sweet to handle even at high speed, and it's jets for me from now on".

Source: NA AIR 2/7070, 177A

Engine Comparison Table

This table is provided to help make sense of the many different engines mentioned in the text. It includes both piston engines and gas turbine engines (turbojets and turboprops). Piston engine powers have been converted to thrust to enable an order of magnitude comparison between the two different types. The rough conversion demonstrates that existing piston engines served as a reference point for early gas turbine engines and that early gas turbine engines did not necessarily outperform piston engines. A variety of other measurements, which are not included here—including particularly fuel consumption—demonstrate the many compromises that occur in engine design and show that a given engine was rarely superior to all other engines in all of its characteristics; different engines were best suited to different applications.

Converting piston engine powers to thrust is an unnatural operation because the two engine types produce thrust in fundamentally different ways. The calculation of equivalent thrust done here is necessarily somewhat arbitrary. A reasonable speed for piston-engined aircraft, 300 mph, was chosen, and a reasonable propeller efficiency (for a variable pitch propeller) of 85 percent was assumed. Piston engine shaft power depends on speed and altitude, and piston engine specifications often list multiple values to reflect the use of superchargers at altitude. The shaft horsepower specifications used below reflect measurements for particular altitudes as indicated. A further complication is the fact that the thrusts and horse powers given are not readily converted into flight speeds; the performance of a piston engine / airframe combination depended on the match between the engine's output and the conditions of flight for which the airframe was designed. The calculations provided here should be viewed with these considerations in mind.

The figures given for engines in this table reflect mostly published figures (where engines became service engines, their model numbers are given in the chart), but output varied drastically for a given prototype over time, between different prototypes and between prototypes and service articles.

Equivalent power was calculated as follows:

Thrust = Power available / V_∞

V_∞ = cruising velocity

Assume V_∞ = 300 mph = 440 ft./sec.

Power available (from a piston engine–propeller combination) = $\eta_{propeller} \cdot P_s$

P_s = shaft power in horsepower

1 HP = 550 ft. $\cdot$ lbs./sec.

Choose (optimistically) $\eta_{propeller}$ = 0.85

Thrust = $\eta_{propeller} \cdot P_s \cdot$ 550 ft. $\cdot$ lbs./sec./HP / V_∞

Thrust = 0.85 $\cdot P_s \cdot$ 550 / 440

Source: John Anderson, *Introduction to Flight* (McGraw-Hill, 2008), 416 and 421–24.

Description of Values Given in the Comparison Table

1. **T**: engine type, TJ = turbojet; TP = turboprop; PE = piston engine
2. **LAYOUT**: 3 columns
   a. For gas turbine engines: compressor (A = axial; CR = contra-rotating; C = centrifugal; number of stages is given where known), combustion system (A = annular, for C = individual can combustors or ST = straight through— individual combustors the number of combustion chambers is given where known), turbine (giving the number of stages and the type, A = axial, C = centrifugal)
   b. For piston engine: number of cylinders and layout (either IL = in-line, R = radial, or V, X, or H), displacement (L), and SC indicates the engine was supercharged
3. **PR**: highest pressure ratio of the supercharger or pressure ratio of the gas turbine

4. **TH (lbs.)**: thrust of gas turbine engines and residual thrust of turboprops
5. **HP**: shaft horsepower for piston engines, a high cruising power was chosen rather than takeoff power
6. **ALT (1,000 ft.)**: altitude where given for piston engine powers in thousands of feet, otherwise assume sea level
7. **EQ TH (lbs.)**: equivalent thrust calculated for piston engines and turboprops, figures for turboprops this figure represents total engine power in thrust
8. **DRY (lbs.)**: dry weight of unit; the weights of engines that drive propellers are generally given without propeller weights
9. **DESIGN / FIRST RUN**: gives the dates when designs were begun and when engines ran for the first time self-sustaining
10. **S**: Sources of data in chart—where no source is given the information was taken from the preceding text
    (A) Paul H. Wilkinson, *Aircraft Engines of the World* (New York: Wilkinson, 1945)
    (B) "British Power Units," *Flight*, September 4, 1947
    (C) Wagner, *Die Ersten Strahlflugzeuge*
    (D) Brooks, *Vikings at Waterloo*
    (E) Kay, *Turbojet History and Development*

Power Jets

	TYPE	LAYOUT	PR	TH (lbs.)	HP (max.)	ALT (ft.)	EQUIV TH (lbs.)	DRY WT (lbs.)	DATE DESIGN /FIRST RUN	NOTES
WU	TJ	1C C 1A							1936/1937	
W1	TJ	1C C 1A							1939/1941	
W1A	TJ	1C C 1A		860			860		1941	(F)
W2	TJ	1C 10C 1A		562			562		1941	(F)
W2B	TJ	1C 10C 1A		1,526			1,526		1941	(F)
W2/500	TJ	1C C 1A		2,485			2,485		1942	(F)
W2/700	TJ	1C C A		2,500			2,500			(F)
LR1	TJ	A C A								
ROVER										
STX/B26	TJ	1C 10ST 1A		1,295			1,295		1941–1942	STX1 2.43, (E)
METROVICK										
F2	TJ	9A A A		1,800			1,800		1941	(F)
F2/4	TJ	A A A		3,850			3,850	1550		(B)
F9	TJ	13A 2A 2A		7,000			7,000		1945	(F)
De Havilland										
GIPSY KING	PE	12IL 18.37 SC			425		452	1,056	/1937	(A)
GIPSY QUEEN 70	PE	6IL 10.18 SC			345	7750	367	660		
H1	TJ	1C C 1A		3,000			3,000			
GOBLIN	TJ	1C 16ST 1A		3,000			3,000	c. 1,550	1941/1942	(B)
GHOST	TJ	1C 10ST 1A		5,000			5,000	2,011	1944	(B)

continued

	TYPE	LAYOUT	PR	TH (lbs.)	HP (max.)	ALT (ft.)	EQUIV TH (lbs.)	DRY WT (lbs.)	DATE DESIGN /FIRST RUN	NOTES
Rolls-Royce										
VULTURE	PE 24X	42.4 SC	7.28		1,845	5,000	1,960	2,450		(A)
MERLIN XX	PE 12V	27 SC	9.49		1,480	6,000	1,573	1,385		(A)
MERLIN 61	PE 12V	27 SC	8.03		1,570	11,500	1,668	1,640		(A)
MERLIN 140	PE 12V	27 SC			1,780	4,500	1,891	1,780		(B)
GRIFFON 121	PE 12V	36.7 2S2S			2,300	15,750	2,444	2,150		(B)
CRECY	PE 12V	26 SC	7		2,000		2,125	2,090		(A)
EAGLE	PE 24H	46 2S			3,500	3,250	3,719	4,220		(B)
WR1	TJ 1C	C 1A		1,984			1,984			(F)
WELLAND (W2B/23)	TJ 1C	C 1A		1,698			1,698			(F)
DERWENT (B/37)	TJ 1C	ST 1A		1,984			1,984			(F)
DERWENT V	TJ 1C	9ST 1A		3,600			3,600	1,280		(B)
RB41 NENE	TJ 1C	9ST 1A		4,500			4,500	1,640		(B)
CR1	TJ 14A	A 14A								
AVON (AJ65)	TJ 15A	C 2A		5,000			5,000			(F)
RB39 CLYDE	TP 1A1C	9 2A		1,225	3,020		4,434	2,800		(B)
DART	TP 2C	7 2A			890		946	847		(F)
RCA1	TJ A	C A								
ASM										
CHEETAH 25	PE 7R	13.67 SC	6.5		405	4,000	430	805		(B)
DEERHOUND	PE 21R				1,115		1,185			
WOLFHOUND	PE 24R				3,000		3,188			
ASX	TJ 14A	C 2A		2,551			2,551			
HEPPNER UNIT	TJ CR								1942/1943	(F)
ASP	TP A	C 2A			3,600		3,825			
PYTHON	TP 14A	11 2A		1,150	3,670		3,899	3,190		(B)

BRISTOL												
HERCULES 120	PE	14R	38.7	SC			1,800	6,000	1,913	2,025		(B)
CENTAURUS XX	PE	2X18R	107.2	SC			5,160	4,000	5,483	C 2,900		(B)
THESEUS I	TP	1A,1C	8			500	1,950		2,572	2,310		(B)
NAPIER												
SABREVII	PE	24 H	36.7	2S			3,055	2,250	3,246	2,540		(B)
NAIAD	TP	A	5			241	1,500		1,835	1,095		(B)
ELAND	TJ	10A	6	3A		825	2,690		3,683	1,661		(C)
HEINKEL												
HeS 1	TJ	1C	A	1C							1936/1937	(F)
HeS 2	TJ	1C	A	1		300			300			(D)
HeS 3b	TJ	1C	A	1A	2.8	1,102			1,102	792	1938/1939	
HeS 8 (109-001)	TJ	1C	A	1A	2.7	1,587			1,587	836	1940/1941	
HeS 30 (1909-006)	TJ	5A	9C	1A	3.2	1,808			1,808	836	1939/1942	
HeS 11	TJ	1D+3A	A	1A	4.4	2,866			2,866	2,090	1941/1945	
JUNKERS												
211-J	PE	12V	35	SC	11.37		1,300	12,500	1,381	1,429		(A)
213-A	PE	12V	36.9	SC			1,700	9,800	1,806	1,543		(A)
109-004A-0	TJ	8A	6	1		1,852			1,852	1,870	1939/1940	(F)
109-004B-1	TJ	8A	6	1	3.14	1,984			1,984	1,639	1945	(D)
109-012	TJ	11A	6	2	5.5	6,614			6,614	3,520	1945	(D)
109-022	TP	11A	A	3	5.5		6,000		6,375	6,600	1945	(D)
BRAMO												
FAFNIR 323-P1	PE	9R	26.8	SC			810	8,500	861	1,320		(A)
329	PE	14R	A				2,000		2,125			
109-002 WEINRICH SIMPLE AXIAL	TJ	CR										
(P.3302)	TJ	6A	A	1A		330	330		330		1938/1940	

continued

	TYPE	LAYOUT			PR	TH (lbs.)	HP (max.)	ALT (ft.)	EQUIV TH (lbs.)	DRY WT (lbs.)	DATE DESIGN /FIRST RUN	NOTES
BMW												
132K	PE	9R	SC	27.7	12.4		960	9,800	1,020	1168		(A)
139	PE	18R					1,500		1,594	1,940		(A)
801	PE	14R	SC	41.8			1,600	13,100	1,700	1,940		(A)
P.3303	TJ	2C	A								1938	(F)
P.3302 (revised)	TJ	6A	1A			1,323			1,323		1940/1942	(F)
109-003A-0	TJ	7A	1A		3.1	1,764			1,764	1,309		(D)
109-003A-2	TJ	7A	1A		3.4	1,984			1,984	1,309		(D)
109-018	TJ	12A	3A		7.1	7,716			7,716	4,840	1945	(D)
109-028	TP	12A	4A		7	2,750	3,100		3,294	7,040	1945	(D)
DAIMLER-BENZ												
601E	PE	12V	SC	33.9	10		1,375	18K	1,461			(A)
603a	PE	12v	SC	44.5	9.22		1,680	18K	1,785			(A)
605-a1	PE	12V	SC	35.7	10		1,350	19.7K	1,434			(A)
109-007	TJ	17CR	1A		8	2,810			2,810	2,860	1939/1943	DUCTED FAN, (F)
109-016	TJ	9A	2A			4,400			4,400		1943	(F)
109-021	TP	1D+3A	2A		4.4	1,741	2,000		3,866	2,860	1944	(D)

Acknowledgments

This book would not have been possible without the generous help of many people whom I would like to mention here.

First of all, my thanks to all of the groups who contributed the funding for this project: Imperial College's Centre for the History of Science, Technology and Medicine; the Royal Air Force Historical Section and Royal Aeronautical Society Historical Group; the Smithsonian Institution's National Air and Space Museum; the Deutscher Akademischer Austausch Dienst; the International Committee for the History of Technology; and the Society for the History of Technology.

During my research, I worked in many archives around the world, and the staff at each was invaluable in helping me find my way around. The many archives cited in this work all have my thanks for enabling the sprawling comparative study that I undertook. I would like to especially thank Dave Piggott (Rolls-Royce Heritage Trust) and Jessika Wichner (Göttingen Archives of the Deutsches Zentrum für Luft- und Raumfahrt) for their ongoing assistance and interest. The Smithsonian National Air and Space Museum and Deutsches Museum were very helpful in opening their archives to me and helping me find relevant material. In addition, the Science Museum Library and Archives has also been an invaluable resource.

For making my research trips to Germany and the United States possible, I want to give special thanks to the Sattler, Heinze, Scholten, and Goldblatt families for generously giving me a place in their homes.

In addition, my thanks to Duncan Denie of Denie Technology for making models for me, although he had never seen a jet engine from the inside before!

I could not have imagined embarking on this book without the help and inspiration of the lively debate, encouragement and draft reading of David Edgerton, who encouraged me to embark on this project and who patiently supported my subsequent efforts to radically rethink the history of the jet. I'd also like to thank particularly Andrew Nahum (Science Museum), Michael Neufeld and Jeremy Kinney (National Air and Space Museum), and Walter Rathjen (Deutsches Museum) for their guidance and encouragement. The university departments that supported me during my work—Imperial College and the University of Utrecht (particularly the members of the Asymmetrical Encounters project)—receive sincere thanks for ongoing support, stimulation, and understanding.

And lastly, I want to thank my family—Rona, David, and Jonathan—for constant help and motivation, particularly when everything took that much longer than expected.

Archives Consulted

AFHR	Air Force Historical Research Agency
BA	Bundesarchiv, Berlin
BA MA	Bundesarchiv-Militärarchiv, Freiburg
BL	British Library
BMW	BMW Corporate Archive, Munich
CAC	Churchill Archives Centre, Churchill College Cambridge
DB	Daimler-Benz Corporate Archive, Stuttgart
DM	Archives of the Deutsches Museum, Munich
GOAR	Archives of the Deutsches Zentrum für Luft- und Raumfahrt, Göttingen
IWM	Archives of the Imperial War Museum
KNA	Kings Norton Archive, Cranfield University
MOSI	Archive of the Manchester Museum of Science and Industry
NA	British National Archives, London
NASM	Archives of the Smithsonian Institution, National Air and Space Museum
RR	Archives of the Rolls-Royce Heritage Trust
WS	Wright State University, Special Collections and Archives

Notes

Introduction

1. Gilfillan, *Inventing the Ship*; Ogburn and Thomas, "Are Inventions Inevitable?"; Usher, *A History of Mechanical Inventions*, 1929 edition, 8–31.

2. Susskind and Zybkow, "The Argument," ix.

3. This can be compared to the "unheroic bipolarity" between inventors and industrial R&D laboratories in the twentieth century described by MacLeod (MacLeod, *Heroes of Invention*, 351).

4. Maurice Holland and Henry F. Pringle, *Industrial Explorers* (New York: Harper and Brothers, 1928), quoted by Gilfillan, *Sociology of Invention*, 54; W. B. Kaempffert, *Invention and Society* (Chicago: American Library Association, 1930), 30, quoted in Jewkes, Sawers, and Stillerman, *Sources of Invention*, 36; Joseph Schumpeter, 1942, quoted in Edgerton, *Industrial Research and Innovation in Business*, xi.

5. MacLeod, *Heroes of Invention*, 5.

6. Jewkes, Sawers, and Stillerman, *Sources of Invention*.

7. Hounshell, "Hughesian History of Technology and Chandlerian Business History," 217; Hughes, *American Genesis*, 15.

8. For example, Jewkes, Sawers, and Stillerman, *Sources of Invention*, 246; Edgerton, *Shock of the Old*, 257–58.

9. Dennis, "Accounting for Research," 505–6; Edgerton, "'The Linear Model' Did Not Exist," 31–57; Edgerton, *Shock of the Old*, 254.

10. For moral divisions, see, for example, Hughes, *American Genesis*, 138, and Mokyr, *Lever of Riches*, 19.

11. White, "Act of Invention: Causes, Contexts, Continuities, and Consequences," 486.

12. Freeman, *Economics of Industrial Innovation* (1974 and 192 editions); Mowery and Rosenberg, *Paths of Innovation*; Friedel, *Culture of Improvement*, 492. Yet Friedel's own work on invention has been some of the best to look outside

the transition narrative. See, for example, Friedel, Israel, and Finn, *Edison's Electric Light*; Friedel, *Zipper*; Friedel, *Pioneer Plastic*.

13. For example, Bijker, Hughes, and Pinch, *Social Construction of Technological Systems*; Bijker, *Of Bicycles, Bakelites, and Bulbs*.

14. Edgerton, *Shock of the Old*.

15. Israel, *From Machine Shop to Industrial Laboratory*; Fox and Guagnini, "Sites of Innovation in Electrical Technology, 1880–1914;" Jakab, *Visions of a Flying Machine*, xvi.

16. Scranton discusses the same activity in the context of the dominance of attention to mass production and the modern corporation in business history (Scranton, "Technology, Science, and American Innovation"); the work of authors on topics less amenable to the transition narrative than nineteenth-century inventors is generally not regarded as part of the literature on invention. For example, Noble, *Forces of Production*; Vincenti, *What Engineers Know and How They Know It*; Schatzberg, *Wings of Wood, Wings of Metal*.

17. Jewkes, Sawers, and Stillerman, *Sources of Invention*, 320–21.

18. Historian of aerospace Richard Hallion described it as a "thoughtful model of technological change"; Schabel referred to it as an "excellent model"; Nahum a "compelling model"; Aitken praised the "challenging quality of its theoretical constructs" (Hallion, "Review: Aerospace Technology"; Schabel, *Illusion der Wunderwaffen*, 35; Nahum, "Two-Stroke or Turbine?"; Aitken, "Reviewed Work(s): Networks of Power: Electrification in Western Society by Thomas P. Hughes"). Hughes identified Constant's "presumptive anomaly" as what he calls a "presumed reverse salient," but distinguished Constant's concept from his own for its (according to him) laudable stress on the role of science (Hughes, "Evolution of Large Technological Systems"). In 1992, Roland included the notion of "presumptive anomaly" in his discussion of models of technological change and grouped it with Hughes's reverse salients (Roland, "Theories and Models of Technological Change").

19. Layton, "Mirror-Image Twins"; Hughes, "Science-Technology Interaction"; Layton, "Scientific Technology, 1845–1900"; Aitken, *Syntony and Spark*; Aitken, *Continuous Wave*; Laudan, *Nature of Technological Knowledge*; Hindle, *Emulation and Invention*.

20. Forman, "Discovery of X-Rays by Crystals," 68.

21. The innovative process has been understood through the "commonly accepted basic model" of invention, development, innovation (Staudenmaier, "Recent Trends in the History of Technology," 717; Edgerton, "'The Linear Model' Did Not Exist").

22. Scranton, "Technology-Led Innovation," 338.

23. As Schlaifer, free of the later assumptions of historians of technology, argued (Schlaifer and Heron, *Development of Aircraft Engines*, 25).

24. Scranton, "Turbulence and Redesign"; Scranton, "Technology, Science, and American Innovation."

25. Schlaifer made a similar argument in his book, *Development of Aircraft Engines.*

26. In a geared supercharger, an extra compressor was driven directly by the piston engine to compress the air going into the pistons, thereby increasing efficiency and power. Turbo-superchargers, in contrast, used a turbine to extract energy from the piston engine's exhaust and to drive the compressor. (See also the glossary.)

27. Other research can be consulted for more information on different aspects of the role of government in jet engine development. For example, Schlaifer and Heron, *Development of Aircraft Engines*; Schabel, *Illusion der Wunderwaffen*; Nahum, *Frank Whittle: Invention of the Jet.*

28. For background on the development of aerodynamics before the first use of jet engines, see Bloor, *The Enigma of the Aerofoil*, and Constant, *Origins of the Turbojet Revolution*, 99–116.

Chapter One

1. Figures for Germany are production until March 1945. Britain produced 745 jet engines and the United States 296 before the end of 1945 (RRHT; Hornby, *Factories and Plant*, 245; Kay, *German Jet Engine and Gas Turbine Development*; IWM FD 5400/45, Jumo Report, TL-Geräte 004 B-1/4 Neubau und Reperatur [Turbojet engines 004 B-1/4 Construction and Repair]; AFHR A2073; AFHR A2072).

2. Pavelec, *Jet Race*, 120 and 155; Schabel, *Illusion der Wunderwaffen*, 12–83; Constant, *Origins of the Turbojet Revolution*, 208.

3. Schlaifer based this conclusion on a discussion of the comparative quality of British and German engines rather than just speed and production numbers (Schlaifer and Heron, *Development of Aircraft Engines*, 429–39). Postwar intelligence reports and testing in Britain confirmed the inferiority of German engines to British engines (NA AVIA 15/2121, June 1945, 205, Edward Pickles (MAP), Note on visit to Germany).

4. Epstein argues similarly in the case of the torpedo that the inferior strength of the American navy as compared to the British, led the two nations to build different research and development organizations and create different weapons designs (Epstein, *Torpedo*, 222–23).

5. Edgerton, *War Machine*, 234.

6. Ibid., 301.

7. Hornby, *Factories and Plant*, 253.

8. Ibid., 265.

9. Factories paid for by the government to expand production were not unique to Britain. German firms, too, for example, were worried about such investments as leading to excess capacity in peacetime (James, *Krupp*).

10. Hornby, *Factories and Plant*, 257.

11. Ibid., 254.

12. NA AVIA 15/2305, February 26, 1940, 5A, Minutes of Meeting, Whittle Jet Propulsion Engine; Tedder quoted in Nahum, "World War to Cold War," 83.

13. Kay, *Turbojet History and Development*, 28.

14. Kershaw, *Jet Pioneers*, 14.

15. The Air Ministry wanted to put the W.1 into a "full-military" fighter airframe, but William Farren insisted that an experimental aircraft be built first (Thomson and Hall, "William Scott Farren, 1892–1970").

16. Ritchie, *Industry and Air Power*, 113–46; Bulman, *An Account of Partnership*, 320.

17. NA AVIA 15/2305, February 17, 1940, 4A, Letter Whyte to Tedder.

18. NA AVIA 15/2305, February 26, 1940, 5A, Minutes of Meeting, Whittle Jet Propulsion Engine; Brooks, *Vikings at Waterloo*, 18.

19. Brooks, *Vikings at Waterloo*, 15.

20. NA AVIA 15/1806, January 5, 1943, 12A, Notes of a discussion, Rolls-Royce–Rover Organisation.

21. Brooks, *Vikings at Waterloo*, 13.

22. NA AVIA 15/1802, December 15, 1942, 5A, Interdepartmental note.

23. Bulman, *An Account of Partnership*, 320.

24. Taylor, *Jane's All the World's Aircraft, 1989–908*, 266.

25. Bulman, *An Account of Partnership*, 320.

26. Such signals were important; as Ricardo complained, Rolls-Royce deemphasized work on the Crecy because its importance was not made clear through the allocation of an airframe to it (Reynolds, *Engines and Enterprise*, 210–12).

27. Thomson and Hall, "William Scott Farren, 1892–1970," 215–41.

28. Bulman, *An Account of Partnership*, 317.

29. NA CAB 66/15/22, March 3, 1941, Memo by Beaverbrook, Minister of Aircraft Production; Edgerton, *Warfare State: Britain*; Peden, *Arms, Economics, and British Strategy*.

30. NA AVIA 46/234, April 18, 1941, Air Supply Board Meeting Summary.

31. NA PREM 3/21/4, July 23, 1941, 184, Memo from Cherwell to Churchill; Clark, *Tizard*, 303.

32. Churchill, quoted in Clark, *Tizard*, 303.

33. Whittle, *Jet: The Story of a Pioneer*, 2nd ed., 168–69.

34. NA AVIA 15/1806, n.d., 24B, Rover Contracts. The engine contract was awarded by October 28, 1941 (RR, A-Numbers List) NA AVIA 15/1509, October 14, 1941, Minute DDGEDP (Bulman) to CRD (Linnell).

35. Brooks, *Vikings at Waterloo*, 50.

36. NA AIR 62/990, May 31, 1942, GTCC Progress Reports.

37. Hooker, *Not Much of an Engineer*, 76.

38. NA AVIA 15/1509, October 14, 1941, Min 8, Minute Bulman to Linnell.

39. Bulman, *An Account of Partnership*, 81.

40. Ibid., 319.

41. Ibid., 331.

42. NA AVIA 15/1614, August 11, 1942, 32A, S. B. Wilks (Rover) to Linnell (MAP).

43. NA AVIA 9/30, April 2, 1942, Linnell to Llewellin.

44. NA AVIA 15/1614, July 26, 1942, 26, Intradepartmental Minute Linnell to B. McEntegart, his deputy.

45. NA AVIA 15/1614, August 11, 1942, 32A, S. B. Wilks (Rover) to Linnell (MAP).

46. NA AVIA 15/1806, December 11, 1942, 1A, Interdepartmental letter to Freeman.

47. NA AIR 62/608, January 5, 1943, Whittle diary, Meeting at Brownsover Hall.

48. Moult, "Development of the Goblin Engine," 654–55.

49. NA AVIA 15/1599, July 18, 1942, 14a, Letter Bulman to de Havilland.

50. Watkins, *De Havilland Vampire*, 4–5; the switch to all metal aircraft in America was more a product of ideology than practical performance, and European manufacturers continued to use wood after American firms had given it up (Schatzberg, *Wings of Wood, Wings of Metal*).

51. Watkins, *De Havilland Vampire*, 3–4.

52. NA AVIA 15/1708, June 21, 1942, Letter Linnell to J. J. Llewellin (Minister); AFHR A2072, October 2, 1942, Doc 2, Report by Kiern.

53. NA AVIA 31/3, March 6, 1943, GTCC Meeting Minutes #8.

54. AFHR A2072, October 2, 1942, Doc 2, Report by Kiern; Shacklady, *Gloster Meteor*, 18–19. Contrary to some accounts, the Ace was designed for the H.1 and was never meant for a single W.2B, which would have been too weak to power it. After the war, three prototypes (now E.1/44) flew with the Rolls-Royce Nene (Shacklady, *Gloster Meteor*, 65).

55. NA AVIA 15/1599, August 8, 1942, Letter Linnell to Controller General.

56. NA AVIA 15/1599, September 1, 1942, 21, Minute Linnell to Bulman.

57. NA AVIA 15/1599, August 24, 1942, 20, Minute Bulman to Linnell.

58. NA AVIA 15/1599, September 15, 1942, 25, Minute Bulman to Linnell.

59. NA AVIA 15/1599, November 3, 1942, 39a, Bulman to Banks.

60. NA AVIA 15/1599, ca. November 1942, 39b, Report on Development Production of H.1 Supercharger Unit for Air Supply Board.

61. NA AVIA 15/1708, November 7, 1942, 15A, Meeting about abandoning W.2B design.

62. NA AVIA 15/1708, November 7, 1942, 15A, Meeting about abandoning W.2B design; November 9, 1942, 17, Minute Linnell to Tizard; AVIA 15/1806, January 27, 1943, 18A, Roxbee Cox to Hives.

63. NA AVIA 15/1708, November 9, 1942, 17, Minute Linnell to Tizard.

64. NA AVIA 15/1708, November 29, 1942, 20, Minute Linnell to Freeman.

65. Furse, *Wilfrid Freeman*, 261–62.

66. NA AIR 62/608, December 4, 1942, Whittle diary.

67. NA AVIA 15/1802, January 7, 1943, 15, Minute Roxbee Cox to Linnell.

68. Brooks, *Vikings at Waterloo*, 72.

69. NA AVIA 15/1806, February 20, 1943, 21A, Letter Hives to Linnell, Roxbee Cox, Freeman.

70. NA AVIA 15/1708, January 31, 1943, 31, Minute Linnell to Freeman.

71. NA AVIA 15/1708, March 30, 1943, 38A, Letter Linnell to Hives.

72. NA AVIA 15/1614, October 15, 1942, 39, Notes of meeting between Linnell and Wilks Brothers (Rover).

73. Postan recorded that a prime ministerial directive was responsible for restoring Meteor production. An order was placed for 120 Meteor I's in mid-1943 (Postan, Hay, and Scott, *Design and Development of Weapons*, 224). Power Jets' leadership believed that the improvement in engine performance that the contract was a response to "owes nothing to RR [Rolls-Royce]" (NA AIR 62/608, May 3, 1943, Whittle diary, Memo by W. E. P. Johnson).

74. Kay, *Turbojet History and Development*, 56.

75. NA AVIA 46/237, July 27, 1944, GTCC Progress Reports; RR HT, Piggott, Development and Production Jets.

76. NA PREM 3/21/4, November 23, 1944, Cripps, Third Two-Monthly Report; Rolls-Royce recorded that they produced ninety-eight Welland engines by the end of November 1944 and an additional twelve before stopping production in October 1945 at Barnoldswick (RR, Piggott, Development and Production Jets).

77. Hooker, *Not Much of an Engineer*, 81.

78. Pavelec, *Jet Race*, 100.

79. Postan, Hay, and Scott, *Design and Development of Weapons*, 225.

80. NA PREM 3/21/4, July 26, 1944, 130, Memo Cherwell to Churchill.

81. The squadron had six Meteors (Golley, Whittle, and Gunston, *Whittle, the True Story*, 222).

82. NA AVIA 15/2101, July 30, 1944, 1A, Letter Churchill to Cripps (Minister of Aircraft Production).

83. Bulman, *An Account of Partnership*, 335; Schlaifer and Heron, *Development of Aircraft Engines*, 435.

84. Schlaifer and Heron, *Development of Aircraft Engines*, 434.

85. The "RB" prefix was already in use when Rover was in charge at Barnoldswick and referred to "Rover Barnoldswick." After Rolls-Royce took over the factory, the prefix came to be understood to mean "Rolls Barnoldswick" (Brooks, *Vikings at Waterloo*).

86. RR, Piggott.

87. The floor space estimate is based on total floor space added between 1943 and 1944, as reported by Eyre, *50 Years with Rolls-Royce*, 87; NA PREM 3/21/4, September 27, 1944, Cripps, Second Two-Monthly Report.

88. NA PREM 3/21/4, November 23, 1944, Cripps, First Third Two-Monthly Report.

89. NA PREM 3/21/4, January 30, 1945, Cripps, Second Third Two-Monthly Report.

90. RR, Piggott.

91. Gunston, *Plane Speaking*, 162; because of Derwent production lagging, the first fifteen Meteor IIIs were fitted with Wellands (Brooks, *Vikings at Waterloo*, 77–78); NA PREM 3/21/4, October 23, 1944, 78, Production of Jet Propelled Aircraft, Supplementary Note.

92. NA AVIA 15/1599, December 31, 1942, 49B, Letter Parkes (de Havilland) to Freeman.

93. Watkins, *De Havilland Vampire*, 6; NA AVIA 31/3, November 20, 1943, GTCC Minutes Meeting #11.

94. NA AVIA 15/161, March 13, 1944, 86A, Interdepartmental Letter.

95. Watkins, *De Havilland Vampire*, 10.

96. NA AVIA 15/1708, November 19, 1942, 19, Minute Linnell to Freeman; RRHT, A-Numbers List; Watkins, *De Havilland Vampire*, 25.

97. NA AVIA 15/1599, July 22, 1944, 55, Minute Watt to Banks.

98. NA AVIA 15/1599, July 25, 1944, 57, Minute Linnell to Banks and Watt; De Havilland, n.d.

99. NA AVIA 31/5, Gas Turbine Collaboration Committee Minutes Meetings #16–18; at the same time, the Rolls-Royce Nene completed a 100-hour development test at 4,000 pounds.

100. De Havilland, "Development of the De Havilland 'Goblin' Jet Propulsion Engine 1940/1945," unpublished, n.d.

101. Hornby, *Factories and Plant*, 254.

102. NA AVIA 15/2101, August 1, 1944, 4, Minute Freeman to Cripps.

103. Watkins, *De Havilland Vampire*, 11.

104. With 247 Squadron (Hassel, "Halford Jets").

105. NA AIR 2/7070, July 27, 1944, 76A, Letter Freeman to Portal; Postan, Hay, and Scott, *Design and Development of Weapons*, 225.

106. Air Force Historical Research Agency (AFHR) A2073, October 14, 1941, Doc 16, Letter H. A. Self (British Air Commission) to H. L. Stimson (Secretary of War).

107. Zimmerman, *Top Secret Exchange*, 94.

108. Ibid., 167–72.

109. NA AVIA 46/234, Monograph by Adderley (MAP).

110. Arnold claimed in his autobiography, *Global Mission*, to have seen the aircraft flying in April 1941, but this was before its first flight (Arnold, *Global Mission*, 242). His diary made no mention of the jet engine during this trip to England. During his trip, however, he toured the British aircraft industry and spoke at length with Portal, Freeman, Beaverbrook, and Tizard, who all knew about the jet.

111. AFHR A2073, July 21, 1941, Doc 1, Cable Brig. Gen. Ralph Royce, Military Attaché London to General Arnold; NA AVIA 46/234, Monograph by Adderley (MAP).

112. AFHR A2073, November 25, 1941, Doc 25, Letter Caroll (Wright Field) to Echols (Materials Division, Washington); AFHR A2072, Doc 1, Report by Kiern (Power Plant Lab, Wright Field); NA AVIA 46/234, Monograph by Adderley (MAP); NA AVIA 46/234, Monograph by Adderley (MAP).

113. AFHR A2073, ca. September 4, 1941, Doc 7, Cable General Arnold to Lt Col J. T. C. Moore-Brabazon, MAP.

114. AFHR A2073, October 14, 1941, Doc 16, Letter H. A. Self (British Air Commission) to H. L Stimson (Secretary of War).

115. Arnold, *Global Mission*, 243.

116. A2073, October 2, 1943, Doc 14, Letter General Arnold to Lyon.

117. NA AVIA 46/234, Monograph by Adderley (MAP); AFHR A2073, September 11, 1941, Doc 9, Cable Brig Gen Ralph Royce, Military Attaché London, to AAF Washington.

118. AFHR A2073 September 22, 1941, Doc 11, Letter British Air Commission to Stimson.

119. AFHR A2073, October 2, 1941, Doc 14, Letter General Arnold to Col A. J. Lyon (Office of Special Army Observer); General Electric, *Seven Decades of Progress*.

120. AFHR A2073.

121. AFHR A2073, September 4, 1941, Doc 6, Memo Col B. W. Chidlaw to Gen Echols (Fighter Br, Engineering Division); AFHR A2065, September 4, 1941, Doc 27, Letter Gen Arnold to American Embassy, London.

122. Library of Congress, Manuscript Division, H. H. Arnold Papers, 471208, July 11, 1941, Letter Bush to Arnold. Quoted in Dawson, *Engines and Innovation*, 50–51.

123. NA AVIA 31/2, December 12, 1942, GTCC Meeting Minutes #7.

124. NA AVIA 31/2, July 18, 1942, GTCC Report Progress in USA.

125. AFHR A2073, April 29, 1943, Doc 56, Progress report for General Arnold Chidlaw.

126. AFHR A2073, May 28, 1943, Doc 60, Letter General Electric to Col D. J. Kiern (Power Plant Lab, Engineering Division, Wright Field).

127. AFHR A2073, July 26, 1943, Doc 69, Letter Chidlaw (Chief, Mat Div, Washington) to General Electric.

128. AFHR A2078, Summary, 1.

129. AFHR A2073, Summary, 11.

130. Ibid., Summary, 33.

131. Ibid., Summary, 15.

132. AFHR A2078, Summary, 3.

133. Ibid., Summary, 4.

134. AFHR A2078, May 20, 1944, Doc 56, Minute Col G. E. Price to Chief Production Division, Wright Field.

135. AFHR A2078, Doc 91, November 26, 1944, Doc 91, Contract W33-038 ac-3849.

136. AFHR A2078, Doc 93, December 6, 1944, Doc 93, Letter Allison Division to Air Technical Service Command Wright Field.

137. AFHR A2078, Doc 100.

138. AFHR A2078 Full Summary, 5; September 18, 1945, Doc 157, Memorandum United States Government, Subject: Termination of General Electric I-40 Jet Engine Procurement.

139. AFHR A2072, Summary.

140. AFHR A2072, Doc 12.

141. AFHR A2073, October 13, 1941, Doc 17, Letter Hunsaker to Arnold.

142. AFHR A2072, Summary, 1–2.

143. Ibid.

144. The contract was for 40 H.1 engines. (AFHR A2072, Summary, 2).

145. AFHR A2072, May 18, 1943, Doc 10, Letter Lt. Col. Thomas Hitchcock (Air Technical Section, London) to Robert A. Lovett (Assistant Secretary of State, Air).

146. AFHR A2072, Summary, 3–4; July 20, 1945, Doc 80.

147. AFHR A2072, May 18, 1943, Doc 10, Letter Lt. Col. Thomas Hitchcock (Air Technical Section, London) to Robert A. Lovett (Assistant Secretary of State, Air).

148. Pavelec, *Jet Race*, 138–41.

149. AFHR A2072, October 12, 1943, Doc 25, Report by Brig. Gen. F. O. Carroll.

150. AFHR A2072, Summary, 3.

151. Watkins, *De Havilland Vampire*, 25–26.

152. Neville and Silsbee, *Jet Propulsion Progress*, 112.

153. BA MA RL 3/13, May 3, 1942, Besprechung bei Milch am April 24, 1942 [Meeting with Milch on April 24, 1942].

154. Schlaifer and Heron, *Development of Aircraft Engines*, 402–4, see especially 403n39.

155. Franz, *From Jets to Tanks*, 18.

156. BA MA RL 3/2187, November 4, 1942, Gemeinschaftssitzung der Arbeitsgemeinschaft für Triebwerksplannung [Meeting of the Association for Engine Planning].

157. BA RL 3/1133, June 11, 1942, Entwicklung von Luftstrahltriebwerken [Development of Jet Engines] (likely based on a report by Schelp report if not actually written by him).

158. Ibid.

159. BA MA RL 3/13, May 3, 1942, GL-Besprechung, Übersicht über die Ent-

wicklungsplannung auf dem Triebwerksgebiet [Overview of engine development planning]; Tooze, *Wages of Destruction*, 581.

160. Tooze, *Wages of Destruction*, 581.

161. Wright, "Factors Affecting the Cost of Airplanes"; Rosenberg, *Inside the Black Box*, 125.

162. BA MA RL 3/2568, March 22, 1945, Letter Speer to Diesing.

163. Tooze, *Wages of Destruction*, 577–81.

164. Ibid., 582–84.

165. Schabel, *Illusion der Wunderwaffen*, 164.

166. BA MA RL 3/51, November 1, 1942, Letter Lahr to Milch.

167. BA MA RL 3/2578, September 22, 1944, Zusammenfassung des Auszuges aus der stenografischen Niederschrift über die Besprechung beim Reichsmarschall [Summary of excerpt from the record of the development meeting with Reichsmarschall Göring].

168. BA MA RL 3/51, November 1, 1942, Letter Lahr to Milch; BA MA RL 3/34, August 28, 1942, Entwicklungsbesprechung im RLM [Development meeting at the RLM].

169. BA MA RL 3/2187, November 4, 1942, Gemeinschaftssitzung der Arbeitsgemeinschaft für Triebwerksplannung [Meeting of the Committee for Engine Planning].

170. Irving, *Rise and Fall of the Luftwaffe*.

171. Tooze, *Wages of Destruction*, 622.

172. BA MA N653/4, January 18, 1944, Generalluftzeugmeister Amtschefbesprechung Flugmotoren-Produktion [RLM Meeting about aircraft production].

173. BA MA N653/4, January 18, 1944, Generalluftzeugmeister Amtschefbesprechung Flugmotoren-Produktion [RLM Meeting about aircraft production]; USSBS, Aircraft Division Industry Report, Second Edition, 99.

174. IMW FD 5400/45, Jumo Report, TL-Geräte 004 B-1/4 Neubau und Reperatur [Turbojet engines 004 B-1/4 Construction and Repair].

175. BA MA N 653/11, January 5, 1944, Flugzeug-Produktion [Aircraft Production].

176. Wagner, *Ersten Strahlflugzeuge der Welt*, 87–90.

177. Government Report July 18, 1944, quoted in Schabel, *Illusion der Wunderwaffen*, 238.

178. Irving, *Rise and Fall of the Luftwaffe*, 233.

179. Tooze, *Wages of Destruction*, 633.

180. BA MA RL 3/1133, June 11, 1942, Entwicklung von Luftstrahltriebwerken [Development of Jet Engines] (likely based on a report by Schelp report if not actually written by him); BA MA N653/4, January 18, 1944, Generalluftzeugmeister Amtschefbesprechung Flugmotoren-Produktion [RLM Meeting about aircraft production].

181. BA MA R 3/1567, May 15, 1944, Jägerstab Besprechung.

182. IMW FD 5400/45, Jumo Report, TL-Geräte 004 B-1/4 Neubau und Reperatur [Turbojet engines 004 B-1/4 Construction and Repair].

183. NA AVIA 15/2121, July 27, 1945, 260, Interrogation of Max Eichler.

184. Kay, *German Jet Engine and Gas Turbine Development*, 78.

185. IWM FD 5400/45, Jumo Report, TL-Geräte 004 B-1/4 Neubau und Reperatur [Turbojet engines 004 B-1/4 Construction and Repair].

186. NA AVIA 15/2121, January 27, 1945, 68, Adderley.

187. IWM FD 5400/45, Jumo Report, TL-Geräte 004 B-1/4 Neubau und Reperatur [Turbojet engines 004 B-1/4 Construction and Repair].

188. Kay, *German Jet Engine and Gas Turbine Development*, 118.

189. NA AVIA 15/2121, July 27, 1945, 260, Interrogation of Max Eichler.

190. BMW Corporate Archive, Munich, FA79, September 2, 1943, Letter Zborowski to Himmler.

191. NA AVIA 15/2121, July 27, 1945, 260, Interrogation of Max Eichler.

192. BA MA NS19/57, September 15, 1943, Letter Jüttner to Himmler.

193. BA MA NS19/57, September 15, 1943, Letter Jüttner to Himmler; Archives of the Deutsches Museum, Munich (DM) FA 001/0323, August 29, 1944, Memo by Meschkat; NA AVIA 15/2121, July 27, 1945, 260, Interrogation of Max Eichler.

194. USSBS, Aircraft Division Industry Report, 32.

195. Irving, *Rise and Fall of the Luftwaffe*, 301–8.

196. Speer, *Inside the Third Reich*, 450; Irving, *Rise and Fall of the Luftwaffe*, 301–3.

197. Allen, *Business of Genocide*, 232–39.

198. Tooze, *Wages of Destruction*, 627.

199. BA MA R 3/1509, June 8, 1944, Führer Besprechung.

200. Tooze, *Wages of Destruction*, 633; in 1944 there was a "drastic shift of production" to single-engined fighter types (USSBS, January 1947, Aircraft Division Industry Report, Second Edition, 99).

201. Irving, *Rise and Fall of the Luftwaffe*, 308–9.

202. CIOS XXXII-17, 26.

203. CIOS XXXII-17; Neufeld, *Rocket and the Reich*, 200–213; Irving, *Mare's Nest*, 166.

204. W. R. J. Cook et al., *Underground Factories in Central Germany* (Combined Intelligence Objectives Sub-Committee File No. XXXII-17, London: HMSO); Neufeld, *Rocket and the Reich*, 200–213.

205. Tooze, *Wages of Destruction*, 627–34.

206. DM FA 001/0832, November 8, 1944, Notes of a meeting on November 7, 1944, between the Ernst Heinkel Aktien Gesellschaft and the Nationalsozialistisches Fliegerkorps; Combined Intelligence Objectives Sub-Committee File No. XXIV-6; Neufeld, *Rocket and the Reich*, 230–45.

207. Neufeld, *Rocket and the Reich*, 200–213; Cook, *Underground Factories*

in Central Germany; USSBS, Aircraft Division Industry Report, Second Edition, 112–14.

208. Jet engine production is mentioned in Allen, *Business of Genocide*; Neufeld, *Rocket and the Reich*; Ordway and Sharpe, *Rocket Team*; Irving, *Mare's Nest*, and in the memoirs of those who worked as slaves on V2 production. See Bülow, "Ein Blick in Die Hölle," 38–41, and Béon and Neufeld, *Planet Dora*. Kay mentions the factory (Kay, *German Jet Engine and Gas Turbine Development*, 78) likely because it appears in Jumo 004 production statistics but in no way makes its importance or nature clear. Green notes the dispersal of Junkers engine production to Nordhausen (Green, *Warplanes of the Third Reich*). Schabel mentions the Mittelwerk only with reference to V2 and planned Me 262 production (Schabel, *Illusion der Wunderwaffen*, 225). No professional history of the jet engine mentions production of the Jumo 004 at Nordhausen.

209. See, for example, Neufeld, *Rocket and the Reich*; Ordway and Sharpe, *Rocket Team*; Bülow, "Ein Blick in Die Hölle," 38–41; and Béon and Neufeld, *Planet Dora*.

210. Combined Intelligence Objectives Sub-Committee File No. XXXII-17; Allen, *Business of Genocide*.

211. Fedden, "German Piston-Engine Progress," 602–4.

212. Werner, *Kriegswirtschaft und Zwangsarbeit bei BMW*, 313.

213. Eyre, *50 Years with Rolls-Royce*, 87.

214. Werner, *Kriegswirtschaft und Zwangsarbeit bei BMW*, 329 and 346.

215. Combined Intelligence Objectives Sub-Committee File No. XXIV-6; NA AVIA 15/2121, July 27, 1945, 260, Interrogation of Max Eichler.

216. Combined Intelligence Objectives Sub-Committee File No. XXIV-6.

217. Schabel, *Illusion der Wunderwaffen*, 225; Tooze, *Wages of Destruction*, 622.

218. DM FA 001/0832, November 8, 1944, Notes of a meeting on November 7, 1944, between Heinkel and the Nationalsozialistisches Fliegerkorps; P. Lloyd, *Gas Turbine Development, B.M.W., Junkers, Daimler Benz* (Combined Intelligence Objectives Sub-Committee Report No. XXIV-6: London, HMSO); Neufeld, *Rocket and the Reich*, 245.

219. USSBS, January 1947, Aircraft Industry Report, Second Edition, 30–31.

220. Combined Intelligence Objectives Sub-Committee File No. XXX-II17.

221. Ibid.

222. USSBS, January 1947, Aircraft Division Industry Report, Second Edition.

223. IWM FD 5400/45, Jumo Report, TL-Geräte 004 B-1/4 Neubau und Reperatur [Turbojet engines 004 B-1/4 Construction and Repair].

224. Kay, *German Jet Engine and Gas Turbine Development*; IWM FD 5400/45, Jumo Report, TL-Geräte 004 B-1/4 Neubau und Reperatur [Turbojet engines 004 B-1/4 Construction and Repair]; USSBS, Aircraft Division Industry Report, Second Edition, 99–100. In December 1944, piston engine production was recovering from a trough in October 1944, but as turbojet engine production continued to

climb until the end of the war, the proportion of turbojet engines likely continued to increase.

225. Milward, *War, Economy, and Society*, 183.

226. BA R 3/100, July 22, 1944, Speer, Anordnung zum Erlaß des Fürhrers über die Konzentration der Rüstungs- und Kriegsproduktion vom 19 June 1944 [Order regarding the Führer decree about the concentration of defense and war manufacture on June 19, 1944].

227. Speer, *Inside the Third Reich*, 549.

228. Schabel, *Illusion der Wunderwaffen*, 251.

229. Tooze, *Wages of Destruction*. The RLM's policy shifted from emphasizing quality, as favored by Göring, toward increasingly emphasizing the production of a large number of aircraft (Overy, *Goering: The "Iron Man,"* 164–204).

230. BA MA RL 3/2578, September 22, 1944, Stenographsiche Niederschrift der Entwicklungsbesprechung beim Reichsmarschall [Stenographic record of the development meeting with Reichsmarschall Göring].

231. Tooze, *Wages of Destruction*, 611–18.

232. Schabel, *Illusion der Wunderwaffen*, 233–37.

233. For example, Constant, *Origins of the Turbojet Revolution*, 211.

234. BA MA RL 3/51, May 20, 1943, Report by Milch.

235. Ibid.

236. BA MA RL 3/51, June 1942, Jumo Report, Verminderung des Sparstoffgehalts beim Sondertriebwerk 109-004 Jumo [Reduction of strategic metal content in the special aero-engine 109-004 Jumo].

237. Franz, *From Jets to Tanks*, 23.

238. Constant, *Origins of the Turbojet Revolution*, 208–11.

239. NA AVIA 15/2121, July 27, 1945, 260, Interrogation of Max Eichler; DM FA 001/0827, September 29, 1944, Francke. Uziel's work is notable for its interest in aviation production methods (see also Uziel, "Between Industrial Revolution and Slavery").

240. NA AVIA 15/2121, July 27, 1945, 260, Interrogation of Max Eichler.

241. Overy, *Goering: The "Iron Man,"* 187.

242. Schlaifer and Heron, *Development of Aircraft Engines*, 434; IWM FD 5399/45.

243. "Packard vs RR."

244. Schlaifer and Heron, *Development of Aircraft Engines*, 416; Kay reported 600 hours (Kay, *German Jet Engine and Gas Turbine Development*, 78).

245. Kay, *German Jet Engine and Gas Turbine Development*, 106; Schlaifer and Heron, *Development of Aircraft Engines*, 416.

246. Schlaifer and Heron, *Development of Aircraft Engines*, 434.

247. Franz, *From Jets to Tanks*, 29; Kay, *German Jet Engine and Gas Turbine Development*.

248. Franz, *From Jets to Tanks*, 29; Jumo's Strasbourg plant, for example, was ar-

ranged to construct engines from subcontracted parts and send them to Muldenstein for testing (NA AVIA 15/2121, January 27, 1945, Adderley).

249. USSBS, 1945, Overall Report, European War, 3944; Schlaifer and Heron,
Development of Aircraft Engines, 416–21; Schabel, *Illusion der Wunderwaffen*, 230;
BA MA RL 3/1133, June 11, 1942, Entwicklung von Luftstrahltriebwerken [De-
velopment of Jet Engines] (likely based on a report by Schelp report if not actually written by him).

250. Schlaifer and Heron, *Development of Aircraft Engines*, 416 and 421.

251. USSBS, 1945, Overall Report, European War, 3944; Schlaifer and Heron,
Development of Aircraft Engines, 416–21; Schabel, *Illusion der Wunderwaffen*, 230.

252. USSBS, 1945, Overall Report, European War, 3944.

253. Hornby, *Factories and Plant*, 258.

254. IWM FD 5400/45, Jumo Report, TL-Geräte 004 B-1/4 Neubau und Reperatur [Turbojet engines 004 B-1/4 Construction and Repair] suggests that many
thousands more engine's worth of parts may have been completed during the war;
NA AVIA 15/2121, July 27, 1945, 260, Interrogation of Max Eichler.

255. BA MA RL 3/3781, June 4, 1945, Informationsbericht über Verbesserun-
gen an der Regelung von Turboluftstrahl-Triebwerken [Report on improvements
in the control of turbojet engines].

256. Pavelec, *Jet Race*, 91.

257. Cairncross, *Planning in Wartime*, 143; Boog, *Germany and the Second
World War*, 345.

258. BA MA RL 3/2578, September 22, 1944, Stenographische Niederschrift
der Entwicklungsbesprechung bei Reichsmarschall [Stenographic record of the
development meeting with Reichsmarschall Göring].

259. IWM FD 4924/45, October 1944, Stellungnahme 162 [Opinion on the 162]
by Messerschmitt.

260. BA R 3/1510, September 24, 1944, Führer Besprechung [Meeting with
Führer] of September 21–23, 1944; BA MA RL 3/2568, February 12, 1945, Arbeit-
stagung in Potsdam 6 Feb., Steigerung der Geschwindigkeit [Workshop in Potsdam
on February 6 about increasing speed].

261. BA MA RL 3/2578, September 22, 1944, Stenographische Niederschrift
der Entwicklungsbesprechung bei Reichsmarschall [Stenographic record of the
development meeting with Reichsmarschall Göring].

262. Ibid.

263. BA MA RL 3/51, January 15, 1942, Letter from Lusser to Milch.

264. Uziel, "Der Volksjäger: Rationalisierung und Rationalität von Deutschlands Letztem Jagdflugzeug im Zweiten Weltkrieg," 63–82. Uziel's work is notable
for its interest in aviation production methods (see also Uziel, "Between Industrial
Revolution and Slavery").

265. Cairncross, *Planning in Wartime*, 142.

266. DM FA 001/0827, September 29, 1944, Francke; BA MA RL 3/2578, Sep-

tember 22, 1944, Stenographische Niederschrift der Entwicklungsbesprechung bei Reichsmarschall [Stenographic record of the development meeting with Reichsmarschall Göring].

267. BA MA RL 3/2568, March 27, 1945, Letter Führer; BA MA RL 3/2568, April 3, 1945, Letter Purucker to Diesing.

268. BA RL 3/2568, March 28, 1945, Letter Koller (OKL); Schabel, *Illusion der Wunderwaffen*, 219.

269. Scranton, "Turbulence and Redesign."

Chapter Two

1. The literature's account of development in Britain is dominated by Whittle's version of the work of his firm—the story of an outsider and a company that never brought an aero-engine to market (Kay, *Turbojet History and Development*, 22–26; Gunston, *World Encyclopaedia of Aero Engines*, 126–28. Whittle's story was reproduced in Golley, Whittle, and Gunston, *Whittle, the True Story*; and Pavelec, *Jet Race*, chapter 2).

2. Basalla, whose interest was in technological change as a continuous process on the level of the artifact, criticized Constant's model and its arbitrary focus on discontinuity (Basalla, *Evolution of Technology*, 28–30).

3. Constant, *Origins of the Turbojet Revolution*, 118.

4. Ibid., 129.

5. Hallion criticized Constant's focus on what came before the turbojet revolution, calling it a "structural unevenness" (Hallion, "Review: Aerospace Technology").

6. Constant, *Origins of the Turbojet Revolution*, 63–98.

7. Important aspects of the firm's earliest work on gas turbines can be found in *Flight*'s account of the Derwent (*Flight*, October 25, 1945).

8. The most recent addition to the literature on Rolls-Royce is unsurprisingly titled *Hives and the Merlin* (Lloyd and Pugh, *Hives and the Merlin*); Pavelec, *Jet Race*, 118–19. The earliest, three-volume history of Rolls-Royce published in the late 1970s, which covered the period from the firm's founding until 1945, pointed to the importance of Merlin development in shaping Rolls-Royce's turbojet work, but its brief account of this work was marred by inaccuracy (Lloyd, *Rolls-Royce: The Merlin at War*, 131–33). In Peter Pugh's more recent two-volume history, the chapters on the early turbojet are mostly a retelling of the Whittle story (Pugh, *Magic of a Name*, 268–80). Pugh reduces the firm's wartime turbojet story to the takeover of Barnoldswick (Pugh, *Magic of a Name*, 1–13).

9. Constant, *Origins of the Turbojet Revolution*, 214; Schlaifer and Heron, *Development of Aircraft Engines*, 355 and 364; neither is in Kay's recent account of British jet engines (Kay, *Turbojet History and Development*).

10. Postan, Hay, and Scott, *Design and Development of Weapons*, 201.

11. Pugh, *Magic of a Name*, 263–67.

12. Wilde, "Dr. Griffith's CR.1," 85; Eyre, *50 Years with Rolls-Royce*, 72–75; Rubbra, "Alan Arnold Griffith"; Lloyd, *Rolls-Royce: The Merlin at War*.

13. Eyre, *50 Years with Rolls-Royce*, 72–73.

14. Bloor, *The Enigma of the Aerofoil*.

15. NA AVIA 46/234, July 7, 1926, Royal Aircraft Establishment report 1111, Griffith, "An Aerodynamic Theory of Turbine Design."

16. Eyre, *50 Years with Rolls-Royce*, 72–73.

17. Baxter, *Professional Aero Engineer Novice Civil Servant*, 144–45.

18. Eyre, *50 Years with Rolls-Royce*, 72.

19. Hooker, *Not Much of an Engineer*, 101.

20. Hawthorne, Cohen, and Howell, "Hayne Constant, 1904–1968," 176–77.

21. Wilde, "Dr. Griffith's CR.1," 87.

22. RR L, December 20, 1944, Letter Bristol to EJW.

23. RR HC, June 14, 1939, Letter A. G. Elliott to Hives, Re Dr. Griffith's department; Hooker, *Not Much of an Engineer*, 29–30 and 35; Wilde, "Dr. Griffith's CR.1"; Mordell, "Better Than a PhD," 9–11.

24. Extrapolated from Eyre's figures (Eyre, *50 Years with Rolls-Royce*, 87).

25. Mordell, "Better Than a PhD," 9; Wilde, "Dr. Griffith's CR.1."

26. RR HC, July 25, 1941, Letter Hives to MAP.

27. RR HC, August 21, 1941, Letter Roxbee Cox to Hives; AVIA 15/1201, September 13, 1941, 15, Minute by Bulman.

28. NA AVIA 15/1201 September 13, 1941, 15, Minute by Bulman; RR HC, September 17, 1941, letter Linnell to Hives.

29. Mordell, "Better Than a PhD," 13.

30. Hooker, *Not Much of an Engineer*, 68–69.

31. NA AVIA 15/1201, January 16, 1943, Minute Watt.

32. NA AVIA 15/1201, October 30, 1942, 26, Minute Pye to Roxbee Cox.

33. NA AVIA 15/1201, January 6, 1943, Letter by Adderley; Wilde, "Dr. Griffith's CR.1."

34. Hooker, *Not Much of an Engineer*, 63.

35. RR HC, May 28, 1941, Letter Hives to Hennesay (MAP); RR HC, February 11, 1942, Letter Hives to S. B. Wilks.

36. Hooker, *Not Much of an Engineer*, 67.

37. RR HC, July 25, 1941, Letter Hives to MAP.

38. RR February 24, 1942, Director's Minute Book 11, ICTs.

39. Whittle, *Jet: The Story of a Pioneer*, 2nd ed., 192.

40. NA AVIA 15/1607, January 27, 1942, Minute Linnell to Pye.

41. *Archive #79*, "LH and His RCA.3."

42. NA AVIA 15/1607, March 17, 1943, Memo by Roxbee Cox.

43. NA AVIA 15/1607, February 9, 1942, 4a, Discussion of Whittle-Rolls-Royce Jet Propulsion unit on February 4, 1942.

44. Whittle, *Jet: The Story of a Pioneer*, 2nd ed., 192.

45. RR HC, April 16, 1942, Letter Rolls-Royce to W. E. P. Johnson.

46. NA AIR 62/990, April 1942, GTCC Progress Report #6.

47. RR HT, February 17, 1943, Excerpt from Report J. P. Herriot to S. B. Wilks (Rover); NA AIR 62/990, November 1942, GTCC Progress Report #13.

48. NA AIR 62/990, 4.42, GTCC Progress Report #6; NA AVIA 13/962, February 23, 1942, 8A, Minute Constant to Roxbee Cox.

49. RR HC, August 21, 1941, Letter Roxbee Cox to Hives.

50. NA AVIA 31/3, July 1, 1947, Terms of Reference of the GTCC.

51. NA AVIA 31/3, November 1, 1941, GTCC Meeting Minutes #1.

52. NA AVIA 46/237, September 1, 1941, 3b, Interdepartmental letter; NA AVIA 46/237, July 31, 1941, 3a, Gas Turbine Collaboration.

53. NA AVIA 13/935, April 20, 1942, 273, Letter Perring to MAP; NA AVIA 43/237, September 2, 1941, 3c, Interdepartmental memo; Bulman, *An Account of Partnership*, 327.

54. RR DMB #11, February 24, 1942.

55. Schlaifer and Heron, *Development of Aircraft Engines*, 345–46.

56. See Johnson's argument below; Bailey and Whittle, "Early Development of the Aircraft Jet Engine," 42–43; interview of James Foulds, March 29, 2010.

57. NA AVIA 13/962, February 12, 1942, Letter Perring to Griffith; February 23, 1942, 8A, Letter Constant to MAP.

58. RR HC, November 17, 1942, Hives to MAP; Lloyd, *Rolls-Royce: The Merlin at War*, 126–33.

59. NA AIR 62/990, November 1942, GTCC Progress Report #13.

60. NA AVIA 31/2, December 12, 1942, GTCC Minutes Meeting #7.

61. NA AIR 62/990, December 1942, GTCC Progress Report #14; February 1943, GTCC Progress Report #16.

62. Roxbee Cox, "British Aircraft Gas Turbines," 68; Postan, Hay, and Scott, *Design and Development of Weapons*, 202; Schlaifer and Heron, *Development of Aircraft Engines*, 364.

63. Haworth, *Rolls-Royce Tyne*, 16.

64. NA AVIA 15/1607, February 10, 1942, Minute Bulman to Linnell.

65. NA AVIA 15/1614, April 2, 1942, 14A, Letter A. G. Elliott to Roxbee Cox; RR HC, March 25, 1942, W. E. P. Johnson to Linnell; April 7, 1942, W. E. P. Johnson to Hives; Pugh, *Magic of a Name*, 299.

66. NA AVIA 15/1806, August 1944, 11A, Draft memo to Churchill.

67. NA AVIA 15/1802, January 7, 1943, 15, Minute Roxbee Cox to Linnell.

68. NA AVIA 15/1708, March 27, 1943, 37A, Letter Hives to Linnell.

69. *Archive* #79, February 5, 1943, Letter Haworth to Constant.

70. NA AVIA 15/1802, February 24, 1943, 38A, Interdepartmental letter Robinson.

71. NA AIR 62/608, April 19, 1943, Whittle diary, Visit with Hives.

72. Hooker, *Not Much of an Engineer*, 80; Hives decided to put Mr. Reid, a

Rolls-Royce production engineer who had been superintending Merlin production at Packard in the United States, in charge of Barnoldswick (NA AVIA 15/1806, 11A; January 5, 1943, 12A, Notes of discussion held at Power Jets Ltd.). Reid, however, never worked at Barnoldswick (Pugh, *Magic of a Name*, 317). Hives also assigned Les Buckler from Derby to be works manager at the turbojet factory, where Buckler made a crucial contribution to the acceleration of work there. Buckler later became manager of Barnoldswick when design and development was moved back to Derby (Pugh, *Magic of a Name*, 296).

73. Interview with James Foulds, March 29, 2010.

74. NA AIR 62/608, May 5, 1943, Whittle diary, Meeting W. E. P. Johnson, J. C. B. Tinling, R. D. Williams, and Hives on April 30, 1943.

75. Pugh, *Magic of a Name*, 291.

76. NA AVIA 15/1607, February 10, 1943, 23, Minute Watt to Roxbee Cox; NA AVIA 15/1607, July 17, 1943, 28a, Letter Rolls-Royce to MAP; NA AIR 62/608, April 19, 1943, Whittle diary, Visit with Hives; NA AIR 62/608, May 13, 1943, Whittle diary, Phone call with Roxbee Cox.

77. Rover's design team at Clitheroe designed the STX or B/26 to ease the production problems expected to arise if the W.2B were to be manufactured on a large scale, using unskilled workers. The STX used the W.2B's compressor-turbine assembly, but had a completely different, straight-through combustion system (NA AVIA 13/937, May 21, 1942, Meeting at Clitheroe).

78. Hooker, *Not Much of an Engineer*, 82–83; RR L December 20, 1944, Letter.

79. RR L December 20, 1944, Letter; Hooker, *Not Much of an Engineer*, 88–89.

80. Hooker, *Not Much of an Engineer*, 104.

81. Haworth, *Rolls-Royce Tyne*, 7.

82. Ibid., 17–18.

83. RR L, December 20, 1944, Letter; Wilde, "Dr. Griffith's CR.1."

84. NA AVIA 15/1960, July 27, 1943, 7A, Letter Roxbee Cox to Farren; NA AVIA 15/1960, February 19, 1943, Letter Roxbee Cox to Haworth.

85. NA AVIA 15/1960, October 7, 1943, 10A, Letter A. G. Elliott to Roxbee Cox.

86. Pugh, *Magic of a Name*, 291.

87. Haworth, *Rolls-Royce Tyne*, 17–18.

88. NA AVIA 15/1960, February 9, 1946, 16, Minute.

89. Haworth, *Rolls-Royce Tyne*, 19; RR HT, March 3, 1911, Comments from Max Alderston.

90. The two were good friends (Hooker, *Not Much of an Engineer*, 101).

91. Gunston, *World Encyclopaedia of Aero Engines*, 162.

92. Heathcote, *Rolls-Royce Dart*, 14; Feilden, "Lionel Haworth," 200; Gunston argues that the Clyde was superior to the axial turboprops that replaced it (Gunston, *World Encyclopaedia of Aero Engines*, 162).

93. RR HC, February 18, 1944, Memo by Griffith.

94. *Archive* #79, Reprint of February 5, 1943, Letter Haworth to Constant.

95. Feilden, "Lionel Haworth," 204.

96. MOSI, November 1945, MAP turbine engine data sheet.

97. NA AVIA 15/1201, March 12, 1945, 48, Minute Watt to Sorley; AVIA 15/1201, March 25, 1945, 49, Minute Sorely to Watt.

98. Hooker, *Not Much of an Engineer*, 107 and 131; Haworth, Nedham, and Wilde, "Robin Ralph Jamison," 179; Feilden, "Lionel Haworth," 197.

99. Hooker, *Not Much of an Engineer*, 120.

100. Hornby, *Factories and Plant*, 253–54.

101. For more detail, see particularly Whitfield, "Metropolitan Vickers, the Gas Turbine, and the State."

102. The official history gives a matter-of-fact account of how MAP recruited all of Britain's aero-engine firms to turbojet work during the war, but has been mostly overlooked (Postan, Hay, and Scott, *Design and Development of Weapons*, 199–227). The first attempt to compile detailed information on the early turbojets of the entire British aero-engine industry was the recent 2007 book by Anthony Kay, in which he extended, through extensive research, the same comprehensive treatment as demonstrated in his 2002 book on German gas turbines to the Britain gas turbine aero-engine field (Kay, *Turbojet History and Development*).

103. Schlaifer and Heron, *Development of Aircraft Engines*, 355–57 and 374; Constant, *Origins of the Turbojet Revolution*, 216–18. Some accounts of these firms don't even mention their wartime turbojet work. James, *Bristol Aeroplane Company*; Gunston's biography of Roy Fedden gave some early details, but Fedden left Bristol in 1942. Gunston, *Fedden*; Lawton, *Parkside*, 65–84; Wilson and Reader, *Men and Machines*, 159–65.

104. Bulman, *An Account of Partnership*, 324; Brodie, "Frank Bernard Halford," 201; Sharp, *D. H.: An Outline of De Havilland History*, 201–4.

105. NA AVIA 13/934, May 23, 1941, Letter E. S. Moult to Hayne Constant; Watkins, *De Havilland Vampire*, 28–29; Sharp, *D. H.: An Outline of De Havilland History*, 207.

106. Taylor, *Boxkite to Jet*, 106–8; Hassel, "Halford Jets"; Moult, "Development of the Goblin Engine," 658–75; Sharp, *D. H.: An Outline of De Havilland History*, 201; NA AVIA 15/1599, January 10, 1942, 1, Minute Bulman to Linnell.

107. NA AVIA 13/934, May 15, 1941, Letter RAE to Halford; July 8, 1941, Letter Moult to RAE.

108. Moult worked closely with Napier on the Sabre and with de Havilland on the H.1 (Brodie, "Frank Bernard Halford," 199; NA AVIA 13/934, May 23, 1941, E. S. Moult to Constant; May 20, 1941, E. S. Moult to Chief Superintendent RAE; De Havilland, "Development of the De Havilland 'Goblin' Jet Propulsion Engine").

109. *Flight*, 1945.

110. Sharp, *D. H.: An Outline of De Havilland History*, 203.

111. *Flight*, 1945.

112. Moult, "Development of the Goblin Engine," 657.

113. On design trade-offs see Pye, *The Nature and Art of Workmanship*.

114. NA AVIA 13/934, September 27, 1941, Comments on De Havilland-Halford Project (requested 6 Aug); De Havilland, "Development of the De Havilland 'Goblin' Jet Propulsion Engine."

115. De Havilland, "Development of the De Havilland 'Goblin' Jet Propulsion Engine."

116. NA AVIA 13/935, August 5, 1942, Letter Farren to Halford; November 10, 1942, Letter Farren to Halford; NA AVIA 46/237, May 22, 1942, 138, Interdepartmental letter; NA AVIA 13/935, April 20, 1942, 273, Letter Perring to Roxbee Cox; NA AVIA 43/237, September 2, 1941, 3c, Letter Roxbee Cox to Pye.

117. Sharp, *D. H.: An Outline of De Havilland History*, 202–3.

118. NA AVIA 15/1708, November 29, 1942, 20, Minute Linnell to Freeman; NA AVIA 15/1599, July 22, 1944, 55, Minute Watt to Banks.

119. Bulman, *An Account of Partnership*, 325 and 350–52.

120. NA AVIA 15/1599, November 16, 1944, 58, Minute Watt to Banks.

121. NA AVIA 15/1599, July 22, 1944, 55, Minute Watt to Banks.

122. NA AVIA 15/1599, January 7, 1943, 44, Minute Roxbee Cox to Linnell.

123. NA AVIA 31/4, November 11, 1944, GTCC Minutes Meeting #15.

124. NA AVIA 15/1599, August 24, 1942, 20, Interdepartmental minute.

125. NA AVIA 15/1599, November 16, 1944, 58, Minute Watt to Banks; PREM 3/21/4, January 30, 1945, 95, Third Two-Monthly Report from Cripps.

126. Brodie, "Frank Bernard Halford," 201.

127. MOSI, September 10, 1942, Eng/2038.R/HC/21, Royal Aircraft Establishment report by Hayne Constant.

128. Hornby, *Factories and Plant*, 259; Gunston, *World Encyclopaedia of Aero Engines*, 19; NA AVIA 15/1509, October 6, 1941, Engine Development Policy ASM; October 3, 1941, 4A, Meeting at MAP; NA AVIA 15/1509, November 20, 1941, 25, Minute Roxbee Cox to Linnell, Banks and Pye. Collaboration with Metro-Vick began almost immediately, and the first meeting of the "Joint Committee on F2 Plant" took place on March 18, 1942 (NA AVIA 15/1509, March 18, 1942, 43A, Minutes Meeting #1).

129. Hawthorne, "Aircraft Propulsion from the Back Room," 96; Constant, "The Development of the Internatl Combustion Turbine"; MOSI, December 16, 1937, Report by Mr. H. G. Rhoden to Mr. K Baumann, "Report on Conference with R.A.E Engineers."

130. Hawthorne, "Aircraft Propulsion from the Back Room," 96; Baxter, *Professional Aero Engineer*, 147.

131. NA AVIA 13/1403, April 20, 1940, 2A, Roxbee Cox on behalf of Chief Superintendent RAE to Air Ministry.

132. MOSI, Artus, July 27, 1940, Letter Smith to RAE.

133. MOSI, Artus, September 15, 1939, Letter RAE to Metrovick.

134. MOSI, Artus, July 30, 1940, Report Smith to Baumann; August 13, 1940, Memo Smith to Chief Draughtsman.

135. Gunston, *World Encyclopaedia of Aero Engines*, 117.

136. NA AVIA 15/1509, October 10, 1941, 5B, Letter Bulman to Armstrong Siddeley Motors with notes of meeting on October 6, 1941; NA AVIA 15/1509, 5A; September 18, 1941, 1A, Letter Perring to Roxbee Cox; NA AVIA 15/1509, October 23, 1941, 17, Minute Bulman to Roxbee Cox; NA AVIA 15/1509, 38A, January 15, 1942.

137. NA AVIA 15/1509, July 30, 1942, 53, Minute Bulman to Roxbee Cox.

138. NA AVIA 15/1509, December 8, 1941, 30, Minute Bulman to Roxbee Cox quoting letter from Spriggs.

139. NA AVIA 15/1509, July 30, 1942, 53, Minute Bulman to Roxbee Cox.

140. NA AVIA 31/2, May 2, 1942, GTCC Meeting Minutes #4.

141. NA AVIA 15/1509, October 14, 1941, 8, Minute Bulman to Linnell.

142. NA AVIA 15/1614, March 9, 1942, 11, Intradepartmental minute.

143. NA AVIA 15/1509, August 7, 1942, 56B, Letter Spriggs to Bulman; AVIA 15/1614, May 12, 1942, Jet Engine Propulsion Panel 5th Meeting and October 8, 1942, 10th meeting; AVIA 31/3, March 6, 1943, GTCC Meeting Minutes #8.

144. Lawton, *Parkside*, 68–70.

145. Ibid., 69; Armstrong Siddeley Motors Ltd and Fritz Albert Max Heppner, "Improvements relating to jet-propelled aircraft." GB577950, 1946.

146. NA AVIA 13/911, December 5, 1939, Interdepartmental letter Roxbee Cox and Hayne Constant.

147. NA AVIA 13/911, June 26, 1941, Hayne Constant for Perring to Roxbee Cox; Bulman and Ross also had a high opinion of Heppner (NA AVIA 15/1509, November 7, 1941, 18, Minute Bulman to Roxbee Cox).

148. NA AVIA 13/911, December 12, 1941, Visit to Armstrong Siddeley Motors, Discussion at Coventry on proposed Heppner Turbine.

149. Hodgson, "Stewart S. Tresilian."

150. NA AVIA 13/911, May 1942, 23, Technical Appreciation of TB Engine (Heppner) from Armstrong Siddeley Motors brochure dated May 5, 1942; NA AVIA 13/911, July 30, 1942, Hayne Constant and William Hawthorne to Roxbee Cox (revised proposal).

151. NA AVIA 13/911, July 1, 1942, 43a, Notes for discussion with Armstrong Siddeley Motors.

152. NA AVIA 15/1509, August 7, 1942, 56B, Letter Spriggs to Bulman.

153. NA AVIA 13/911, August 14, 1942, Armstrong-Siddeley Motors visit to Royal Aircraft Establishment; NA AVIA 13/911, July 1, 1942, 43a, Notes for discussion with Armstrong Siddeley Motors.

154. NA AVIA 13/911, August 14, 1942, Armstrong Siddeley Motors visit to Royal Aircraft Establishment; NA AVIA 13/911, May 1942, Eng/2038-11/SJH/6S, Royal Aircraft Establishment report.

155. NA AVIA 13/911, August 14, 1942, Armstrong Siddeley Motors visit to the Royal Aircraft Establishment.

156. NA AVIA 13/911, August 15, 1942, 55a, Letter Hayne Constant to Roxbee Cox; Constant, "The Early History of the Axial Type of Gas Turbine Engine," 421; Lawton, *Parkside*, 71.

157. Kay, *Turbojet History and Development*, 118; Gunston, *World Encyclopaedia of Aero Engines*, 20; Lawton, *Parkside*, 71.

158. NA AVIA 13/911, June 18, 1942, Hayne Constant to Armstrong-Siddeley Motors.

159. Lawton, *Parkside*, 68–71.

160. NA AVIA 13/911, October 6, 1942, Hayne Constant to Roxbee Cox.

161. Gunston, *World Encyclopaedia of Aero Engines*, 19.

162. NA AVIA 31/3, May 15, 1943, GTCC Minutes Meeting #9; Brodie, "Frank Bernard Halford," 200; Lawton, *Parkside*, 71.

163. NA AVIA 31/3, August 28, 1943, GTCC Minutes Meeting #10.

164. NA AVIA 31/5, April 20–21, 1945, GTCC Minutes Meeting #17.

165. NA AVIA 31/5, October 12, 1945, GTCC Minutes Meeting #18.

166. Gunston, *World Encyclopaedia of Aero Engines*, 19.

167. Whitfield, "Metropolitan Vickers, the Gas Turbine, and the State," 127.

168. The piston engine design, development, and production capability of Rolls-Royce and Bristol were by far the most significant during World War II (Postan, Hay, and Scott, *Design and Development of Weapons*, 26).

169. Owner, "9th Barnwell Memorial Lecture"; NA AVIA 10/261, October 20, 1941, Letter Fedden to Linnell.

170. Owner joined Bristol in 1922 as the company's first university graduate. While at Manchester University in 1920, he had designed a gas turbine (Gunston, *Fedden*, 86–90).

171. Gunston, *Fedden*, 90.

172. Owner, "9th Barnwell Memorial Lecture"; Gunston, *Fedden*, 244–45 (reproduction of Whittle's letter, 245).

173. MOSI, September 10, 1942, Eng/2038.R/HC/21, Royal Aircraft Establishment report by Hayne Constant.

174. NA AVIA 10/261, December 4, 1941, 14a, Brochure; AVIA 10/261, March 2, 1942, 16A, Letter Linnell to Fedden; October 2, 1942, 26, Minute Rowe to Linnell.

175. NA AVIA 10/261, November 4, 1941, 11B, Letter Fedden to Linnell; March 2, 1942, 16A, Letter Linnell to Fedden.

176. Owner, "9th Barnwell Memorial Lecture"; Gunston states that Fedden initiated work on a turboprop in May 1941 (Gunston, *Fedden*, 252).

177. Owner explained to Roxbee Cox that "various non-technical circumstances" had prevented him from presenting the proposal sooner (NA AVIA 15/1911, April 8, 1943, 2B, Report on Visit to Dr. Roxbee Cox). Bristol's reputation

was built on Fedden's design skill, and it suffered after his departure (Gunston, *Fedden*, 86). Fedden became an advisor to MAP in various capacities throughout the rest of the war (Gunston, *Fedden*, 259 and 266–94).

178. Owner, "9th Barnwell Memorial Lecture."

179. Ibid.

180. Smith, *Gas Turbines and Jet Propulsion for Aircraft*, 93–94; "Theseus Air Testing," *Flight*, March 27, 1947.

181. NA AVIA 15/1911, April 8, 1943, 2B, Report on Visit to Dr. Roxbee Cox on 7th April 1943.

182. "Theseus Air Testing," *Flight*, March 27, 1947.

183. NA AVIA 15/1911, April 8, 1943, 2B, Report on Visit to Dr. Roxbee Cox.

184. NA AVIA 15/1911, August 9, 1943, Meeting on Gas Turbine policy held at Bristol, August 5–6, 1943.

185. Bristol, "Short History of Bristol Gas Turbine Project," 1943.

186. At first he recruited men who had just finished work on the design of Fedden's last piston engine, the Orion (NA AVIA 15/1911, April 8, 1943, 2B, Report on Visit to Dr. Roxbee Cox).

187. Owner, "9th Barnwell Memorial Lecture"; Fedden became well known for his opposition to the new gas turbine engines and his defence of the piston engine, although he supported the engines after their worth had been proven (Gunston, *Fedden*, 244 and 251–52).

188. NA AVIA 15/1911, January 22, 1944, 29A, Letter Roxbee Cox to Jelfs.

189. Owner, "9th Barnwell Memorial Lecture"; the lack of detail designers on the project not only slowed down the design work on the Theseus but also caused problems to appear later in development because the early designs were not as thorough as they might have been (Banks, *I Kept No Diary*, 145; NA AVIA 15/1911, July 7, 1945, 41a, Meeting at Bristol on July 6, 1945).

190. NA AVIA 31/4, September 27, 1944, GTCC Minutes Meeting #14. The engine's impeller was set up for test on the Northampton test rig and by February 17, 1945; it had been tested for twenty-two hours (NA AVIA 31/5, February 17, 1945, GTCC Minutes Meeting #16).

191. Owner, "9th Barnwell Memorial Lecture"; Gunston, *World Encyclopaedia of Aero Engines*, 36.

192. Owner, "9th Barnwell Memorial Lecture"; Kay, *Turbojet History and Development*, 139.

193. NA AVIA 15/1911, February 14, 1945, Minute Watt to Banks; Banks, *I Kept No Diary*, 145.

194. NA AVIA 15/1911, July 25, 1945, Notes.

195. NA AVIA 15/1911, July 7, 1945, 41A, Interdepartmental note, Meeting at Bristol on July 6, 1945.

196. Owner, "9th Barnwell Memorial Lecture."

197. Bulman, *An Account of Partnership*, 345.

198. Banks, *I Kept No Diary*, 144; NA AVIA 10/222, May 2, 1945, 64, Note by F. R. Banks, Development Plant for Napier E.124/125 Engines.

199. NA AVIA 15/2106, August 21, 1944, Specification.

200. Wilson and Reader, *Men and Machines*, 163.

201. NA AVIA 10/222, 1945, 42a, note by DED.

202. NA AVIA 10/222, September 26, 1945, note by DCRF and DED for the Air Supply Board.

203. By 1946, the specification had been reduced to 3,000 horsepower (Wilson and Reader, *Men and Machines*, 163).

204. Quote from NA AVIA 10/222, May 2, 1945, Document 64; NA AVIA 15/2106, August 24, 1944, 2, Minute Banks to Sorley; NA AVIA 15/2106, January 13, 1945, Letter Curzon to Trend.

205. 0.34 lb/BHP/hr (AVIA 15/2106, August 24, 1944, 2, Minute Banks to Sorley).

206. NA AVIA 15/2106, September 8, 1944, 4, Minute Sorley to Banks.

207. NA AVIA 31/4, November 11, 1944, GTCC Meeting Minutes #15.

208. NA AVIA 15/2106, August 11, 1944, Conference; August 22, 1944, 1a, Letter Napier to Banks.

209. Wilson and Reader, *Men and Machines*, 162; NA AVIA 10/222, May 2, 1945, 64, Napier development plant for Napier E.124/125 Engines; May 4, 1945, 64a, Extracts from Air Supply Board Meeting.

210. Banks, *I Kept No Diary*, 172.

211. Schelp's testimony (Ermenc, *Interviews with German Contributors to Aviation History*, 113).

212. Schlaifer and Heron, *Development of Aircraft Engines*, 387.

213. Schlaifer contradicted the common misconception that the use of axial compressors was dictated by the German government, which has nevertheless continued to circulate (Schlaifer and Heron, *Development of Aircraft Engines*, 435–36).

214. Hirschel, Prem, and Madelung, *Aeronautical Research in Germany*, 79–81; in contrast to the work of another research establishment, the Deutsche Versuchsanstalt für Luftfahrt, exclusively on centrifugal compressors (Ermenc, *Interviews with German Contributors to Aviation History*, 111).

215. Two historians who mention Bramo's work: Schlaifer and Heron, *Development of Aircraft Engines*, 390; Constant, *Origins of the Turbojet Revolution*, 212.

216. Constant, *Origins of the Turbojet Revolution*, 213; Schlaifer and Heron, *Development of Aircraft Engines*, 402.

217. Franz, *From Jets to Tanks*, 11; DM NL172/2, August 1981, Wagner, Bemerkungen zu Heinkels Darstellung [Remarks on Heinkel's Account].

218. Combined Intelligence Objectives Sub-Committee File No. XXIV-6, Gas Turbine Development, Franz interrogation.

219. Franz, *From Jets to Tanks*, 16; Wagner and Cox, *First Jet Aircraft*, 235–36; Hirschel, Prem, and Madelung, *Aeronautical Research in Germany*, 232.

220. Franz, *From Jets to Tanks*, 11.

221. Schlaifer and Heron, *Development of Aircraft Engines*, 426.

222. Some authors have claimed that Jumo's axial engine came from the work of a turbojet design team led by Müller who worked at the Junkers Airframe Company between the mid-1930s and 1939 (for example: Constant, *Origins of the Turbojet Revolution*, 202). Franz consistently contradicted this allegation (Franz, "Development of the 'Jumo 004' Turbojet Engine"; Franz, *From Jets to Tanks*, 13).

223. DM NL 172/2 Franz, Vortrag auf der DGLR-Jahrestagung 1981; Schlaifer and Heron, *Development of Aircraft Engines*, 396.

224. DM NL 172/2.

225. The 109- prefix and numbering system was an RLM system that was modified during the war.

226. Franz, *From Jets to Tanks*, 26–27.

227. NASM VO, Box 16, 463; Wagner and Cox, *First Jet Aircraft*, 31.

228. Franz, *From Jets to Tanks*, 18–20; Combined Intelligence Objectives Sub-Committee File No. XXIV-6, Gas Turbine Development.

229. Franz, *From Jets to Tanks*, 21–22.

230. Ibid., 16.

231. Ibid., 26–27.

232. Ibid., 17.

233. Ibid., 17; Schlaifer and Heron, *Development of Aircraft Engines*, 419.

234. Franz, *From Jets to Tanks*, 17. Franz, like Hooker at Rolls-Royce and Oestrich at BMW, was most important as a technically adept manager who guided Jumo's work.

235. IWM FD 5400/45, December 31, 44, Jumo Report, Gefolgschafts = Einsatz nach Motortypen [Employment by engine type]; Vajda and Dancey used the same source, but leaves out foreign workers (Vajda and Dancey, *German Aircraft Industry and Production*, 242–43).

236. Homze, *Arming the Luftwaffe*, 186.

237. Franz, *From Jets to Tanks*, 13.

238. Bentele, *Engine Revolutions*, 45–46.

239. Franz, *From Jets to Tanks*, 38.

240. Budraß, "Hans Joachim Pabst von Ohain"; Mönnich, Bastow, and Henson, *The BMW Story*, 232–34. The purchase was encouraged by the RLM as part of its efforts to rationalize aero-engine development (Homze, *Arming the Luftwaffe*, 27).

241. In late 1938, BMW and Bramo's development sections had about 1,794 workers together (1,053 at Munich and 741 at Spandau) against 757 at Junkers Dessau and 441 at Daimler-Benz Stuttgart-Untertürkheim. In 1938, Bramo, which had been taken into government control in 1936, was the smallest of the four major German manufacturers by both employment and market share (12.4 percent). BMW was just ahead, representing 13.8 percent of the aero-engine industry (Homze, *Arming the Luftwaffe*, 186–88).

242. Schlaifer and Heron, *Development of Aircraft Engines*, 388–90 and 411.

243. Wagner and Cox, *First Jet Aircraft*, 244.

244. Ibid., 237–38.

245. Schlaifer and Heron, *Development of Aircraft Engines*, 389–90.

246. Its contra-rotating blades were mounted half on an internal and half on an external cylinder. After his collaboration with Bramo, Weinrich became a consultant for Brückner-Kanis, a steam turbine company that worked on a marine gas turbine based on Weinrich's ideas. His career indicated the wide interest in gas turbines for air, land, or sea traction in Germany, which did not arise in Britain or the United States (Kay, *German Jet Engine and Gas Turbine Development*, 97 and 179).

247. Gunston, *World Encyclopaedia of Aero Engines*, 30.

248. Wagner and Cox, *First Jet Aircraft*, 239.

249. Gersdorff and Grasmann, *Flugmotoren und Strahltriebwerke*, 1985 edition, 195–97.

250. Schlaifer and Heron, *Development of Aircraft Engines*, 389.

251. When he left BMW in 1944, Müller began promoting the use of gas turbines for land traction, especially in tanks (Kay, *German Jet Engine and Gas Turbine Development*, 156–57).

252. Wagner and Cox, *First Jet Aircraft*, 237; Schlaifer and Heron, *Development of Aircraft Engines*, 411n47.

253. Budraß, "Hans Joachim Pabst von Ohain"; Schlaifer and Heron, *Development of Aircraft Engines*, 401.

254. Combined Intelligence Objectives Sub-Committee File No. XXIV-6, Gas Turbine Development; Schlaifer and Heron, *Development of Aircraft Engines*, 401.

255. Wagner and Cox, *First Jet Aircraft*, 239.

256. Gersdorff and Grasmann, *Flugmotoren und Strahltriebwerke*, 1985 edition, 197.

257. Wagner and Cox, *First Jet Aircraft*, 239.

258. Combined Intelligence Objectives Sub-Committee File No. XXIV-6, Gas Turbine Development.

259. Wagner and Cox, *First Jet Aircraft*, 239.

260. Gersdorff and Grasmann, *Flugmotoren und Strahltriebwerke*, 1985 edition, 200.

261. Schlaifer and Heron, *Development of Aircraft Engines*, 413–15; Kay, *German Jet Engine and Gas Turbine Development*, 107.

262. Kay, *German Jet Engine and Gas Turbine Development*, 97.

263. Combined Intelligence Objectives Sub-Committee File No. XXIV-6, Gas Turbine Development.

264. Combined Intelligence Objectives Sub-Committee File No. XXIV-6, Gas Turbine Development; Werner, *Kriegswirtschaft und Zwangsarbeit bei BMW*, 347.

265. The 109-018 was a large turbojet with a twelve-stage axial compressor, annular combustion system with twenty-four burners, and three turbine stages. In

addition, the RLM had contracted for a 018 turbojet outfitted with contra-rotating propeller driven by an extra turbine stage (requiring two spools—the country's first compound gas turbine), known as the 109-028 (Oestrich, "Die Entwicklung der Fluggasturbine bei den Bayrischen Motorenwerken").

266. Kay, *German Jet Engine and Gas Turbine Development*, 97–98.

267. Wagner and Cox, *First Jet Aircraft*, 239.

268. Werner, *Kriegswirtschaft und Zwangsarbeit bei BMW*, 68; Mönnich, Bastow, and Henson, *The BMW Story*, 234.

269. A significant number of people offered "miscellaneous help" in the engineering department; a number that more than doubled from 200 in 1942 to 500–600 in 1945 (Wagner and Cox, *First Jet Aircraft*, 241; Schlaifer and Heron, *Development of Aircraft Engines*, 415–16).

270. Homze, *Arming the Luftwaffe*, 186.

271. Gersdorff and Grasmann, *Flugmotoren und Strahltriebwerke*, 1985 edition, 151.

272. Ermenc, *Interviews with German Contributors to Aviation History*, 111–12; Hirschel, Prem, and Madelung, *Aeronautical Research in Germany*, 244–45.

273. A third possibility, that of cooling the turbine's surface, was used by Sanford Moss at General Electric (Gersdorff and Grasmann, *Flugmotoren und Strahltriebwerke*, 1985 edition, 154–55.)

274. Combined Intelligence Objectives Sub-Committee File No. XXIV-6, Gas Turbine Development; BA MA RL 3/4399, October 5, 1943, Minutes ZTL-Entwicklungssitzung [Bypass-Turbojet Development meeting].

275. GOAR 3619, October 6, 1939, Letter Daimler-Benz to Betz; October 30, 1939, Letter Daimler-Benz to Encke.

276. Combined Intelligence Objectives Sub-Committee File No. XXIV-6, Gas Turbine Development.

277. Schlaifer and Heron, *Development of Aircraft Engines*, 387; Kay, *German Jet Engine and Gas Turbine Development*, 145; Gersdorff and Grasmann, *Flugmotoren und Strahltriebwerke*, 1985 edition, 204–5; Niemann, Feldenkirchen, and Herman, *Gasturbinen und Flugtriebwerke der Daimler-Benz AG*, 50–52.

278. DB Haspel 8.9, October 23, 1943, Zeitlicher Ablauf der Entwicklung des ZTL-Triebwerkes [Bypass-Turbojet-Power Plant].

279. GOAR 3619, June 14, 1940, Daimler-Benz Auftrag; GOAR 3619, April 1, 1941, Daimler-Benz Memo.

280. GOAR 3619, June 28, 1940, Letter Betz to Daimler-Benz; GOAR 3619, June 24, 1940, Letter Daimler-Benz an Bevollmächtigten des RLM für das Luftfahrt-Industriepersonal.

281. GOAR 3619, April 1, 1941, Daimler-Benz Memo.

282. Although none had yet been received, the contract was extended to five compressors on December 18, 1941 (DB Haspel 8.9, June 23, 1942, Contract between Daimler-Benz and J. M. Voith).

283. DB Haspel 8.9, July 20, 1943, Letter Daimler-Benz to Voith (Heidenheim).

284. DB Haspel 8.9, October 23, 1943, Zeitlicher Ablauf der Entwicklung des ZTL-Triebwerkes [Bypass-Turbojet-Power Plant].

285. DB Haspel 8.9, September 21, 1944, Minute Nallinger to Haspel.

286. DB Haspel 8.9, October 23, 1943, Zeitlicher Ablauf der Entwicklung des ZTL-Triebwerkes [Bypass-Turbojet-Power Plant].

287. Schlaifer and Heron, *Development of Aircraft Engines*, 403.

288. Niemann, Feldenkirchen, and Herman, *Gasturbinen und Flugtriebwerke der Daimler-Benz AG*, 53.

289. BA MA RL 3/4399, September 28, 1943, ZTL-Entwicklungssitzung [Bypass-Turbojet Development meeting]; the engine was never built (Kay, *German Jet Engine and Gas Turbine Development*, 150).

290. BA MA RL 3/4399, November 9, 1943, Besprechung im Forschungsinstitut für Kraftfahrwesen und Fahrzeugmotoren Stuttgart; Niemann, Feldenkirchen, and Herman, *Gasturbinen und Flugtriebwerke der Daimler-Benz AG*, 44–45; BA MA RL 3/4399, October 8, 1943, Company Minutes 60KR-Notiz.

291. Haspel 8.9, ca. October 20, 1943.

292. Niemann, Feldenkirchen, and Herman, *Gasturbinen und Flugtriebwerke der Daimler-Benz AG*, 42.

293. DB Ordner XVIII, June 1, 1944, Contract between Ernst Heinkel Aktien Gesellschaft and Daimler-Benz.

294. The firm stood to lose around 800 unskilled, foreign workers who could not be used for development (DM FA 001/0323, February 10, 1945, Letter Schif to Frydag).

295. DB Haspel 8.1, February 9, 1944, Letter Haspel, Nallinger to Frydag; DB Haspel 6.11, Produktionsstatistiken.

296. Gersdorff and Grasmann, *Flugmotoren und Strahltriebwerke*, 1985 edition, 205; Kay, *German Jet Engine and Gas Turbine Development*, 150; Gregor, *Daimler-Benz in the Third Reich*.

297. DB Haspel 8.1, February 3, 1944, Memo Betr. Besprechung in Untertürkheim am 22 January 1944 [Memo regarding meeting in Untertürkheim on January 22, 1944].

298. DB Haspel 8.9, September 22, 1944, Minutes; this agrees with Schlaifer's assertion that the firm was "extremely anxious to catch up in the new field" (Schlaifer and Heron, *Development of Aircraft Engines*, 405).

299. Gersdorff and Grasmann, *Flugmotoren und Strahltriebwerke*, 1985 edition, 205–206; DB Haspel 6.11, Produktionsstatistiken.

300. DB Haspel 8.9, September 21, 1944, Nallinger to Haspel.

301. "75 years: MTU Aero Engines celebrates anniversary—An established player in the engine industry for decades."

302. *Flight International*, October 2, 1969.

303. Craven and Cate, *Army Air Forces in World War II*, 252–53; various authors have blamed different parties for American failure: the Air Corps' leadership's lack of expertise in science and engineering and the decision to pursue

piston engines (Holley, "Jet Lag in the Army Air Corps"); General Electric's "persistent myopia" (Constant, *Origins of the Turbojet Revolution*, 211); or the short sightedness of the National Advisory Committee for Aeronautics (NACA), the federal agency responsible for aeronautical research in the United States (Roland, *Model Research*, 187–89).

304. St. Peter, "The History of Aircraft Gas Turbine Development in the US."

305. Roland, *Model Research*, 186.

306. NACA-TR159, 1922, Edgar Buckingham, "Jet Propulsion for Airplanes" (Bureau of Standards), 86. Buckingham was careful to distinguish his jet from a rocket, in that it takes in air from the atmosphere ("Jet Propulsion for Airplanes," 76). In 1920, there was still little distinction made between rockets and air-breathing jet propulsion in the United States (Roland, *Model Research*, 187).

307. St. Peter, "The History of Aircraft Gas Turbine Development in the US"; Roland, *Model Research*; Hansen, *Engineer in Charge*.

308. Roland, *Model Research*, 188; Hansen, *Engineer in Charge*, 224.

309. NACA-TR159, 1922, Edgar Buckingham, "Jet Propulsion for Airplanes" (Bureau of Standards), 86.

310. These technical reports give accounts of earlier research carried out at the NACA: NACA TR-802, 1943, Macon C. Ellis and Clinton E. Brown, "NACA Investigation of a Jet-Propulsion System Applicable to Flight"; NACA TR-758, 1944, John T. Sinnette, Oscar W. Schey, and J. Austin King, "Performance of NACA Eight-Stage Axial-Flow Compressor Designed on the Basis of Airfoil Theory."

311. Hansen, *Engineer in Charge*, 221–22.

312. NACA TR-758, 1944, John T. Sinnette, Oscar W. Schey, and J. Austin King, "Performance of NACA Eight-Stage Axial-Flow Compressor Designed on the Basis of Airfoil Theory."

313. Hansen, *Engineer in Charge*, 226.

314. NACA TR-802, 1943, Macon C. Ellis and Clinton E. Brown, "NACA Investigation of a Jet-Propulsion System Applicable to Flight."

315. Dawson, *Engines and Innovation*, 58; Nichelson, "Early Jet Engines and the Transition from Centrifugal to Axial Compressors."

316. Nichelson, "Early Jet Engines and the Transition from Centrifugal to Axial Compressors," 136–62; Daso, *Hap Arnold and the Evolution of American Airpower*, 194. Schlaifer dismissed the lack of tactical utility for a short-range aircraft as a reason for the lack of jet engine development, citing the American Air Forces' development of jet aircraft as disproving this assertion (Schlaifer and Heron, *Development of Aircraft Engines*, 483).

317. Constant, *Origins of the Turbojet Revolution*, 218–20.

318. Nichelson, "Early Jet Engines and the Transition from Centrifugal to Axial Compressors," 29–30.

319. Constant, for example, blames Moss's conservatism (*Origins of the Turbojet Revolution*, 221).

320. Constant, *Origins of the Turbojet Revolution*, 7–21.

321. Dawson and Hansen argue that Arnold's decision to write to Bush as head of the NDRC indicated his distrust for the NACA. It was Bush's decision to place the new project with an NACA committee (Hansen, *Engineer in Charge*, 230–31).

322. Hansen, *Engineer in Charge*, 230–31. Arnold insisted on the exclusion of piston-aero-engine manufacturers "because he feared they would oppose any radical new departures in engine development; later it was claimed that they were excluded because their 'energies' were judged to be 'completely absorbed in production problems.'" Roland, *Model Research*, 189.

323. AHFR A2073, October 13, 1941, Doc 17, Letter Hunsaker to Arnold.

324. Dawson, *Engines and Innovation*, 47.

325. NACA TR-802, 1943, Macon C. Ellis and Clinton E. Brown, "NACA Investigation of a Jet-Propulsion System Applicable to Flight," 491–92.

326. A2073 September 22, 1941, Doc 11, Letter British Air Commission to Stimson.

327. Ibid.

328. AFHR A2073, September 20, 1941, Doc 10, Letter Lyon to Arnold.

329. AFHR A2073, October 2, 1941, Doc 14, Letter Arnold to Lyon.

330. Dawson, *Engines and Innovation*, 51.

331. Hansen, *Engineer in Charge*, 234.

332. Pavelec, "The Development of Turbojet Aircraft in Germany, Britain, and the United States."

333. Roland, *Model Research*, 191–92.

334. A2073 June 9, 1943, Doc 61, Letter Headquarters Material Command Wright Field to Commanding General Army Air Forces AAF Materials Division Washington D.C.

335. A2073, September 4, 1941, Doc 7, Draft Cable, Arnold to Moore-Brabazon.

336. A2073, September 11, 1941, Doc 9, Cable.

337. Dawson, *Engines and Innovation*, 42; NA AIR 62/953.

338. AIR 62/876, 13 March 1942, HTMRC Special Mtg.

339. AVIA 31/2, May 2, 1942, GTCC Meeting #4.

340. AVIA 46/237 February 27, 1942, 122a Minute Watt to Roxbee Cox; AVIA 31/2, May 2, 1942, GTCC Meeting #4; NA AVIA 15/1509, February 16, 1943, Minute by Watt.

341. AVIA 31/2, February 14, 1942, GTCC Meeting Minutes #3 (Meeting 3 minutes discuss meeting 2 minutes regarding Hastelloy).

342. AVIA 31/2, May 2, 1942, GTCC Meeting #4; Accession file: Hastelloy Turbine in Type I-A in Smithsonian.

343. AVIA 31/2, July 18, 1942, GTCC Meeting #5.

344. AVIA 31/2, May 2, 1942, GTCC Meeting #4; AVIA 31/2, July 18, 1942, GTCC Meeting #5.

345. AVIA 31/2, May 2, 1942, GTCC Meeting #4.

346. AVIA 13/937, March 14, 1942 Wilks from RAE.

347. AVIA 31/3, March 6, 1943, GTCC Meeting #8.

348. Mr. Vickers of Rolls-Royce on blade casting in the United States (AVIA 31/2) and Whittle on Vitallium.

349. See, for example, Whittle's report: AVIA 31/2, July 18, 1942, GTCC Meeting #5 in which he discussed General Electric's axial engine and its Whittle engine.

350. NA AVIA 31/6, February 1, 1946, GTCC Meeting #19; NA AVIA 31/6, June 21, 1946; GTCC Meeting #20.

351. NASM Registrar Files, Inventory number A19500082000, June 15, 1949, "Background to Presentation of Historic Engines."

352. A2073, August 1943, Doc 67.

353. NA AIR 62/990, December 25, 1943, GTCC Reports.

354. Berry, "R-R W2B."

355. Zimmerman, *Top Secret Exchange*, 120.

356. Ibid.; Phelps, *The Tizard Mission*.

357. Schlaifer and Heron, *Development of Aircraft Engines*, 97.

358. Lagasse, "The Westinghouse Aviation Gas Turbine Division."

359. General Electric, *Seven Decades of Progress*, 59.

360. Schlaifer and Heron, *Development of Aircraft Engines*, 97.

361. Hounshell, *From the American System to Mass Production*.

362. Scranton, "Technology, Science and American Innovation."

363. Schlaifer and Heron, *Development of Aircraft Engines*, 92–96 and 493.

364. Ibid., 97–98.

365. Scranton, "Technology, Science and American Innovation."

Chapter Three

1. Ferguson, *Engineering and the Mind's Eye*, 15.

2. Postan, Hay, and Scott, *Design and Development of Weapons*; Boyne and Lopez, *Jet Age*; Constant, *Origins of the Turbojet Revolution*; Golley, Whittle, and Gunston, *Whittle, the True Story*; Nahum, *Frank Whittle*; Pavelec, *Jet Race and the Second World War*; Kay, *Turbojet History and Development*.

3. Frank Whittle, "Future Developments in Aircraft Design" (PhD thesis, Cranwell, 1928) (held by the Science Museum Library in London).

4. GB#347206A.

5. GB#347766A.

6. While at Felixstowe he filed a patent with Herbert McCarthy Reynolds relating to supercharging piston aero-engines in July 1931. It was granted in June 1932 (GB#375104A).

7. Whittle, *Jet: The Story of a Pioneer*, 2nd ed., 50–51; Hall and Morgan, "Bennett Melvill Jones."

8. Whittle, *Jet: The Story of a Pioneer*, 2nd ed., 314.

9. Quoted from Tizard's letter to Whyte (Whyte, *Focus and Diversions*, 144).

10. The exact date is unclear, but von Ohain's HeS 1 ran in late February or early March (Constant, *Origins of the Turbojet Revolution*, 198).

11. Although he recalls patenting a turbojet idea in 1935, von Ohain's first traceable patent in Germany on the topic of "Strahltriebwerk, insbesondere fuer Luftfahrzeuge" ["Jet engine, particularly for aircraft"] (DE#767258A) was applied for in September 1939. It was granted in 1952. Margaret Conner alleges that von Ohain had an earlier patent number 317/38, but this is untraceable (Conner, *Elegance in Flight*, 34).

12. Wagner and Cox, *First Jet Aircraft*, 230 and 239.

13. Gunston. *World Encyclopaedia of Aero Engines*, 117.

14. Schlaifer and Heron, *Development of Aircraft Engines*, 336–48; Whittle's memoir commented on the firm's labor and space problems and the lack of machine tools, but Whittle denied any real knowledge of the firm's finances or administration (Whittle, *Jet: The Story of a Pioneer*, 2nd ed.).

15. Constant, *Origins of the Turbojet Revolution*, 187–88; Pavelec, *Jet Race*, 45–46; Kay, *Turbojet History and Development*, 21–22. The official history contrasts Power Jets' hardship to the supposedly easy financial situation of the Heinkel Aircraft Company (Postan, Hay, and Scott, *Design and Development of Weapons*, 191–92). Similarly, Kay, *Turbojet History and Development*, 21–22, and Golley, Whittle, and Gunston, *Whittle, the True Story*.

16. Schlaifer and Heron, *Development of Aircraft Engines*, 359; Kay, *Turbojet History and Development*, 59; Pavelec, *Jet Race*, 58 and 114; Postan, Hay, and Scott, *Design and Development of Weapons*, 215–17; Whittle, *Jet: The Story of a Pioneer*, 2nd ed.

17. Golley used a quote from Whittle originally made in reference to the conversion of the government company into a research establishment in his section on the nationalisation of Power Jets (Golley, Whittle, and Gunston, *Whittle, the True Story*, 217; original quote in Whittle, *Jet: The Story of a Pioneer*, 2nd ed., 283).

18. NASM T, Box 1, Folder 8, Agreement.

19. Schlaifer and Heron, *Development of Aircraft Engines*, 339.

20. GB#457972A applied for on 09.05.1935, granted on December 9, 1936, "Improvements relating to the electrical transmission of mechanical power" [dynamos]; GB#456976A, applied 16.05.1935, November 16, 1936, "Improvements relating to centrifugal compressors"; GB#461887A applied 25.07.1935, granted February 25, 1937, "Improvements relating to internal combustion turbines"; GB#471368A applied 04.03.1936, granted September 3, 1937, "Improvements relating to the propulsion of aircraft."

21. Schlaifer and Heron, *Development of Aircraft Engines*, 337.

22. BL Add 61931, Board Meeting March 26, 1936; Certificate of Incorporation No 211791.

23. The report is reprinted in Bailey and Whittle, "The Early Development of the Aircraft Jet Engine," appendix.

24. Whittle, *Jet: The Story of a Pioneer*, 2nd cd., 45; Schlaifer and Heron, *Development of Aircraft Engines*, 337; BL Add 61931, January 25, 1940.

25. BL Add 61931, March 26, 1936.

26. Schlaifer and Heron, *Development of Aircraft Engines*, 337.

27. Whyte, *Focus and Diversions*, 143–44.

28. Reader, Weir, and Thomson, *Weir Group*, 117–19; BL Add 61931.

29. BL Add 61931; Schlaifer and Heron, *Development of Aircraft Engines*, 344–45.

30. Schlaifer and Heron, *Development of Aircraft Engines*, 345–46; BL Add 61931, December 20, 1938.

31. Schlaifer and Heron, *Development of Aircraft Engines*, 341.

32. Nahum, *Frank Whittle*, 31–33.

33. Whittle, *Jet: The Story of a Pioneer*, 2nd ed., 98.

34. Ibid., 120; NA AVIA 15/2305, March 13, 1940, Memo of discussion.

35. Postan, Hay, and Scott, *Design and Development of Weapons*, 191.

36. Whittle, *Jet: The Story of a Pioneer*, 2nd ed., 56.

37. Ibid., 66.

38. Ibid., 57.

39. Whittle, *Jet: The Story of a Pioneer*, 2nd ed., 52–62.

40. BL Add 61931, January 10, 1938.

41. BL Add 61931, ecember 2, 1937.

42. Schlaifer and Heron, *Development of Aircraft Engines*, 344.

43. Two night watchmen, an office boy, Mary Phillips (Whittle's secretary), and Victor Crompton (Whittle, *Jet: The Story of a Pioneer*, 2nd ed., 80).

44. Gunston, *Development of Jet and Turbine Aero Engines*, 123.

45. Whittle, *Jet: The Story of a Pioneer*, 2nd ed., 50–51.

46. Gunston, *Development of Jet and Turbine Aero Engines*, 123; after leaving Cambridge, Hall embarked on a distinguished career in aeronautics (Tucker, "Sir Arnold Hall").

47. NA AVIA 46/174, Scott Narrative—the National Gas Turbine Establishment, §28, Interview with Constant, December 14, 1949; the reports by Constant and Griffith, RAE Note #E3546 and Report #3545, can be found in NA AVIA 46/234.

48. NA AVIA 46/174, Scott Narrative—the National Gas Turbine Establishment, §28, Interview with Constan December 14, 1949; NA AIR 62/10, December 22, 1937, Whittle diary, Long "phone call to Dr. Griffith."

49. Whittle, *Jet: The Story of a Pioneer*, 2nd ed., 128–29.

50. Ibid., 135–36.

51. Clark, *Tizard*, 299–300; Nahum, *Frank Whittle*, 170.

52. NA AVIA 15/2987, April 23, 1941, MAP meeting, discussion on PJ and Rover.

53. NA AVIA 15/2305, February 17, 1940, 4A, Letter Whyte to Tedder.

54. Hooker, *Not Much of an Engineer*, 63–67.

55. Whittle, *Jet: The Story of a Pioneer*, 2nd ed., 122–23.

56. Ibid., 146.

57. Ibid., 124.

58. NA AVIA 15/1806, 24B, Rover Contracts (C/Eng/523).

59. Brooks, *Vikings at Waterloo*, 47.

60. Whittle, *Jet: The Story of a Pioneer*, 2nd ed., 88.

61. Ibid., 92.

62. NA AVIA 46/174, Scott Narrative—the National Gas Turbine Establishment, §13.

63. Bloor, *Enigma of the Aerofoil*.

64. NA AVIA 46/174, Scott Narrative—the National Gas Turbine Establishment, §12.

65. Bulman, *An Account of Partnership*, 319.

66. NA AVIA 31/2, July 18, 1942, GTCC Meeting Minutes #15.

67. Whittle, *Jet: The Story of a Pioneer*, 2nd ed., 117; Bulman, *An Account of Partnership*, 319.

68. Hawthorne, "The Early History of the Aircraft Gas Turbine in Britain," 99.

69. Whittle, *Jet: The Story of a Pioneer*, 2nd ed., 120.

70. Whyte, *Focus and Diversions*, 145.

71. Farren quoted in Whittle, *Jet: The Story of a Pioneer*, 2nd ed., 101.

72. BL Add 16931, April 13, 1937.

73. Whittle, *Jet: The Story of a Pioneer*, 2nd ed., 104.

74. Ibid., 96–113.

75. Ibid., 105 and 126.

76. NA AVIA 46/174, Scott Narrative—the National Gas Turbine Establishment; to its never-ending dismay, MAP gave Rover direct contracts without an agreement on patents being reached between Power Jets and Rover (NA AVIA 10/412, February 24, 1944, Tinling's talk); NA AIR 62/2, Whittle's comments on Keppel's narrative.

77. NA AVIA 22/1540, November 4, 1939, 1a, Air Ministry (Supply).

78. NA AVIA 15/1614, October 15, 1942, 39, Minute Linnell.

79. Postan, Hay, and Scott, *Design and Development of Weapons*, 205–6.

80. NA AVIA 31/1, November 1, 1941, GTCC Meeting Minutes #1.

81. NA AVIA 46/237, Extracts from GTCC File; November 11, 1941, Letter Roxbee Cox to Ricardo.

82. NA AVIA 46/237, Extracts from GTCC File, 167a, Letter from Roxbee Cox to Admiral Hoare.

83. Whittle and von Ohain, *An Encounter between the Jet Engine Inventors*, 97. Schlaifer argued that the GTCC had contributed little to generating new ideas, which generally came from the work of individuals; the only exception was perhaps in combustion (Schlaifer and Heron, *Development of Aircraft Engines*, 436–38).

84. BL Add 61931, January 8, 1942; NA AIR 62/1006, January 23, 1942, Whittle, Gas Turbine Collaboration Committee.

85. NA AIR 62/1006, January 23, 1942, Whittle, Gas Turbine Collaboration Committee.

86. This included Hawthorne's loan to Power Jets (NA AIR 62/1006, January 23, 1942, Whittle, Gas Turbine Collaboration Committee).

87. NA AIR 62/994, December 18, 1941, Letter W. E. P. Johnson to Watt.

88. NA AIR 62/608, April 21, 1943, Whittle diary, Note W. E. P. Johnson.

89. NA AVIA 15/1806, May 10, 1943, Letter Tinling to Hives.

90. Rover's B/26 entirely changed the architecture of the W.2B engine by replacing the reverse-flow combustion system of the W.2B (used in all Power Jets engines) with a straight-through system. Rather than derivative, the B/26 had been seen by some in MAP as a completely new private venture (NA AVIA 13/937, May 8, 1942, Rovers Brochure; NA AVIA 15/1614, April 2, 1942, 15, Minute Linnell). In defense of its claim to the engine, Power Jets emphasized that it had previously designed a straight-through engine, the W.3X. Yet Rover had built one without seeing Power Jets' drawings (NA AVIA 13/937, May 19, 1942, Whittle's Comments).

91. NA AVIA 15/1806, May 10, 1943, Tinling to Hives.

92. NA AVIA 15/1806, February 28, 1943 30B, Letter Rowlands to Sidgreaves.

93. Whittle, *Jet: The Story of a Pioneer*, 2nd ed., 104.

94. NA AVIA 46/174, Scott Narrative—the National Gas Turbine Establishment, §10.

95. Ibid.

96. Schlaifer and Heron, *Development of Aircraft Engines*, 346. The Treasury happily abandoned the pay-as-you-go scheme in 1944 when the firm was nationalized (NA T 161/1240).

97. Whittle, *Jet: The Story of a Pioneer*, 2nd ed., 93; Whyte, "Whittle and the Jet Adventure," 147.

98. Bulman, *An Account of Partnership*, 319.

99. NA AVIA 15/2987, April 11, 1941, copy of March 29, 1941, L. L. Whyte, Power Jets Ltd. Schedule of Factory Space.

100. NA AVIA 15/2987, July 4, 1941, Roxbee Cox to Pye.

101. NA AVIA 15/2987, September 3, 1941, Letter Walker to Roxbee Cox.

102. NA AVIA 15/2987, July 28, 1941, R. D. Williams, Proposed new works for Power Jets.

103. Power Jets had been open to the state purchasing a bigger share in the company earlier, but earlier negotiations in July 1941 had foundered on the asking price—£300,000. Power Jets offered the state a share of up to 51 percent, including its existing 12 percent stake. The Treasury objected to MAP's involvement in a private business (NA T 161/1240, July 17, 1941, Letter to Gilbert).

104. NA AVIA 46/237, July 31, 1941, Doc 3a, Gas Turbine Collaboration.

105. NA AVIA 15/2987, April 23, 1941, MAP meeting, Discussion on Power Jets and Rover.

106. NA AVIA 15/2987, October 17, 1941, Air Supply Board Meeting Conclusions.

107. Estimate from second half of 1943 (NA T 161/1240).

108. For the year to June 30, 1942, and June 30, 1943, the Treasury offered Power Jets fixed profits according to the shadow scheme of £4,500. The firm's management declined the profits, offering to work without fees during the war (NA T 161/1240). Power Jets had negotiated profits of £1,000 (roughly 5 percent on capital employed) and £1,500 for the year to June 30, 1940, and June 30, 1941 (NA T 161/1240, January 16, 1941, Letter Graham to Fletcher).

109. Whittle, *Jet: The Story of a Pioneer*, 2nd ed., 237.

110. NA AVIA 15/2987, July 28, 1941, Conference at MAP.

111. Whittle, *Jet: The Story of a Pioneer*, 2nd ed., 253–54.

112. Ibid., 252.

113. NA AVIA 46/174, Scott Narrative—the National Gas Turbine Establishment, §49; Postan, Hay, and Scott, *Design and Development of Weapons*, 228–29; Nahum, "World War to Cold War," 99–100.

114. Whittle, *Jet: The Story of a Pioneer*, 2nd ed., 258.

115. Ibid., 216–17; NA AIR 62/2, Whittle's comments on Keppel's narrative.

116. NA AVIA 15/1806, January 5, 1943, 12A, Notes of discussion held at Power Jets.

117. RRHT, A-Numbers List.

118. Of the thirty-two development engines built by Rover at Barnoldswick, over twenty were used on development work at Barnoldswick and Clitheroe; Power Jets received just four of these engines (Whittle, *Jet: The Story of a Pioneer*, 2nd ed.).

119. Ibid., 227.

120. Whittle deployed many of the same arguments that he had used earlier against Rover's STX or B/26 design (NA AIR 62/608, August 19, 1943, Whittle diary, Letter Whittle to Roxbee Cox).

121. NA AVIA 46/174, Scott Narrative—the National Gas Turbine Establishment, §50.

122. Schlaifer and Heron, *Development of Aircraft Engines*, 346. The contemporary Treasury figures are approximate and cover different time periods. The Treasury calculated that total expenditure on Power Jets had been about £650,000 in the four years to June 30, 1943, plus an additional £325,000 spent on Whetstone (including plant) (NA T 161/1240).

123. Nahum, *Frank Whittle*, 112; Cripps's plans for a "National Aeronautical Establishment" were realized, with Farren's active support, as RAE Bedford. See Nahum, *Frank Whittle*, 108–24 on Cripps's vision for British aviation.

124. NA T 161/1240, December 13, 1943, Letter Cripps to Tobin; Clarke, *Cripps Version*, 34–37.

125. NA T 161/1240, November 29, 1943, Note on Power Jets.

126. NA AVIA 46/174, Scott Narrative—the National Gas Turbine Establishment, §52.

127. Postan, Hay, and Scott, *Design and Development of Weapons*, 229–30.

128. NA AVIA 10/412, "A suitable mandate for Power Jets (R&D) Ltd."; NA AVIA 46/174, Scott Narrative—the National Gas Turbine Establishment, §57.

129. NA T 161/1240, Note on the Future of Power Jets (Research and Development) Ltd. These were some of the very same reasons that Tizard had seen Power Jets' establishment as a private company advantageous despite the fact that the government would inevitably have to support it (Nahum, *Frank Whittle*, 31–34).

130. Cf. Edgerton, "Technical Innovation, Industrial Capacity, and Efficiency"; NASM T, Box 1, Folder 1, April 6, 1944, Letter Sidgreaves to Tinling.

131. NA T 161/1240, April 11, 1944, Letter Weir to Cripps; similar questions were raised by the nationalization of Shorts Brothers in 1943 (Edgerton, "Technical Innovation, Industrial Capacity, and Efficiency").

132. NASM T, Box 1, Folder 1, April 10, 1944, Letter Mrs. Lubbock to Bonham-Carter; NASM T, Box 1, Folder 1, April 13, 1944, Letter Tinling to Pettitt.

133. NA T 161/1240, April 11, 1944, Letter Weir to Cripps.

134. NA T 161/1240, February 10, 1944, Letter Lee to Tucker.

135. Edgerton, "Technical Innovation, Industrial Capacity and Efficiency."

136. PREM 3/21/4, June 2, 1944, 143, Memo Cherwell to Churchill.

137. (NA AVIA 46/174, Scott Narrative—the National Gas Turbine Establishment, §41); Bulman had earlier argued that the company was in effect operating like a government research establishment and should be brought under MAP control (Nahum, *Frank Whittle*, 110).

138. Ibid., 124.

139. Postan, Hay, and Scott, *Design and Development of Weapons*, 207; NA AVIA 46/174, Scott Narrative—the National Gas Turbine Establishment, §50.

140. NA T 161/1240; Schlaifer and Heron, *Development of Aircraft Engines*, 346; Nahum, "World War to Cold War," 108.

141. Nahum, *Frank Whittle*, 124.

142. Whittle, *Jet: The Story of a Pioneer*, 2nd ed., 295.

143. During 1944, Whittle was in the hospital for six months (Golley, Whittle, and Gunston, *Whittle, the True Story*, 221–29).

144. NA AVIA 46/234, Griffith, Royal Aircraft Establishment report No H.1111; Constant, "Early History of the Axial Type of Gas Turbine Engine," 411.

145. Whittle, *Jet: The Story of a Pioneer*, 56.

146. Hawthorne, "Aircraft Propulsion from the Back Room," 96.

147. NA AVIA 46/234, 87th Report of Engine Sub-Committee.

148. Roxbee Cox, "British Aircraft Gas Turbines," 55; NA AVIA 46/234, February 1937, Griffith, Report #3545 on Whittle Jet Propulsion System; NA AVIA 46/234, March 1937, Constant, RAE Note E #3546 "The ICT as a power plant for aircraft." The thoroughness of Constant's report contrasts with the report prepared by Bramson for O. T. Falk in 1935 (Reprinted in Kay, *Turbojet History and Development*).

149. Nahum, "Two-Stroke or Turbine?" 315–19; Power Jets was promised government support in August 1937 and received its first payment in May 1938 (Schlaifer and Heron, *Development of Aircraft Engines*, 434–35).

150. Constant, "Development of the Internal Combustion Turbine," 409.

151. NA AVIA 46/237, July 31, 1941, 3a, Gas Turbine Collaboration.

152. Hawthorne, "Aircraft Propulsion from the Back Room," 98; NA T 161/1240, December 7, 1943, Letter from Trend to Lee.

153. Thomson and Hall, "William Scott Farren," 229–30.

154. Ibid., 230; Postan, Hay, and Scott, *Design and Development of Weapons*, 107n1; Constant, "Early History of the Axial Type of Gas Turbine Engine," 412.

155. Hawthorne, "Early History of the Aircraft Gas Turbine in Britain," 104.

156. Resistance to the RAE's growing influence came from George P. Bulman, who was head of engine development at MAP. Ever since he had taken charge of engine development in the Air Ministry at the beginning of 1928, Bulman had fought to curb the influence that the RAE had gained in aero-engine design during World War I (Bulman, *An Account of Partnership*, 54–57).

157. Thomson and Hall, "William Scott Farren," 218–26.

158. NA AVIA 15/1417, January 27, 1942, 7A, Letter Farren to Pye.

159. NA AVIA 15/1417, February 22, 1942, 11, Minute Pye to Bulman and Rowe.

160. Hawthorne, "Aircraft Propulsion from the Back Room," 98; Halford's suspicion about the RAE's input on the H1 demonstrated that the Establishment's work in the gas turbine field went beyond its role in piston engine work (NA AVIA 13/935, August 5, 1942, Letter Farren to Halford; November 10, 1942, Letter Farren to Halford).

161. NA AVIA 15/1708, November 2, 1942, 15A, Meeting in Linnell's Office about "the future of the W.2B design."

162. Hawthorne, Cohen, and Howell, "Hayne Constant," 275; NA AVIA 46/174, Scott Narrative—the National Gas Turbine Establishment, §30; Postan, Hay, and Scott, *Design and Development of Weapons*, 207.

163. Hawthorne, "Aircraft Propulsion from the Back Room," 100; Hawthorne, Cohen, and Howell, "Hayne Constant," 276; Postan, Hay, and Scott, *Design and Development of Weapons*, 230.

164. NA PREM 3/21/4, May 26, 1944, 147, Memo Cherwell to Churchill.

165. Postan, Hay, and Scott, *Design and Development of Weapons*, 230.

166. Hawthorne, "Aircraft Propulsion from the Back Room," 100.

167. NA AVIA 46/174, Scott Narrative—the National Gas Turbine Establishment, §54.

168. KNA, File 2000.443, April 30, 1944, Letter Roxbee Cox to Mother; in 1948, Roxbee Cox finished negotiations for a new site north of the RAE turbine section's existing site (Roxbee Cox, "British Aircraft Gas Turbines"). The test facilities of the former Power Jets Whetstone factory were not fully combined with those of the original RAE turbine section until 1955.

169. RR HC, April 17, 1945, Commentary by James E. Ellor.

170. Whittle, *Jet: The Story of a Pioneer*, 2nd ed., 282–84.

171. Ibid., 108.

172. NASM T, Power Jets (R&D) Technical Report May–June 1945.

173. Whittle, *Jet: The Story of a Pioneer*, 2nd ed., 287.

174. Postan, Hay, and Scott, *Design and Development of Weapons*, 231.

175. Whittle, *Jet: The Story of a Pioneer*, 2nd ed., 283–84.

176. NA T 161/1240, Note on the future of Power Jets (Research and Development) Ltd.

177. NA T 161/1240, December 21, 1945, Letter Flett to Bailey.

178. Whittle, *Jet: The Story of a Pioneer*, 2nd ed., 304–5.

179. Quoted in Whittle, *Jet: The Story of a Pioneer*, 2nd ed., 296–98.

180. Emphasis in the original. Whittle, *Jet: The Story of a Pioneer*, 2nd ed., 283.

181. Ibid., 301.

182. Emphasis in the original. Ibid., 302.

183. "Jet-Propulsion Development Sixteen Engineers Resign," *The Times*, April 15, 1946; Whittle, *Jet: The Story of a Pioneer*, 2nd ed., 300; NA T 161/1240, March 19, 1946, Letter Newton to Curtis.

184. KNA, File 2000.467, April 14, 1946, Letter Roxbee Cox to Mother; NA AVIA 46/234, April 15, 1946, Extract from Parliamentary Debates; *The Times*, April 15, 1946.

185. NA T 161/1240, March 19, 1946, Letter Newton to Curtis.

186. The Treasury described the terms of his engagement as "at the extreme limit, but not over the limit of what is reasonable" (Nahum, *Frank Whittle*, 131; KNA, File 2000.443, April 30, 1944, Letter Roxbee Cox to Mother; NA T 161/1240, February 7, 1944, Letter Forde to Smith).

187. *The Times*, November 9, 1949; the amount originally suggested had been halved in light of payments individual American firms had already made for manufacturing licenses ("U.S. Use of JET Patents Claim Settled for $4,000,000," *The Times*, October 13, 1951).

188. Homze, *Arming the Luftwaffe*, 185.

189. There have been several book-length studies of Heinkel and his firm by German authors, in which the jet features as one of the firm's many pioneering innovations. Nowarra, *Heinkel und seine Flugzeuge*; Köhler, *Ernst Heinkel: Pionier der Schnellflugzeuge*; Koos, *Ernst Heinkel: Vom Doppeldecker zum Strahltriebwerk*.

190. Heinkel and Thorwald, *Stürmisches Leben* and *He 1000*; Neufeld, "Rocket Aircraft and the 'Turbojet Revolution'"; NASM VO, Box 4, January 16, 1981, Letter Schelp to von Ohain.

191. Postan, Hay, and Scott, *Design and Development of Weapons*, 190–92; Postan does not give an adequate account of the Heinkel Aircraft Company (Postan, Hay, and Scott, *Design and Development of Weapons*, 187–94).

192. Constant, *Origins of the Turbojet Revolution*, 200–201.

193. Conner gives a detailed account of von Ohain's early work at Heinkel,

drawing on many sources; the details of his later work are somewhat confused (Conner, *Elegance in Flight*).

194. Köhler, *Ernst Heinkel: Pionier Der Schnellflugzeuge*, 178 and 198.

195. Schlaifer and Heron, *Development of Aircraft Engines*, 406–10.

196. Koos, *Ernst Heinkel: Vom Doppeldecker zum Strahltriebwerk*, 165.

197. Schlaifer and Heron, *Development of Aircraft Engines*, 410; Whittle and von Ohain, *An Encounter between the Jet Engine Inventors*, 85.

198. Koos, *Ernst Heinkel: Vom Doppeldecker zum Strahltriebwerk*, 132–35.

199. DM FA 001/0258, August 4, 1938, Letter Heinkel to Udet; the Heinkel Company was not the only German aviation firm to make a virtue of its design rather than production capability (Overy, *Air War*, 179). The huge number of German development projects is often interpreted as proof of German technical superiority, but they reflected rather an unfocused procurement system.

200. Koos, *Ernst Heinkel: Vom Doppeldecker zum Strahltriebwerk*, 141.

201. Köhler, *Ernst Heinkel: Pionier Der Schnellflugzeuge*, 165.

202. Whittle and von Ohain, *An Encounter between the Jet Engine Inventors*, 16.

203. NASM VO, Box 22, April 3, 1936, Copy of contract signed by von Ohain and Heinkel.

204. Whittle and von Ohain, *An Encounter between the Jet Engine Inventors*, 15.

205. Gundermann, "Zur Geschichte des Strahltriebwerks."

206. Including the HeS 1, 2, 3, 6, 8 and 10 (Combined Intelligence Objectives Sub-Committee File No. XXIII-14, May 1945, Turbine Activity at EHAG [Ernst Heinkel AG, Founded 1 April 1943]).

207. Using hydrogen gas, which was easy to burn, avoided the problems of liquid fuels, which had somehow to be mixed with the incoming airstream (Whittle and von Ohain, *An Encounter between the Jet Engine Inventors*, 15).

208. Like Whittle, von Ohain, was at first only able to test complete engines (Ermenc, *Interviews with German Contributors*, 34–39; Whittle and von Ohain, *An Encounter between the Jet Engine Inventors*, 93; Postan, Hay, and Scott, *Design and Development of Weapons*, 190–91).

209. Hirschel, Prem, and Madelung, *Luftfahrtforschung in Deutschland*, 236.

210. Conner, *Elegance in Flight*, 80.

211. Von Ohain quoted in Ermenc, *Interviews with German Contributors*, 37.

212. Gundermann, "Zur Geschichte des Strahltriebwerks."

213. Koos, *Ernst Heinkel: Vom Doppeldecker zum Strahltriebwerk*, 143–45.

214. Whittle and von Ohain, *An Encounter between the Jet Engine Inventors*, 95; he did however receive important help in learning how the Heinkel Company worked from Wilhelm Gundermann (Conner, *Elegance in Flight*, 256).

215. Wagner and Cox, *First Jet Aircraft*, 52n12; NASM VO.

216. Koos, *Ernst Heinkel: Vom Doppeldecker zum Strahltriebwerk*, 140; DM FA 001/0258, February 6, 1939 Besprechung in Rostock January 23, 1949; Reprint of Protokoll Nr S-017, end July/start August 1938 in Wagner and Cox, *First Jet Aircraft*, 17.

217. Emphasis in the original. Whittle and von Ohain, *An Encounter between the Jet Engine Inventors*, 16–17.

218. Conner, *Elegance in Flight*, 257.

219. Reproduction of Heinkel's museum directive in Conner, *Elegance in Flight*, 85.

220. Combined Intelligence Objectives Sub-Committee File No. XXIII-14, 5.45, Turbine Engine activity at EHAG [Ernst Heinkel AG, Founded 1 April 1943].

221. Ermenc, *Interviews with German Contributors*, 42.

222. Wagner and Cox, *First Jet Aircraft*, 30.

223. Schlaifer and Heron, *Development of Aircraft Engines*, 406; Koos, *Ernst Heinkel: Vom Doppeldecker zum Strahltriebwerk*, 156.

224. Schlaifer and Heron, *Development of Aircraft Engines*, 406–7.

225. Wagner and Cox, *First Jet Aircraft*, 31; NASM VO, Box 16, 463; Köhler, *Ernst Heinkel: Pionier Der Schnellflugzeuge*, 198; Koos, *Ernst Heinkel: Vom Doppeldecker zum Strahltriebwerk*, 154.

226. Koos, *Ernst Heinkel: Vom Doppeldecker zum Strahltriebwerk*, 155.

227. Budraß, "Hans Joachim Pabst von Ohain."

228. NASM VO, Box 16, 461–464; DM NL172/9, August 7, 1983, Letter von Ohain to Knausenberger; Koos claims it was eighteen engineers (Koos, *Ernst Heinkel: Vom Doppeldecker zum Strahltriebwerk*, 155).

229. Ermenc, *Interviews with German Contributors*, 47; Kay gives a figure of 120 for turbojet airframe and engine groups at the end of 1939, which implies that the group working on the airframe was far larger (Kay, *Turbojet History and Development*, 178).

230. NASM VO, Copy of chapter for DGLR [German Society for Aeronautics and Astronautics] book by von Ohain; Schlaifer and Heron, *Development of Aircraft Engines*, 402.

231. NA AVIA 15/2121, May 25, 1945, 230, Report by Helmut Schelp.

232. Wagner and Cox, *First Jet Aircraft*, 30–41; Kay, *German Jet Engine and Gas Turbine Development*, 35–37.

233. NA AVIA 15/2121, May 25, 1945, 230, Report by Helmut Schelp.

234. Köhler, *Ernst Heinkel: Pionier Der Schnellflugzeuge*, 204; the exception was the HeS 50 motorjet engine, which Bentele continued work on at Hirth (Bentele, *Engine Revolutions*, 63–64).

235. DM FA 001/0314, June 12, 1940, Pfistermeister to Heinkel; Bentele of HMZ claimed to have received visits with regard to the sale from Daimler-Benz, Jumo, BMW, Messerschmitt, Heinkel, Dornier, and several smaller aviation companies (Bentele, *Engine Revolutions*, 33–34).

236. DM FA 001/0314, June 11, 1940, Heinkel to Udet.

237. Homze, *Arming the Luftwaffe*, 186.

238. Bentele, *Engine Revolutions*, 50.

239. Whittle and von Ohain, *An Encounter between the Jet Engine Inventors*, 18.

240. NA AVIA 15/2121, May 25, 1945, 230, Report by Helmut Schelp.

241. DM FA 001/0314, August 8, 1948, Letter Heinkel to Herr Baron von le Fort.

242. Conner, *Elegance in Flight*, 257.

243. NA AVIA 15/2121, May 25, 1945, 230, Report by Helmut Schelp.

244. Letter von Ohain to Günther, July 1, 1967 quoted in Wagner and Cox, *First Jet Aircraft*, 33.

245. DM FA 001/0314, June 27, 1941, Betriebsappell anlässlich der Übernahme von Hirth [Talk to the factory on the occasion of the takeover of Hirth].

246. DM FA 001/0314, June 12, 1940, Letter Pfistermeister to Heinkel.

247. Koos, *Ernst Heinkel: Vom Doppeldecker zum Strahltriebwerk*, 156; DM FA 001/0314, March 10, 1941, Memo.

248. DM FA 001/0314, March 10, 1941, Memo. The papers weren't signed by Heinkel until June 4, 1941 (DM FA 001/0314, June 4, 1941, Letter Heinkel to Steiger).

249. Ermenc, *Interviews with German Contributors*, 116.

250. Schlaifer and Heron, *Development of Aircraft Engines*, 398.

251. The Heinkel Aircraft Company was well known for being more like a research facility than a production company (Schlaifer and Heron, *Development of Aircraft Engines*, 410; Koos, *Ernst Heinkel: Vom Doppeldecker zum Strahltriebwerk*, 165); Heinkel bragged about the fact (DM FA 001/0314, June 11, 1940, Letter Heinkel to Udet).

252. Ermenc, *Interviews with German Contributors*, 116.

253. NA AVIA 15/2121, May 25, 1945, 230, Report by Helmut Schelp.

254. Schlep remembered that the contract was granted in late 1941 (NA AVIA 15/2121, May 25, 1945, Report by Helmut Schelp); Heinkel recorded that it was granted in late 1942 (DM FA 001/050, HeS 11 contract).

255. Wagner and Cox, *First Jet Aircraft*, 36; NA AVIA 15/2121, May 25, 1945, 230, Report by Helmut Schelp.

256. Letter von Ohain to Günther, July 1, 1967, quoted in Wagner and Cox, *First Jet Aircraft*, 33.

257. DM FA 001/0322, April 1943, Draft letter to Milch.

258. Koos, *Ernst Heinkel: Vom Doppeldecker zum Strahltriebwerk*, 160.

259. DM FA 001/0322, March 15, 1943, Letter Heinkel to Vorstand der Ernst Heinkel AG; Müller continued to correspond with Heinkel after leaving the firm (DM FA 001/0322, January 3, 1943, Letter Müller to Heinkel); DM FA 001/0319, July 7, 1941, Müller; Bentele, *Engine Revolutions*, 44.

260. Bentele, *Engine Revolutions*, 52.

261. Koos, *Ernst Heinkel: Vom Doppeldecker zum Strahltriebwerk*, 162.

262. DM FA 001/0322, January 24, 1943, Letter Wolff to Heinkel.

263. DM FA 001/0322, March 25, 1943, Letter Milch to Wolff.

264. Schlaifer and Heron, *Development of Aircraft Engines*, 410.

265. Ermenc, *Interviews with German Contributors*, 46; Whittle and von Ohain,

An Encounter between the Jet Engine Inventors, 18; Bentele, *Engine Revolutions*, 51.

266. DM FA 001/0322, January 22, 1943, Letter Heinkel to Wolff.

267. For example, engine lubrication (Conner, *Elegance in Flight*, 257).

268. Bentele, *Engine Revolutions*, 57.

269. DM FA 001/0323, November 9, 1944, Letter Heinkel to Frydag; Koos, *Ernst Heinkel: Vom Doppeldecker zum Strahltriebwerk*, 187. Accusations of delay were often based on the testimony of lay persons and so may not reflect the engine's actual status (see, for example, BA MA NS19/3652, September 5, 1944, Letter Himmler to Speer).

270. DM FA 001/0323, November 9, 1944, Letter Heinkel to Frydag.

271. NA AVIA 15/2121, May 25, 1945, 230, Report by Helmut Schlep.

272. Von Ohain was very proud that HMZ produced twenty-three engines between April 1944 and April 1945 (Köhler, *Ernst Heinkel: Pionier Der Schnellflugzeuge*, 203); Budraß argues that Wolff may intentionally have sabotaged production work on the HeS11, in part by encouraging von Ohain's scientific interests, but it seems unlikely that the engine could have successfully reached production in so short a time (Budraß, "Hans Joachim Pabst Von Ohain").

273. Köhler, *Ernst Heinkel: Pionier Der Schnellflugzeuge*, 203.

274. Bentele, *Engine Revolutions*, 59.

275. Ziegler, "Germany's Aviation Industry: A Personal Report, part 1" and "Germany's Aviation Industry: A Personal Report, part 2."

276. Wagner and Cox, *First Jet Aircraft*, 37; "Aero-Engines 1956," *Flight*, May 11, 1956.

277. Nahum, *Frank Whittle*, 128.

Chapter Four

1. Cf. Dennis, "Historiography of Science, an American Perspective."

2. MacLeod, "Concepts of Invention and the Patent Controversy in Victorian Britain"; in her subsequent study, Christine MacLeod examined the "popular memory of the inventor" in the second half of the nineteenth century in Britain. She describes her study as a "social history of remembering" (MacLeod, *Heroes of Invention*, 6–8).

3. Neufeld, "The Nazi Aerospace Exodus."

4. Edgerton, *Shock of the Old*, 103–8.

5. This interpretation is made particularly clear in Michael Pavelec's recent book, *Jet Race*.

6. NA AIR 2/7070, 22 November 22, 1943, 1A, Letter Freeman to Evill.

7. Ibid.

8. NA AIR 20/7070, November 25, 1943, Letter Evill to Walsh.

9. NA AIR 2/7070, December 22, 1943, 3D, Cypher from RAFDEL [Royal Air Force Delegation] to Air Ministry; 12A, Cypher Portal to RAFDEL; November 22, 1943, 1A, Letter Freeman to Evill.

10. NA AIR 2/7070, December 28, 1943, Note by Caines; this same argument resurfaced later in the discussion of Power Jets' nationalization and in the question of Whittle's right to an award for his work.

11. The PR biography was supplemented by Roxbee Cox in consultation with Whittle himself (NA AIR 2/7070, December 30, 1943, 18B, Meeting about release).

12. NA AIR 2/7070, December 22, 1943, RAFDEL [Royal Air Force Delegation]to Air Ministry (for Evill from Welsh).

13. NA AIR 62/1023, January 7, 1944, Extract from BBC News Bulletin, 7 am, and Home Service, 9 pm.

14. NA AIR 62/1023, January 11, 1944, Whittle address to Power Jets.

15. NA AIR 2/7070, January 8, 1944, 29A, Letter Sporborg to Cripps.

16. NA AIR 2/7070, January 17, 1944, Interdepartmental letter; Maurice Bonham Carter, "Jet Propulsion (Letters to the Editor)," *The Times*, January 15, 1944; "B. T. H. [British Thomson Houston Company] and the Jet: British Manufacturer's Share in Development," *Flight*, February 24, 1945.

17. NA AIR 2/7070, November 22, 1943, 1A, Letter Freeman to Evill; the rivalry in aviation technology remained intense in the early postwar period (Engel, *Cold War at 30,000 Feet*).

18. NA AIR 2/7070, January 21–31, 1944, 36A–36D.

19. NA AIR 2/7070, March 31, 1944, 48A, Circular from J. E. Keel (MAP).

20. NA AIR 2/7070, April 19, 1944, 54A, Letter Cripps to Portal.

21. NA AIR 2/7070, April 19, 1944, 54A, Letter Cripps to Portal; April 21, 1944, Interdepartmental memo.

22. NA AIR 2/7070, March 6, 1944, Letter Sorley; May 5, 1944, 61a, Letter Freeman to Evill; April 19, 1944, 54A, Letter Cripps to Portal.

23. NA AIR 2/7070, July 27, 1944, 76A, Freeman to Portal.

24. NA AIR 2/7070, August 16, 1944, 81A, Letter Welsh to Evill.

25. NA AIR 2/7070, August 20, 1944, 85A, Cypher Portal to Arnold; NA AIR 8/785, August 14, 1944, Cypher from Arnold.

26. NA AIR 2/7070, September 11, 1944, Extract from War Cabinet, 122nd conclusions.

27. NA AIR 2/7070, September 27, 1944, 108a, Air Ministry, "Progress of Jet-Propelled Aircraft."

28. NA AIR 8/785, September 27, 1944, 109c, Cypher Portal to Welsh; September 27, 1944, 109d, Cypher Welsh to Portal.

29. NA AIR 2/7070, October 25, 1944, 123a, Extract from report by S/Ldr. Bown (RAFDEL).

30. NA AIR 2/7070, August 20, 1944, 85A, Cypher Portal to Arnold; NA AIR 8/785, August 14, 1944, Cypher from Arnold.

31. NA AIR 20/4235, 34-36A, September 29, 1944–October 5, 1944.

32. *Aviation News* was published from August 2, 1943, to June 30, 1947; NA AIR 2/7070, Excerpt at 124B.

33. NA AIR 2/7070, November 14, 1944, 126A, Letter Henderson to Air Ministry.

34. NA AIR 2/7070, February 1, 1945, 148A, DCS Evill to Freeman.

35. NA AIR 2/7070, February 20, 1945, 159B, Cypher Air Ministry to RAFDEL; February 25, 1945, 168A, Cypher Air Ministry to RAFDEL.

36. NA AIR 2/7070, February 22, 1945, 166A, RAFDEL to Air Ministry.

37. NA AIR 2/7070, February 28, 1945, 177a, "RAF [Royal Air Force] 'Jet' Fighters in Action."

38. "British Achievement in Jet Propulsion," *The Times*, April 19, 1945.

39. *Flight*, October 25, 1945.

40. NA AIR 2/7070, June 26, 1945, 191A, Memo from Joint Security Control.

41. NA AIR 2/7070, January 18, 1946, 192A, Letter from Griffiths.

42. "British Jet-Engine Development; Rivalry with Co-Operation," *The Times*, December 27, 1950.

43. In 1944 he also received in May the 1943 James Alfred Ewing Medal from the Institutions of Civil and Mechanical Engineering, in October the Freedom of Leamington (where he grew up), and lastly the Gold Medal of the Royal Aeronautical Society (Feilden and Hawthorne, "Sir Frank Whittle"; "News in Brief," *The Times*, January 22, 1944; May 9, 1944; October 30, 1944). In 1946, he was awarded the 1945 Clayton Prize (£1,000) by the Institution of Mechanical Engineers, and granted an honorary doctorate by Manchester University ("News in Brief," *The Times*, March 23, 1946; "University Education," *The Times*, May 16, 1946).

44. Feilden and Hawthorne, "Sir Frank Whittle."

45. "The Birthday Honours," *The Times*, June 10, 1948. In 1948 he received the Kelvin Medal from the Institution of Electrical Engineers ("The 600th Anniversary Of Caius College," *The Times*, October 8, 1948). He was granted an honorary doctorate at Oxford on the same day as Sir Stafford Cripps ("Encaenia at Oxford Making Latin Fit the Atomic Age," *The Times*, June 23, 1949). Also in 1949, he was presented the Melchett Medal for 1949 by the Institution of Fuel ("News in Brief," *The Times*, October 14, 1949). In 1953, he was awarded the Royal Society of Art's Albert Medal (Feilden and Hawthorne, "Sir Frank Whittle").

46. The award was followed by a lecture tour, after which Whittle publicly reported that American engine development lagged two years behind that in Britain ("American Honour for Air Commodore Whittle," *The Times*, August 19, 1946; "Jet-Propelled Aircraft Pioneer on American Development," *The Times*, September 19, 1946).

47. Feilden and Hawthorne, "Sir Frank Whittle."

48. NA AVIA 10/412, January 11, 1944, Letter Whittle to Cripps.

49. "News in Brief," *The Times*, June 6, 1946.

50. Golley, Whittle, and Gunston, *Whittle, the True Story*, 266.

51. NA AVIA 22/1540, November 4, 1939, 1A, Air Ministry (Supply)'s view on need for tribunal.

52. NA AVIA 22/1540 and NA AVIA 65/446; *Nature* 158, no. 443 (September 28, 1946); "Award to JET Engine Inventor; Sir Frank Whittle's Explanation," *The Times*, January 16, 1952.

53. April 5, 1945, Letter Sir Harold Scott, Secretary of the Ministry of Supply, to Whittle, reprinted in Whittle, *Jet: The Story of a Pioneer*, 2nd ed., 294.

54. NA T 166, Royal Commission on Awards to Inventors (Cohen Commission).

55. The first Royal Commission on Awards to Inventors was established in 1919. Interestingly, a similar number of cases were heard by the first and second Royal Commissions (444 of 1,834 claims over almost fifteen years versus 448 cases in about eight years), yet significantly less money was paid out by the second commission. The first commission awarded about £1,500,000 total ("Progress on Awards to Inventors Commission Nearing End of Work," *The Times*, January 12, 1954).

56. NA T 166/5, Pickles' case to the Royal Commission on Awards to Inventors, and Whittle, *Jet: The Story of a Pioneer*, 2nd ed., 276.

57. "Award to JET Engine Inventor; Sir Frank Whittle's Explanation," *The Times*, January 16, 1952; Whittle, *Jet: The Story of a Pioneer*, 2nd ed., 308.

58. NA AVIA 65/446, Royal Commission on Awards to Inventors Progress report; "£94,600 Awards for Radar; Sir R. Watson-Watt to Get £50,000," *The Times*, January 15, 1952.

59. Pickles had been director of contracts in the MAP during the war and had served as a chairman of Power Jets (R&D). In 1948, he was an undersecretary at the Ministry of Supply (NA T 166/5, May 26, 1948, Transcript of Royal Commission on Awards to Inventors hearing in the matter of "Inventions made by Frank Whittle").

60. NA T 166/5, Capt. J. H. Noad, Transcripts of proceedings and confidential reports.

61. NA T 166/5, May 26, 1948, Transcript of Royal Commission on Awards to Inventors hearing in the matter of "Inventions made by Frank Whittle."

62. Figures around £100,000 had been earlier proposed within the Ministry of Supply (NA T 225/28, Whittle Case); "Award to JET Engine Inventor; Sir Frank Whittle's Explanation," *The Times*, January 16, 1952.

63. Whittle, *Jet: The Story of a Pioneer*, 2nd ed., 304–6.

64. "£100,000 for Air Cdre. Whittle JET Engine Inventor (News)," *The Times*, May 28, 1948.

65. Whittle made the official endorsement of this interpretation a requirement of giving up his shares to avoid "the risk of going down in history as a fool" (Whittle, *Jet: The Story of a Pioneer*, 2nd ed., 274).

66. Whittle later emphasized that he had had not applied to the Royal Commission for an award and that the award was not meant as reimbursement for his shares (Whittle, *Jet: The Story of a Pioneer*, 2nd ed., 306; "Award to JET Engine Inventor; Sir Frank Whittle's Explanation," *The Times*, January 16, 1952).

67. "11 Inventors Claim Tanks: Royal Commission in England Will Decide Who Was First," *New York Times*, October 7, 1919.

68. The notes of the Royal Commission on Awards to Inventors on invention, arising from their work, are lengthy (Royal Commission on Awards to Inventors Report, HMSO, 1953).

69. "£100,000 for Air Cdre. Whittle JET Engine Inventor," *The Times*, May 28, 1948; "£94,600 Awards for Radar Sir R. Watson-Watt to Get £50,000," *The Times*, January 15, 1952; Pickles equated Whittle's achievement with that of Stephenson, Parsons, and Watt (NA T 166/5, May 26, 1948, Transcript of Royal Commission on Awards to Inventors hearing in the matter of "Inventions made by Frank Whittle").

70. "Progress on Awards to Inventors Commission Nearing End of Work," *The Times*, January 12, 1954; in addition to Watson-Watt's award for the development of radar, almost £50,000 was granted to various other men for work on radar installations, bringing the total to almost £95,000 (NA AVIA 65/446, Cases submitted to Investigating Committee of Royal Commission on Awards to Inventors).

71. "£94,600 Awards for Radar Sir R. Watson-Watt to Get £50,000," *The Times*, January 15, 1952.

72. NA T 225/28, June 18, 1948, letter from WC Pearce; reply from Glenvil Hall, MP, Treasury.

73. NA CAB 103/260, October 22, 1947, Letter Williams-Thompson (CIO, the Ministry of Supply) to Luke (Cabinet Office).

74. INF 6/408, February 8, 1950, Crown Job No. 118.

75. Ibid.

76. The Unit was not a government department itself, although it was sponsored by several. BFI Screenonline, "Crown Film Unit."

77. Whittle, *Jet: The Story of a Pioneer*, 2nd ed., 61.

78. NA CAB 102/393, Keppel Narrative—Development of Jet Propulsion and Gas Turbine Engines in the UK.

79. NA CAB 103/260, October 6, 1947, Hubback, Note for Record.

80. Postan edited Keppel's case history, removing its most judgmental passages, before its eventual publication. A draft of Keppel's account can be found in NA CAB 102/393.

81. Interview with Keppel, March 13, 2009.

82. Postan, Hay, and Scott, *Design and Development of Weapons*.

83. INF 6/408, February 8, 1950, Crown Job No. 118.

84. "First Jet Engine for U.S. Museum Gift Handed Over," *The Times*, November 9, 1949.

85. "The First Jet Engines: Gifts to Britain and the United States," *The Times*, October 27, 1949.

86. FO 371/74280, March 10, 1949, Letter Johnson to Secretary of State for Foreign Affairs (USA Division).

87. NASM Registrar Files, Accession file A19500082000, Presentation at Ceremonies Attending the Presentation of the W1X Engine; the engine was gifted by the British government, not Power Jets ("First turbojet for the nation—and the W1X for the American Smithsonian Institution," *Flight*, November 3, 1949).

88. Saxon, "Robert O. Schlaifer"; Feinberg, "The Early Statistical Years."

89. Schlaifer and Heron, *Development of Aircraft Engines*.

90. Ibid., vii–vii, sponsors' foreword.

91. NA CAB 103/260, Preparation of a narrative on the development of jet propulsion and gas turbines in the UK.

92. Schlaifer interviewed Whittle "at some length" in late 1947. Schlaifer's drafts were commented on by W. E. P. Johnson, Lees, Tinling, and Roxbee Cox (NASMT, Box 1, Folder 12). Whittle was unhappy about an earlier draft, but positive about the published version: "may I congratulate you on your success in sorting out the wheat from so much chaff" (CAC WHTL A.320, January 9, 1950, Letter Whittle to Schlaifer). Schlaifer's German sources included at least Hans A. Mauch, ex-German Air Ministry, whom he interviewed in Dayton (DM NL172/2, June 29, 1980, Conversation with Wagner and Mauch).

93. For example, Schlaifer used the case of Power Jets to demonstrate the need for government funding in the early stages of an innovation. He argued that Falk and Co. would have balked at funding the project if they'd better understood it. Nahum has persuasively argued that the project was a surrogate official project from the start (Nahum, *Frank Whittle*, 33).

94. Including Constant, *Origins of the Turbojet Revolution*; Schabel, *Illusion der Wunderwaffen*; Nahum, "World War to Cold War"; and Pavelec, "The Development of Turbojet Aircraft in Germany, Britain, and the United States."

95. Whittle, *Jet: The Story of a Pioneer*, 2nd ed., preface.

96. Ibid., back cover.

97. Jones, *Jet Pioneers: The Birth of Jet-Powered Flight*, used similar tropes to the author's earlier work on Barnes-Wallace (Zaidi, "Janus-Face of Techno-Nationalism").

98. "Forgotten jet pioneer," *Guardian*, December 24, 2007; see also Dyson, "Britain Can Lead in Technology, but It Must Change."

99. Golley and Gunston, *Jet: Frank Whittle and the Invention of the Jet Engine*, 5th impression; Golley's book was initially published after the fiftieth anniversary of the running of Whittle's first engine, in 1987. Another edition appeared after his death in 1996.

100. A review of current practice was given immediately after the war through a series of articles on gas turbine aero-engines in War Emergency Issue 12 of

the *Proceedings of the Institution of Mechanical Engineers*; other early contributions included Whittle, "Early History of the Whittle Jet Propulsion Gas Turbine"; Smith, "Development of an Axial Flow Gas Turbine for Jet Propulsion"; Moult, "The Development of the Goblin Engine." Such work continued to be published throughout the twentieth century: Hawthorne, "Aircraft Propulsion from the Back Room"; Roxbee Cox, "Beginnings of Jet Propulsion"; Hawthorne, "Early History of the Aircraft Gas Turbine in Britain"; Windley, "Gas Turbine Combustion: A Personal Perspective."

Public lectures included: Roxbee Cox, "British Aircraft Gas Turbines: The Ninth Wright Brothers Lecture"; Hooker, "Cantor Lectures: 'High Speed Flight'"; Halford, "Jet Propulsion"; MOSI MS0539/39, January 9, 1963, D. M. Smith, "The Development of the Jet Engine," lecture to the Women's Engineering Society, Manchester Branch.

Early textbooks included: Keenan, *Elementary Theory of Gas Turbines and Jet Propulsion*; Smith, *Gas Turbines and Jet Propulsion for Aircraft*.

101. Biographical memoirs appeared for engineers who were made Fellows of the Royal Society: Rubbra, "Alan Arnold Griffith"; Hawthorne, Cohen, and Howell, "Hayne Constant"; Saunders, "David Macleish Smith"; Feilden and Hawthorne, "Sir Frank Whittle, O.M., K.B.E"; Feilden, "Lionel Haworth." Book-length biographies were written of Whittle (Rowland, The Jet Man; Golley, Whittle, and Gunston, Whittle: The True Story; Golley and Gunston, Genesis of the Jet) and von Ohain (Conner, Elegance in Flight). See also Taylor, *Boxkite to Jet*.

The most influential autobiographical accounts included: Grierson, *Jet Flight*; Whittle, *Jet: The Story of a Pioneer*, 2nd ed.; Whyte, "Whittle and the Jet Adventure," and *Focus and Diversions*; Hooker, *Not Much of an Engineer*, and Bulman, *An Account of Partnership*. Such stories often derive their historical authority from the technical expertise of their authors and generally dwell only briefly on personal details, similar to the "failed autobiographies" described by Oldenziel (Oldenziel, *Making Technology Masculine*, 92–123). Boyne and Lopez, *The Jet Age*, includes articles by Frank Whittle and Hans von Ohain among other pioneers.

Articles on the jet appeared again in Britain on commemorations of important jet dates: Armstrong, "The Aero Engine and Its Progress—Fifty Years after Griffith"; Howell, "Griffith's Early Ideas on Turbomachinery Aerodynamics"; Nahum, "Origins of the British Jet Engine"; Hawthorne, "Early History of the Aircraft Gas Turbine in Britain"; Golley, "Whittle Revolution"; Golley, "Working with a Genius"; Jones, "Jubilee for the Jet Engine"; Hawthorne, "Early History of the Aircraft Gas Turbine in Britain." The magazine of the Rolls-Royce Heritage Trust, *Archive*, is a good source for numerous articles on engines and individuals.

102. Taylor, *Boxkite to Jet*.

103. Roxbee Cox, "British Aircraft Gas Turbines," 53.

104. "Designs of a Decade," *Flight*, May 11, 1951.

105. Dennis, *Farnborough's Jets*.

106. Hawthorne, "Aircraft Propulsion from the Back Room."

107. Crowther, *Six Great Inventors: Watt, Stephenson, Edison, Marconi, Wright Brothers, Whittle*.

108. Bulman, *An Account of Partnership*, 306.

109. For example, the high altitude wind tunnel at BMW in Munich; Neufeld, "Nazi Aerospace Exodus."

110. A good listing of German wartime reports as well as Allied intelligence documents, both sources of information drawn from interrogations, was given in Gersdorff and Grasmann, *Flugmotoren und Strahltriebwerke*.

111. German work was also reported on by British visitors, often members of the British aero-engine industry. Notable among these was the aero-engine designer Roy Fedden, whose reflections were publicized in the trade journal *Flight* in 1945 (Fedden, "Inquest on Chaos" and "German Jet Development").

112. Edgerton, *Shock of the Old*, 146.

113. Ermenc, *Interviews with German Contributors*, 112. The Air Ministry's release on the Meteor in January 1945 claimed that the E.28/39 was "the first turbine jet aircraft in the world to fly (in May 1941)" (NA AIR 2/7070, February 28, 1945, 177a, "RAF [Royal Air Force] 'jet' fighters in action").

114. The Allies captured their first Me 262 in September 1944, and various British companies and establishments had carried out extensive tests on the aircraft's two Jumo 004 engines, which showed them to be inferior to the British jet engines, as the third release made known (NA AVIA 15/2121, September 13, 1944, 2A, Report on crashed enemy aircraft, first example of Me 262).

115. Whittle, *Jet: The Story of a Pioneer*, 2nd ed., 57.

116. Ibid., 289.

117. Conner, *Elegance in Flight*, 159.

118. *Jet Power*, 1952. Produced by General Electric Film. NASM Archive VE00932.

119. Neville and Silsbee, *Jet Propulsion Progress*.

120. Heiman, *Jet Pioneers*. A similar narrative can be found in the more recent Pavelec, *Jet Race*.

121. Conner, *Elegance in Flight*.

122. Ibid., 251.

123. https://www.aiaa.org/HonorsAndAwardsRecipientsList.aspx?awardId= 9e15dd84-e9ac-48b0-9849-9c7575ecc05c Accessed on December 2, 2015.

124. NASM Photograph, SI 74-11297-07, October 24, 1974.

125. A transcript of their public, joint interview is given in Whittle and von Ohain, *An Encounter between the Jet Engine Inventors*. The two men also appeared together the following day at the US Air Force Museum in Dayton, Ohio. On November 19, 1986, the two inventors reprised their roles as part of the US Air Force Museum's guest lecture series with a lecture titled "The Birth of the Jet Engine in Britain and the Development of the First Turbo Engine in Germany."

126. The proceedings were published, edited by Walter Boyne and Donald Lopez, in *The Jet Age*.

127. The unedited film footage of the event is available in the NASM archive. An edited version was released publicly.

128. Conner, *Elegance in Flight*, 253. According to the website of the National Academy of Engineering, the joint award was given "for their independent development of the turbojet engine" ("NAE Website—Previous Recipients of the Charles Stark Draper Prize").

129. Conner, *Elegance in Flight*, 261.

130. For more on the gallery and its political ramifications, see Giffard, "Politics of Donating Technological Artifacts," 61–82.

131. Giffard, "Politics of Donating Technological Artifacts."

132. The diplomatic importance of the engine's journey was clear in the carefully choreographed events that accompanied the engine's departure from London and arrival at the Smithsonian. At the ceremony that sent off the W1X in London, Power Jets Ltd. presented the W1, the company's first flight engine, to the Science Museum in London (NASM Registrar Files, Accession file A19500082000, Copy of letter from Johnson to Smithsonian, 59).

133. "First Jet Engine for U.S. Museum Gift Handed Over," *The Times*, November 9, 1949.

134. NASM Registrar Files, Accession file A19500082000; NASM Photograph SI-75-16332, ca. October 24, 1974.

135. WS MS-335, Box 4, File 1, August 25, 1972 Letter Meyer to Morrison.

136. Conner, *Elegance in Flight*, 254.

137. Giffard, "Politics of Donating Technological Artifacts."

138. Conner, *Elegance in Flight*, 235.

139. Oestrich, "Die Entwicklung der Fluggasturbine bei den Bayrischen"; Decher, "Die Entwicklung des Triebwerkes Jumo 004 bei den Junkers Flugzeug- und Motorenwerke in Dessau"; Gundermann, "Das Erste Düsenflugzeug der Welt; Heinkel and Thorwald, *Stürmisches Leben* and *He 1000*; H. Scherenberg, "Die Flugmotorenentwicklung bei Daimler-Benz"; Münzberg, "Nachruf Auf Hermann Oestrich"; Franz, "Development of the 'Jumo 004' Turbojet Engine," and *From Jets to Tanks*; Bentele, *Engine Revolutions*. Bentele remarked that two versions of the jet history were created after the war. The fuller one circulated among professionals and jet engineers but did not reach the public (Bentele, *Engine Revolutions*, 65).

140. Lipstadt, *Eichmann Trial*.

141. Rieger, *People's Car*.

142. See the many volumes of the book series Der Deutschen Luftfahrt, Bernard & Graefe Verlag, Bonn; Giffard, "Politics of Donating Technological Artifacts."

143. The Deutsches Museum embraced this notion in its appeal to NASM for

any "important, original artifacts accredited to German designers in excess of NASM needs and for which original examples no longer exist in Germany," which might be displayed in Munich (NASM Inventory number: A19810039000, December 9, 1980, Letter Pancoast to Meyer).

144. The fourth edition of Gersdorff and Grasmann's *Flugmotoren und Strahltriebwerke* was published in 2006.

145. Kay, *Turbojet History and Development*. His two volumes treat (1) Great Britain and Germany and (2) the USSR, USA, Japan, France, Canada, Sweden, Switzerland, Italy, and Hungary. Both cover 1930–60 and include both aircraft and engines.

146. DM FA 001/1072, *Göttenberger Tagblatt*, May 1981; *Frankfurter Allgemeine*, December 14, 1981; Janda, "Oma finanzierte den ersten Jet."

147. Von Ohain's speech was published with a brief introduction by Walter Rathjen in the Deutsches Museum publication, *Technik & Kultur*, in 1981 (Rathjen, "Weniger Gerausch, Keine Vibration!"). Rathjen's introduction made clear that a copy of the engine would sit beside Whittle's engine in the Smithsonian Institution's National Air and Space Museum. In fact, von Ohain had already redrawn the engine—in part at his own expense—for an earlier, similar plan of the Smithsonian Institution, which was never realized (Conner, *Elegance in Flight*, 254).

148. In the 1970s and 1980s, von Ohain received inquiries from the authors Gersdorff, Köhler, and Wagner, all of whom he continued to correspond with from the United States and whom he met later in Germany (NASM VO Box 3, 4, and 12).

149. Von Ohain's correspondents included Schelp and Wagner (NASM VO Box 4, 11 and 16).

150. Constant's book was published in 1980. It was based on his 1977 dissertation from Northwestern University. On May 10, 1970, as a graduate student in history at Tulane, he wrote to Whittle asking for information for his "history of the development of the turbo-jet engine" (CAC WHTL A.325, May 10, 1970, Letter Edward Constant to Sir Frank Whittle). He received responses from Whittle (May 20), von Ohain (July 31), and Wagner (8September 18 and October 26) during 1970 (Constant, *Origins of the Turbojet Revolution*, 296–97).

151. Constant, *Origins of the Turbojet Revolution*, 114 and 218.

152. Schabel, *Illusion der Wunderwaffen*.

153. Ibid., 289–90. In fact, Göring intentionally promoted the split (Overy, *Goering: The "Iron Man,"* 164–204).

154. Nahum, *Frank Whittle*; this book was written for a popular audience but was based on a chapter of his thesis (Nahum, "World War to Cold War," 54–121).

155. Janda, "Oma finanzierte den ersten Jet."

156. Heiman, *Jet Pioneers*.

Conclusion

1. Hughes, *American Genesis*, 381–421; Edgerton, *Shock of the Old*, 260.

2. Kay, *Turbojet History and Development*, 2: 228–49.

3. Edgerton, *Warfare State*; Peden, *Arms, Economics, and British Strategy*; Edgerton, *England and the Aeroplane*; Tooze, *Wages of Destruction*; Rieger, *The People's Car*, 14.

4. Robert Schlaifer argued already in 1950 that for this reason, the aero-engine industry had played the central role in jet engine development. This book has gone further to show in detail how well-known centrifugal compressors, exhaust nozzles, high temperature alloys, fuel systems, turbines, and (in Germany) turbine blade cooling from piston engine practice, as well as development methods more generally, were in important cases usefully transferred from piston engine to turbojet practice by aero-engine firms. Cf. Constant, *Origins of the Turbojet Revolution*, 18. I agree with Constant that "there was [no] simple one-to-one transfer from prior technology to the turbojet," but I interpret the notion of relevant "prior technologies" more broadly and disagree that these transfers tended to "only exacerbate an abstruse decision process," arguing that they notably eased it.

5. Scranton, "Technology, Science, and American Innovation."

6. Pestre, "Commemorative Practices at CERN."

7. Scranton, "Technology-Led Innovation"; Scranton, "Technology, Science, and American Innovation"; see also Nichelson, "Early Jet Engines and the Transition from Centrifugal to Axial Compressors."

8. Vincenti, *What Engineers Know and How They Know It*; Nelson, Peck, and Kalachek, *Technology, Economic Growth, and Public Policy*, 37–43.

9. Fox and Guagnini, *Laboratories, Workshops, and Sites*.

Glossary

Axial Combustion Chamber: A combustion chamber that is a single, radial unit capable of taking the (compressed) airflow of an entire engine. Axial chambers were used rather than separate combustion chambers arrayed around the central axle, each of which takes a proportional part of the incoming air. Smaller chambers were easier to test than larger ones because they required smaller streams of air, and it was difficult to produce a large enough stream of air for a single, axial combustion chamber outside of an engine.

Axial Compressor: A rotating compressor in which the individual blades act like airfoils (in the best case) on the incoming stream of gas (air); these compressors are known as "axial" because the gas flows primarily parallel to the axis of rotation. In an axial compressor, the energy and thus the pressure of the gas increases as the air moves through the compressor due to the action of the rotor blades, typically made up of alternating rows of moving and stationary rotors with increasingly smaller spacing and radius (as the air takes up less volume). Axial flow compressors produce a continuous flow of compressed gas. In the 1930s, they theoretically offered higher efficiency and larger mass flow rates than centrifugal compressors (which also had larger radiuses, causing greater air resistance). The fact that these compressors require several rows of airfoils to impart a large pressure rise makes these compressors complex and expensive relative to centrifugal compressors.

Blade: The part of a compressor or turbine on which the air stream impinges. A blade can either extract energy to drive the rotor or shaft, as in a turbine, or do work on the air, compressing it to higher pressures.

Bypass Turbojet: An engine in which part of the incoming air that enters the engine bypasses the gas turbine, and is mixed directly with the gas turbine's exhaust gases without being compressed or heated. The turbine in these engines tends to be hotter than gas turbine engines without a bypass flow.

Centrifugal Supercharger: Centrifugal refers to the compressor used in the supercharger; generally, early superchargers all used centrifugal compressors (see figure 2.7).

Combustion Chamber: The part of a gas turbine engine where fuel (often kerosene in early jet engines) is mixed with air and burned.

Compound Engine: An engine that has more than one stage for recovering energy from the same working fluid (air stream). The stages may be of differing or similar technologies. The Napier Nomad, for example, coupled a piston engine to a turbine to drive a single propeller.

Compound Jet Engine: An engine that uses multiple turbines to extract energy from an exhaust stream. Such turbines could be used to drive different stages of a compressor at different speeds, thus increasing efficiency, but requiring manufacturers to build shafts that passed concentrically through other shafts.

Compressor: A mechanical device that increases the pressure of a gas (or other fluid) by reducing its volume; similar to pumps, both of which increase the pressure of a fluid and can transport it through a pipe, but because gases are more compressible than liquids, compressors also reduce gas volume.

Compressor Blade Cooling: A method of cooling turbine blades, which because they are located directly downstream of the combustion chamber in a gas turbine, operate in the hottest part of the engine. Cooling allows metals to operate for longer periods of time at high temperatures without losing their shape.

Compressor Stage: A single rotor of a compressor, a disc carrying a row of blades—axial or centrifugal. Each stage of a compressor (multiple in axial compressors) might be powered by an independent turbine or they might all be powered by a single turbine. Low-pressure stages of a compressor are located closer to the engine's air intake, high-pressure stages come directly before the combustion chamber, when the incoming air is already compressed.

Compressor Stall or Surge: A disruption of airflow through a compressor, can range from a momentary power drop to a complete loss of compression (a surge)—a big problem in early jet engine designs, which had manual or mechanical fuel control units.

Contra-Flow Engine: Engine layout in which the air enters at the back of the engine, goes through the compressor, reverses direction in the combustion chamber (at the front of the engine), and flows backward through the turbine, exiting at the back of the engine.

Contra-Rotating Compressor: An axial compressor design in which every other compressor stage rotates in the opposite direction to the one before/after it. A contra-rotating compressor has no stators. This is theoretically advantageous because it avoids stalling.

Contra-Rotating Propeller: Propellers are a type of fan that converts rotational motion into thrust; contra-rotating propellers are a pair of propellers driven in opposite directions about the same axis; the two propellers are arranged one behind the other, on a single shaft. The advantage of contra-rotating propellers is that they do not impart a rotation to an aircraft, they offer increased efficiency because energy is not lost tangentially as with a single propeller.

Ducted Fan: A fan mounted in a shroud or duct, located before the compressor; the duct reduces losses from the tips of the fan blades while the duct's cross section can be varied in order to change the velocity and pressure of the incoming airflow, thereby increasing the efficiency of the compressor.

Fir Tree Blade Root: First developed by Power Jets, the so-called fir blade root allows the base of a turbine blade to expand as it heats to operating temperature. When expanded, the fir tree profile (see figure 2.2) allows for a better transfer of power and heat between the blade and the rotor than other types of blade root, like a bulb root, for example.

Free Turbine: A turbine that extracts energy from a fluid flow produced by a machine, a piston engine for example, independent of (so not linked to) the turbine (as in most jet engines).

Frontal Area: The area that an aero-engine presents to the incoming air current, related to the engine's aerodynamic drag. In general, a smaller frontal area implies less drag. Compressors with a smaller cross-sectional area (axial), thus, would allow the construction of engines with smaller cross-section engines and thus less drag.

ICT, Internal Combustion Turbine: Another name for a gas turbine.

Jet Engine or Gas Turbine: A reaction engine that generates thrust via jet propulsion, generally combustion engines containing a compressor, combustion chamber, and turbine. Gas turbines are characterized by axial symmetry; they work continuously.

Jet Propulsion: Thrust produced by passing a jet of matter (typically air or water) in the opposite direction to the direction of motion. In the case of aero-propulsion, Newton's third law implies that propelling a jet of air will produce an equal and opposite force on the engine, or the plane.

Piston Engine or Reciprocating Engine: An engine that converts pressure into rotational motive power using pistons; each piston takes in air, combusts it, and expels it. Useful power is produced only during the stroke following combustion.

Power-to-Weight Ratio: A common measure for aero-engines expressing the ratio of the power produced by the engine to the weight of the engine. Weight is of crucial importance in aviation manufacturing.

Pressure Ratio: The comparison of the pressure to which a compressor or engine compresses air to the atmospheric air pressure at which the air enters said device.

Radial or Centrifugal Compressor: A compressor in which compression occurs by forcing gas outward. The gas stream leaving a centrifugal compressor has a velocity with a "radial component." The compressor achieves a pressure rise by adding kinetic energy to the gas flowing through the "impeller" rotor, which is then converted into increased pressure by slowing the flow through a "diffuser," the pressure rise in the impeller is generally almost equal to that in the diffuser; the pressure rise achievable by a centrifugal compressor increases with its size.

Radial Turbine: A turbine in which the flow of the gas is radial to the shaft. Whereas for an axial turbine the rotor is impacted by the airflow coming from the compressor, in a radial compressor the air flow drives the turbine in the same way as water drives a watermill, resulting in less mechanical and thermal stress, which enables a radial turbine to be simpler, more robust, and more efficient than an axial; in high power ranges, however, radial turbines no longer have advantages over axial turbines.

Reverse-Flow: Connotes that air enters at the back of the component and moves forward rather than entering at the front.

Stall: A sudden stopping of an engine or compressor, usually accidentally, often caused in turbojets by a lack of airflow.

Stator: Compressor stages that are stationary. Stators do work on air by compressing it; they slow the incoming gas stream, turning the circumferential component of flow into pressure. Also appears as "stator blade," meaning a blade on a stationary rotor.

Supercharger: A device that compresses air before it enters an internal combustion engine, giving the engine more oxygen, which allows it to burn more fuel and do more work, increasing its power and efficiency. Superchargers are particularly advantageous for aero-engines at higher altitudes, where atmospheric oxygen is less concentrated than at sea level. At that elevation, a supercharger provides an aero-engine with enough oxygen to continue to efficiently carry out combustion at altitude. The main component of a supercharger is an air compressor, which can be powered by a turbine or mechanically driven by the engine (often through a belt connected to the crankshaft). Superchargers also improve performance at lower altitudes because there too they can force additional oxygen into an engine's combustion chamber, thus allowing the engine to burn more fuel than it would be able to if relying on atmospheric pressure alone.

Surge: A complete loss of compression in a turbojet—a big problem in early jet engine designs.

Thrust: A turbojet produces thrust by moving a mass of air backward; the thrust varies with speed and altitude.

Turbine: Takes power out of an engine's exhaust stream, often in order to power a compressor or propeller.

Turbo Machinery: A machine that uses a rotor to extract or impart energy from or to a fluid flow (in this case air). This term can refer to a compressor, a turbine, or a machine that includes one of the two or both.

Turbo Supercharger: Also known as turbochargers, a supercharger that is powered by a turbine located in an engine's exhaust stream, which contains useful energy that can be used to do work.

Turboprop: An aero-engine in which a gas turbine produces thrust by driving a propeller, these engines perform better at slower speeds than pure jet engines and offer higher fuel efficiency. In a turboprop, the propeller is powered by a turbine in the gas turbine unit's exhaust stream, a reduction gear is used to convert the high rotational, low torque output of the turbine to the lower rotation, high torque output suited to the propeller.

Bibliography

Archives and Papers

Air Force Historical Research Agency, Maxwell Air Force Base, Alabama
    AFHR A2065: Case History of XP-59, XP-59A, YP-59A, P-59A, P-59B, and
        XP-59B Airplanes
    AFHR A2072: Case History of Turbojet Engine Halford (De Havilland
        H-1)
    AFHR A2073: Case History of Whittle Turbojet Engine (and Supplement)
    AFHR A2078: Case History of Turbojet Engine J33 Series
Archives of the Deutsches Museum, Munich
    DM FA001: Firmenarchiv Heinkel Werke
    DM NL172: Nachlass Herbert Wagner (1900–1982)
Archives of the Deutsches Zentrum für Luft- und Raumfahrt (German Aerospace
    Center), Göttingen
Archives of the Manchester Museum of Science and Industry
Archives of the Rolls-Royce Heritage Trust
    RR DMB: Directors Minute Books, Derby
    RR HC: Hives Correspondence, Derby
    RR L: Lappin files, Bristol
Archives of the Smithsonian Institution, National Air and Space Museum
    NASM R: Registrar's Files
    NASM T: Tinling Papers
    NASM VO: Von Ohain Papers
British Library
    BL Add 61931: Power Jets Limited, Minutes of directors' and sharehold-
        ers' meetings
BMW Corporate Archive, Munich
Churchill Archives Centre, Churchill College Cambridge
    CAC WHTL: Papers of Sir Frank Whittle

Daimler-Benz Corporate Archive, Stuttgart
Das Bundesarchiv (German National Archives)
 Bundesarchiv, Lichterfelde
 BA RL3: Generalluftzeugmeister
 Bundesarchiv-Militärarchiv, Freiburg
 BA MA N653: Nachlass Hertel, Walter (1898–1983)
 BA MA NS19: Persönlicher Stab Reichsführer-SS
 BA MA R3: Reichsministerium für Rüstung und Kriegsproduktion
Imperial War Museum
 IWM FD: Speer / Foreign Documents Collection
Kings Norton Archive, Cranfield University
National Archives, London
 NA AIR: Records created or inherited by the Air Ministry, the Royal Air
  Force, and related bodies
 NA AIR 62: Air Commodore Sir Frank Whittle: Papers
 NA AVIA: Records created or inherited by the Ministry of Aviation and
  successors, the Air Registration Board, and related bodies
 NA AVIA 13: Royal Aircraft Establishment Registered Files
 NA AVIA 15: Ministry of Aircraft Production and predecessor and succes-
  sors: Registered Files
 NA AVIA 31: Records of the National Gas Turbine Establishment and pre-
  decessor
 NA CAB: Records of the Cabinet Office
 NA FO: Records created and inherited by the Foreign Office
 NA INF: Records created or inherited by the Central Office of Information
 NA PREM: Records of the Prime Minister's Office
 NA T: Records created and inherited by HM Treasury
National Archives and Records Administration, College Park, Maryland
 NARA RG 18, Box 42, Case History of Turbojet Engine Halford (De Havil-
  land H-1)
 NARA RG 18, Box 42, Case History of Turbojet Engine J33 Series
 NARA RG 18, Entry 22, Boxes 42-48, Materiel Division R&D
 NARA RG 18, Entry 22, Box 49, Case History of Whittle Turbojet Engine
  (and Supplement)
 NARA RG 255, Entry 8, Box 119, Papers of the Office of Scientific Re-
  search and Development, Special Sub-Committee on Jet Propulsion
 NARA RG 227-130-22, Box 83, Papers of the National Advisory Commit-
  tee for Aeronautics, Travel Files, Eastman Jacobs
Wright State University, Special Collections and Archives
 WS MS-335 Dr. Hans von Ohain Papers

Newspapers and Periodicals

Archive, published by the Rolls-Royce Heritage Trust
Biographical Memoirs of Fellows of the Royal Society
Flight
Flight International
Proceedings of the Institution of Mechanical Engineers
The Times

Government Documents

USSBS	United States Strategic Bombing Survey Reports, European Theater
CIOS	Combined Intelligence Objectives Sub-Committee Reports, HMSO
HANSARD	Daily Debates, House of Commons and House of Lords
RCAI	Final Report of the Royal Commission on Awards to Inventors, HMSO
NACA TR	National Advisory Committee for Aeronautics, Technical Reports

Interviews

Cynthia Keppel, March 13, 2009, research assistant to the official historian M. M. Postan during World War II
James Foulds, March 29, 2010, Barnoldswick employee beginning on September 21, 1942

Other Sources

"75 Years: MTU Aero Engines Celebrates Anniversary—An Established Player in the Engine Industry for Decades." Accessed May 21, 2012, www.gbjyearbook .com/yb_news_story.html?release=4013.
"Aircraft Engine History." Accessed August 29, 2012, www.geaviation.com /aboutgeae/history.html#section3.
Aitken, Hugh G. J. *The Continuous Wave: Technology and American Radio, 1900– 1932*. Princeton, NJ: Princeton University Press, 1985.
———. "Reviewed Work: Networks of Power: Electrification in Western Society by Thomas P. Hughes." *Journal of Interdisciplinary History* 15, no. 2 (1984): 350–51.

————. *Syntony and Spark: The Origins of Radio*. New York: Wiley, 1976.

Allen, Michael Thad. *The Business of Genocide: The SS, Slave Labor, and the Concentration Camps*. Chapel Hill: University of North Carolina Press, 2002.

Anderson, John. *Introduction to Flight*. New York: McGraw-Hill, 2008.

Armstrong, F. W. "The Aero Engine and Its Progress—Fifty Years after Griffith." *Aeronautical Journal of the Royal Aeronautical Society* 80 (1976): 499–520.

Arnold, Henry H. *Global Mission*. Blue Ridge Summit, PA: Tab Books, 1989.

Ashworth, William. *Contracts and Finance*. History of the Second World War, United Kingdom Civil Series. London: HMSO, 1953.

Bailey, Bill, and Ian Whittle. "The Early Development of the Aircraft Jet Engine." Privately published, 2004. Churchill College Archives Centre, Cambridge.

Baldwin, Neil. "The Lesser Known Edison: In Addition to His Famous Inventions, Thomas Edison's Fertile Imagination Gave the World a Host of Little Known Technologies, from Talking Dolls to Poured-Concrete House." *Scientific American* 276, no. 2 (1997): 62–67.

Banks, F. R. *I Kept No Diary: 60 Years with Marine Diesels, Automobile and Aero Engines*. Shrewsbury: Airlife Publications, 1978.

Basalla, George. *The Evolution of Technology*. Cambridge: Cambridge University Press, 1988.

Baxter, A. D. *Professional Aero Engineer, Novice Civil Servant*. Sussex: Book Guild, 1988.

Bentele, Max. *Engine Revolutions: The Autobiography of Max Bentele*. Warrendale, PA: Society of Automotive Engineers, 1991.

Béon, Yves, and Michael J. Neufeld. *Planet Dora: A Memoir of the Holocaust and the Birth of the Space Age*. Boulder, CO: Westview Press, 1997.

Berry, Peter. "R-R W2B." Accessed September 13, 2010, http://www.enginehistory.org/r-r_w2b.htm.

BFI Screenonline. "Crown Film Unit." Accessed December 2, 2007, http://www.screenonline.org.uk/film/id/469778/.

Bijker, Wiebe E. *Of Bicycles, Bakelites, and Bulbs: Toward a Theory of Sociotechnical Change*. Inside Technology. Cambridge, MA: MIT Press, 1995.

————. "The Social Construction of Bakelite: Toward a Theory of Invention." In *The Social Construction of Technological Systems: New Directions in the Sociology and History of Technology*, edited by Wiebe Bijker, Thomas Hughes, and Trevor Pinch, 159–87. Cambridge, MA: MIT Press, 1987.

Bijker, Wiebe, Thomas Hughes, and Trevor Pinch. *The Social Construction of Technological Systems: New Directions in the Sociology and History of Technology*. Cambridge, MA: MIT Press, 2012.

Birtles, Philip. *De Havilland*. London: Jane's, 1984.

Bloor, David. *The Enigma of the Aerofoil: Rival Theories in Aerodynamics, 1909–1930*. Chicago: University of Chicago Press, 2011.

Boog, Horst. *Germany and the Second World War: The Strategic Air War in Europe*. Vol. 7. Oxford: Clarendon Press, 2006.

Bowker, Geof. "What's in a Patent?" In *Shaping Technology / Building Society*, edited by Wiebe Bijker and John Law. Cambridge, MA: MIT Press, 1992.

Boyne, Walter J., and Donald S. Lopez, eds. *The Jet Age: Forty Years of Jet Aviation.* Washington, DC: Smithsonian Institution Press, 1979.

Brodie, John L. P. "Frank Bernard Halford 1897–1955." *Journal of the Royal Aeronautical Society* 63 (1959): 194–205.

Brooks, David S. *Vikings at Waterloo: The Wartime Work on the Whittle Jet Engine by the Rover Company.* Historical Series 22. Derby, England: Rolls-Royce Heritage Trust, 1996.

Brown, Louis. *A Radar History of World War II: Technical and Military Imperatives.* Philadelphia: Institute of Physics Publications, 1999.

Bryant, Lynwood. "The Development of the Diesel Engine." *Technology and Culture* 17, no. 3 (1976): 432–46.

Budraß, Lutz. *Flugzeugindustrie und Luftrüstung in Deutschland 1918–1945.* Droste, 1998.

———. "Hans Joachim Pabst Von Ohain. Neue Erkenntnisse Zu Seiner Rolle in Der Nationalsozialistischen Rüstung." In *Friedrich-Ebert-Stiftung, Landesbüro Mecklenburg-Vorpommern (Hg.): Technikgeschichte Kontrovers: Zur Geschichte des Fliegens und des Flugzeugbaus in Mecklenburg-Vorpommern*, 52–69. Schwerin, 2007.

Bulman, George Purvis. *An Account of Partnership-Industry, Government, and the Aero Engine: The Memoirs of George Purvis Bulman.* Edited by M. C. Neale. Derby: Rolls-Royce Heritage Trust, 2002.

Bülow, Ralf. "Ein Blick in die Hölle: Die Aufzeichnungen von Georg Schmertz über die Kriegswaffen-Produktionsstatte Mittelwerk GmbH." *Kultur und Technik Zeitschrift des Deutschen Museums München* 4, no. 19 (1994): 38–41.

Cairncross, Alec. *Planning in Wartime: Aircraft Production in Britain, Germany, and the USA.* Oxford: Macmillan in association with St. Anthony's College, 1991.

Carlson, W. Bernard. "Invention and Evolution: The Case of Edison's Sketches of the Telephone." In *Technological Innovation as an Evolutionary Process*, edited by. J. M. Ziman, 137–58. Cambridge: Cambridge University Press, 2000.

———. "Thomas Edison as a Manager of R&D: The Case of the Alkaline Storage Battery, 1895–1915." *Technology and Society Magazine, IEEE* 7, no. 4 (1988): 4–12.

Carlson, W. Bernard, and Michael E. Gorman. "Interpreting Invention as a Cognitive Process: The Case of Alexander Graham Bell, Thomas Edison, and the Telephone." *Science, Technology, and Human Values* 15, no. 2 (1990): 131–64.

———. "Understanding Invention as a Cognitive Process: The Case of Thomas Edison and Early Motion Pictures, 1888–1891." *Social Studies of Science* 20 (1990): 387–430.

Cheshire, L. J. "The Design and Development of Centrifugal Compressors for Aircraft Gas Turbines." *Proceedings of the Institution of Mechanical Engineers* 153, no. 12 (1945): 426–40.

Clark, Ronald W. *Tizard*. London: Methuen, 1965.

Clarke, P. F. *The Cripps Version: The Life of Sir Stafford Cripps, 1889–1952*. London: Allen Lane, 2002.

Collins, Harry M. "What Is Tacit Knowledge." In *The Practice Turn in Contemporary Theory*, edited by Theodore R. Schatzki, K. Knorr-Cetina, and Eike von Savigny. Abingdon: Routledge, 2001.

Conner, Margaret. *Hans von Ohain: Elegance in Flight*. Reston, VA: American Institute of Aeronautics and Astronautics, 2001.

Connors, Jack. *The Engines of Pratt and Whitney: A Technical History*. Library of Flight. Reston, VA: American Institute of Aeronautics and Astronautics, 2009.

Constant, Edward W. "On the Diversity and Co-Evolution of Technological Multiples: Steam Turbines and Pelton Water Wheels." *Social Studies of Science* 8 (1978): 183–210.

———. *The Origins of the Turbojet Revolution*. Johns Hopkins Studies in the History of Technology, New Series, 5. Baltimore: Johns Hopkins University Press, 1980.

Constant, Hayne. "The Development of the Internal Combustion Turbine." *Proceedings of the Institution of Mechanical Engineers* 153, no. 12 (1945): 409–10.

———. "The Early History of the Axial Type of Gas Turbine Engine." *Proceedings of the Institution of Mechanical Engineers* 153, no. 12 (1945): 411–25.

Craven, Wesley Frank, and James Lea Cate, eds. *The Army Air Forces in World War II*. Vol. 6. Washington, DC: Office of Air Force History, 1983.

Crowther, J. G. *Six Great Inventors: Watt, Stephenson, Edison, Marconi, Wright Brothers, Whittle*. London: H. Hamilton, 1954.

Daso, Dik Alan. *Hap Arnold and the Evolution of American Airpower*. Washington, DC: Smithsonian Institution Press, 2000.

Dawson, Virginia. *Engines and Innovation: Lewis Laboratory and American Propulsion Technology*. Washington, DC: National Aeronautics and Space Administration, Office of Management, Scientific and Technical Information Division, 1991.

Decher, S. "Die Entwicklung des Tribewerkes Jumo 004 bei den Junkers Flugzeug- und Motorenwerke in Dessau." *Flugwelt* 2 (1951): 41–46.

De Havilland. "Development of the De Havilland 'Goblin' Jet Propulsion Engine 1940/1945." n.d.

Dennis, Michael Aaron. "Accounting for Research: New Histories of Corporate Laboratories and the Social History of American Science." *Social Studies of Science* 17, no. 3 (1987): 479–518.

———. "Historiography of Science: An American Perspective." In *Companion to Science in the Twentieth Century*, edited by John Krige and Dominique Pestre, 1–26. London: Routledge, 2003.

Dennis, R. *Farnborough's Jets*. Fleet, Hants: Footmark Publications, 1999.

Dyson, James. "Britain Can Lead in Technology, but It Must Change." *Telegraph .co.uk*, 2010.

Ebert, Hans J., Johann B. Kaiser, and Klaus Peters. *Willy Messerschmitt: Pioneer der Luftfahrt und des Leichtbaues*. Die Deutsche Luftfahrt 17. Koblenz: Bernard and Graefe, 1992.

Edgerton, David. *England and the Aeroplane: An Essay on a Militant and Technological Nation*. Basingstoke: Macmillan in association with the Centre for the History of Science, Technology and Medicine, University of Manchester, 1991.

———. *Industrial Research and Innovation in Business*. Cheltenham: Edward Elgar, 1996.

———. "'The Linear Model' Did Not Exist—Reflections on the History and Historiography of Science and Research in Industry in the Twentieth Century." In *The Science-Industry Nexus: History, Policy, Implications*, edited by K. Grandin, N. Wormbs, and S. Widmalm, 31–57. New York: Watson, 2004.

———. *The Shock of the Old: Technology and Global History since 1900*. London: Profile, 2008.

———. "Technical Innovation, Industrial Capacity, and Efficiency: Public Ownership and the British Military Aircraft Industry, 1935–1948." *Business History* 26, no. 3 (1984): 247–79.

———. *Warfare State: Britain, 1920–1970*. Cambridge: Cambridge University Press, 2006.

Eltscher, Louis R., and Edward M. Young. *Curtiss-Wright: Greatness and Decline*. New York: Twayne; London: Prentice-Hall International, 1998.

Engel, Jeffrey A. *Cold War at 30,000 Feet: The Anglo-American Fight for Aviation Supremacy*. Cambridge, MA: Harvard University Press, 2007.

Epstein, Katherine C. *Torpedo: Inventing the Military-Industrial Complex in the United States and Great Britain*. Cambridge, MA: Harvard University Press, 2014.

Ermenc, Joseph J. *Interviews with German Contributors to Aviation History*. Westport, CT: Meckler, 1990.

Eyre, Donald. *50 Years with Rolls-Royce: My Reminiscences*. Historical Series. Derby: Rolls-Royce Heritage Trust, 2005.

Fedden, Roy. "German Jet Development." *Flight* (December 13, 1945): 626–29.

———. "German Piston-Engine Progress." *Flight* (December 6, 1945): 602–4.

———. "Inquest on Chaos." *Flight* (November 19, 1945): 575–78.

Feilden, G. B. R. "Lionel Haworth." *Biographical Memoirs of Fellows of the Royal Society* 51 (2005): 196–220.

Feilden, G. B. R., and William Hawthorne. "Sir Frank Whittle, O.M., K.B.E., 1 June 1907–9 August 1996." *Biographical Memoirs of Fellows of the Royal Society* 44 (1998): 434–52.

Feinberg, Stephen E. "The Early Statistical Years: 1947–1967; A Conversation with Howard Raiffa." *Statistical Science* 23, no. 1 (2008): 136–49.

Ferguson, Eugene S. *Engineering and the Mind's Eye*. Cambridge, MA: MIT Press, 1994.

Forman, Paul. "The Discovery of X-Rays by Crystals: A Critique of the Myths." *Archive for History of Exact Sciences* 6 (1969): 38–71.

Fox, Robert, and Anna Guagnini. *Laboratories, Workshops, and Sites: Concepts and Practices of Research in Industrial Europe, 1800–1914*. Berkeley Papers in History of Science 18. Berkeley: Office for History of Science and Technology, University of California, Berkeley, 1999.

———. "Sites of Innovation in Electrical Technology, 1880–1914." *Annales historiques de l'électricité* 2 (2004): 159–72.

Franz, Anselm. "Development of the 'Jumo 004' Turbojet Engine." In *The Jet Age: Forty Years of Jet Aviation*, edited by Walter J. Boyne and Donald S. Lopez, 69–74. Washington, DC: National Air and Space Museum, Smithsonian Institution, 1979.

———. *From Jets to Tanks: My Contribution to the Turbine Age*. Stratford, CT: Avco Lycoming, 1985.

Freeman, Christopher. *The Economics of Industrial Innovation*. Harmondsworth: Penguin, 1974.

———. *The Economics of Industrial Innovation*. 2nd ed. Cambridge, MA: MIT Press, 1982.

Freeze, Karen Johnson. "Innovation and Technology Transfer during the Cold War: The Case of the Open-End Spinning Machine from Communist Czechoslovakia." *Technology and Culture* 48, no. 2 (2007): 249–85.

Friedel, Robert. *A Culture of Improvement: Technology and the Western Millennium*. Cambridge, MA: MIT Press, 2007.

———. *Pioneer Plastic: The Making and Selling of Celluloid*. Madison: University of Wisconsin Press, 1983.

———. *Zipper: An Exploration in Novelty*. New York: Norton, 1994.

Friedel, Robert, Paul Israel, and Bernard Finn. *Edison's Electric Light: Biography of an Invention*. New Brunswick, NJ: Rutgers University Press, 1986.

Furse, Anthony. *Wilfrid Freeman: The Genius behind Allied Survival and Air Supremacy, 1939 to 1945*. Staplehurst: Spellmount, 2000.

General Electric. *Seven Decades of Progress*. Fallbrook, CA: Aero Publishers, 1979.

Gersdorff, Kyrill von, and Kurt Grasmann. *Flugmotoren und Strahltriebwerke: Entwicklungsgeschichte der Deutschen Luftfahrtantriebe von den Anfängen bis zu den Europäischen Gemeinschaftsentwicklungen*. Die Deutsche Luftfahrt 2. Munich: Bernard and Graefe, 1981.

———. *Flugmotoren und Strahltriebwerke: Entwicklungsgeschichte der Deutschen Luftfahrtantriebe von den Anfängen bis zu den Internationalen Gemeinschaftsentwicklungen*. Rev. ed. Die Deutsche Luftfahrt 2. Koblenz: Bernard and Graefe, 1985.

Giffard, Hermione. "Engines of Desperation: Jet Engines, Production, and New

Weapons in the Third Reich." *Journal of Contemporary History* 48, no. 4 (2013): 821–44.

———. "The Politics of Donating Technological Artifacts: Techno-Nationalism and the Donations of the World's First Jet Engines." *History and Technology* 30 (2014): 61–82.

Gilfillan, S. Colum. *Inventing the Ship: A Study of the Inventions Made in Her History between Floating Log and Rotorship; A Self-Contained but Companion Volume to the Author's "Sociology of Invention"; With 80 Illustrations, Bibliographies, Notes, and Index.* Chicago: Follett, 1935.

———. *The Sociology of Invention: An Essay in the Social Causes of Technic Invention and Some of Its Social Results; Especially as Demonstrated in the History of the Ship.* Chicago: Follett, 1935.

Gispen, Kees. *Poems in Steel: National Socialism and the Politics of Inventing from Weimar to Bonn.* New York: Berghahn Books, 2002.

Golley, John. "The Whittle Revolution." *Aeroplane Monthly* 19, no. 6 (1991): 346–51.

———. "Working with a Genius: Engine Pioneer Sir Frank Whittle, Who Died Two Years Ago." *Aeroplane Monthly* 26, no. 10 (1998): 18–23.

Golley, John, and Bill Gunston. *Genesis of the Jet: Frank Whittle and the Invention of the Jet Engine.* Shrewsbury: Airlife, 1996.

———. *Jet: Frank Whittle and the Invention of the Jet Engine.* 5th impression. Fulham: Datum Publishing Limited, 2009.

Golley, John, Frank Whittle, and Bill Gunston. *Whittle, the True Story.* Washington, DC: Smithsonian Institution Press, 1987.

Green, William. *The Warplanes of the Third Reich.* London: Macdonald, 1970.

Gregor, Neil. *Daimler-Benz in the Third Reich.* New Haven, CT: Yale University Press, 1998.

Grierson, John. *Jet Flight.* London: Sampson Low Marston, 1946.

Gundermann, Wilhelm. "Das Erste Düsenflugzeug der Welt." *Zeitschrift Verein Deutscher Ingenieure* 20 (1952): 696–95.

———. "Zur Geschichte des Strahltriebwerks—Die Entwicklungen bei Heinkel von 1936 bis 1939." *DGLR Vortrag* 81-074.

Gunston, Bill. *The Development of Jet and Turbine Aero Engines.* Sparkford: Patrick Stephens, 1995.

———. *The Development of Piston Aero Engines: From the Wrights to Microlights—A Century of Evolution and Still a Power to Be Reckoned With.* Sparkford: Patrick Stephens, 1993.

———. *Fedden—the Life of Sir Roy Fedden.* Historical Series. London: Rolls-Royce Heritage Trust, 1998.

———. *Plane Speaking: A Personal View of Aviation History.* Sparkford: Patrick Stephens, 1991.

———. *World Encyclopaedia of Aero Engines: All Major Aircraft Power Plants,*

from the Wright Brothers to the Present Day. 4th ed. Sparkford: Patrick Stephens, 1998.

Halford, Frank B. "Jet Propulsion." *Royal Society of Arts Journal* 94, no. 4724
(1946): 576–90.

Hall, Arnold, and Morien Morgan. "Bennett Melvill Jones: 28 January 1887–31 October 1975." *Biographical Memoirs of Fellows of the Royal Society* 23 (1977):
253–82.

Hallion, Richard P. "Review: Aerospace Technology." *Science* 215, no. 4529 (1982):
155–56.

Hansen, James R. *Engineer in Charge: A History of the Langley Aeronautical Laboratory, 1917–1958.* National Aeronautics and Space Administration History
Series. Washington, DC: Scientific and Technical Information Office, NASA,
1987.

Hassel, Pat. "The Halford Jets." Unpublished, n.d.

Haworth, Lionel. *The Rolls-Royce Tyne.* Historical Series. Derby: Rolls-Royce
Heritage Trust, 2000.

Haworth, L., M. A. Nedham, and G. L. Wilde. "Robin Ralph Jamison: 12 July 1912–
18 March 1991." *Biographical Memoirs of Fellows of the Royal Society* 40
(1994): 173–94.

Hawthorne, William. "Aircraft Propulsion from the Back Room." *Aeronautical
Journal of the Royal Aeronautical Society* 82 (1978): 93–108.

———. "The Early History of the Aircraft Gas Turbine in Britain." In *50 Jahre
Turbostrahlflug.* Munich: Deutsche Gesellschaft für Luft- und Raumfahrt,
1989.

———. "The Early History of the Aircraft Gas Turbine in Britain." *Notes and
Records of the Royal Society of London* 45 (1991): 79–108.

Hawthorne, William, H. Cohen, and A. R. Howell. "Hayne Constant, 1904–1968."
Biographical Memoirs of Fellows of the Royal Society 19 (1973): 268–79.

Heathcote, Roy. *The Rolls-Royce Dart: Pioneering Turboprop.* Derby: Rolls-Royce
Heritage Trust, 1992.

Heiman, Grover. *Jet Pioneers.* New York: Duell Sloan and Pearce, 1963.

Heinkel, Ernst, and Jürgen Thorwald. *He 1000.* London: Hutchinson, 1956.

———. *Stürmisches Leben.* Stuttgart: Mundus-Verlag, 1953.

Hindle, Brooke. *Emulation and Invention.* New York: New York University Press,
1981.

———. *The Pursuit of Science in Revolutionary America, 1735–1789.* Edited by
Williamsburg Institute of Early American History and Culture, VA. Chapel
Hill: University of North Carolina Press, 1956.

Hintz, Eric S. "Independent Inventors in an Era of Burgeoning Research and Development." *Business and Economic History Online* 5 (2007).

Hirschel, Ernst-Heinrich, Horst Prem, and Gero Madelung. *Aeronautical Research
in Germany: From Lilienthal until Today.* Berlin: Springer, 2004.

———. *Luftfahrtforschung in Deutschland.* Die Deutsche Luftfahrt 30. Koblenz: Bernard and Graefe, 2001.

Hodgson, Richard. "Stewart S. Tresilian—a C.V." Accessed January 17, 2011, http://www.designchambers.com/wolfhound/TresilianCV.htm.

Holley, I. B. "Jet Lag in the Army Air Corps." In *Military Planning in the Twentieth Century: Proceedings of the Eleventh Military History Symposium, 10–12 October 1984,* edited by Harry R. Borowski. Washington, DC: Office of Air Force History, 1984.

Homze, Edward. *Arming the Luftwaffe: The Reich Air Ministry and the German Aircraft Industry, 1919–39.* Lincoln: University of Nebraska Press, 1976.

Hooker, S. G. "Cantor Lectures: 'High Speed Flight.'" *Royal Society of Arts Journal* 94, no. 4714 (1946): 266–96.

Hooker, Stanley. *Not Much of an Engineer: An Autobiography.* Shrewsbury: Airlife, 1984.

Hornby, William. *Factories and Plant.* History of the Second World War. United Kingdom Civil Series. London: HMSO, 1958.

Hounshell, David. *From the American System to Mass Production, 1800–1932: The Development of Manufacturing Technology in the United States.* Studies in Industry and Society. Baltimore: Johns Hopkins University Press, 1985.

———. "Hughesian History of Technology and Chandlerian Business History: Parallel, Departures, and Critics." *History and Technology* 12 (1995): 205–24.

Hounshell, David, and John Kenly Smith. *Science and Corporate Strategy: DuPont R&D, 1902–1980.* Studies in Economic History and Policy: The United States in the Twentieth Century. Cambridge: Cambridge University Press, 1988.

Howell, A. R. "Griffith's Early Ideas on Turbomachinery Aerodynamics." *Aeronautical Journal* (1976): 521–29.

Hughes, Thomas P. *American Genesis: A Century of Invention and Technological Enthusiasm, 1870–1970.* New York: Viking, 1989.

———. *Elmer Sperry: Inventor and Engineer.* Baltimore: Johns Hopkins Press, 1971.

———. "Emerging Themes in the History of Technology." *Technology and Culture* 20, no. 4 (1979): 697–711.

———. "The Evolution of Large Technological Systems." In *The Social Construction of Technological Systems: New Directions in the Sociology and History of Technology,* edited by Wiebe Bijker, Thomas Hughes, and Trevor Pinch, 51–82. Cambridge, MA: MIT Press, 2012.

———. "Inventors: The Problems They Choose, the Ideas They Have, the Inventions They Make." In *Technological Innovation: A Critical Review of Current Knowledge,* edited by Patrick Kelly and Melvin Kranzberg, 166–82. San Francisco: San Francisco Press, 1978.

———. *Networks of Power: Electrification in Western Society, 1880–1930.* Baltimore: Johns Hopkins University Press, 1983.

————. "The Science-Technology Interaction: The Case of High-Voltage Power Transmission Systems." *Technology and Culture* 17 (1976): 646–62.

————. *Thomas Edison: Professional Inventor*. Edited by Science Museum. London: HMSO, 1976.

Irving, David. *Mare's Nest*. London: Panther Books, 1985. Accessed August 14, 2010, www.fpp.co.uk/books/MaresNest/index.html.

————. *The Rise and Fall of the Luftwaffe: The Life of Field Marshal Erhard Milch*. Parcelforce UK, 2002. Accessed August 12, 2010, http://www.fpp.co.uk/books /Milch/Milch.pdf.

Israel, Paul. *Edison: A Life of Invention*. New York: John Wiley, 1998.

————. *From Machine Shop to Industrial Laboratory: Telegraphy and the Changing Context of American Invention, 1830–1920*. Baltimore: Johns Hopkins University Press, 1992.

————. "Inventing Industrial Research Article: Thomas Edison and the Menlo Park Laboratory." *Endeavour* 26 (2002): 48–54.

Jakab, Peter L. *Visions of a Flying Machine: The Wright Brothers and the Process of Invention*. Smithsonian History of Aviation Series. Washington, DC: Smithsonian Institution Press, 1990.

James, Derek N. *Bristol Aeroplane Company*. Archive Photographs Series. Stroud: Chalford, 1996.

James, Harold. *Krupp: A History of the Legendary German Firm*. Princeton, NJ: Princeton University Press, 2012.

Jamison, Andrew. "Technology's Theorists: Conceptions of Innovation in Relation to Science and Technology Policy." *Technology and Culture* 30, no. 3 (1989): 505–33.

Janda, Fritz. "Oma finanzierte den ersten Jet." *Abendzeitung* (February 14, 1989).

Jenkins, Reese. "Elements of Style: Continuities in Edison's Thinking." In *Bridge to the Future: A Centennial Celebration of the Brooklyn Bridge*, edited by Margaret Webb Latimer, 149–81. New York: New York Academy of Sciences, 1984.

————. *Images and Enterprise: Technology and the American Photographic Industry, 1839 to 1925*. Johns Hopkins Studies in the History of Technology. Baltimore: Johns Hopkins University Press, 1975.

Jewkes, John, David Sawers, and Richard Stillerman. *The Sources of Invention*. London: Macmillan, 1958.

————. *The Sources of Invention*. 2nd ed. London: Macmillan, 1969.

Jones, Barry. *Gloster Meteor*. Wiltshire: Crowood Press, 1998.

Jones, Glyn. *The Jet Pioneers: The Birth of Jet-Powered Flight*. London: Methuen, 1989.

————. "Jubilee for the Jet Engine: Frank Whittle and the History of the British Jet Industry." *New Scientist* 130, no. 1768 (1991): 32–35.

Kaempffert, Waldemar. *Invention and Society*. Reading with a Purpose 56. Chicago: American Library Association, 1930.

Kay, Antony L. *German Jet Engine and Gas Turbine Development, 1930–45*. Shrewsbury: Airlife, 2002.

———. *Turbojet History and Development*. 2 vols. Wiltshire: Crowood Press, 2007.

Keenan, J. G. *Elementary Theory of Gas Turbines and Jet Propulsion*. London: Oxford University Press, 1946.

Kershaw, Tim. *Jet Pioneers: Gloster and the Birth of the Jet Age*. Stroud: Sutton, 2004.

Köhler, H. Dieter. *Ernst Heinkel: Pionier der Schnellflugzeuge, Eine Biographie*. Die Deutsche Luftfahrt 5. Koblenz: Bernard and Graefe, 1983.

———. *Ernst Heinkel: Pionier der Schnellflugzeuge, Eine Biographie*: Koblenz: Bernard and Graefe, 1999.

Koos, Volker. *Ernst Heinkel: Vom Doppeldecker zum Strahltriebwerk*. Bielefeld: Delius Klasing, 2007.

Lagasse, Paul. "The Westinghouse Aviation Gas Turbine Division, 1950–1960: A Case Study in the Role of Failure in Technology and Business." Master's thesis, University of Maryland, 1997.

Laudan, Rachel. *The Nature of Technological Knowledge: Are Models of Scientific Change Relevant?* Sociology of the Sciences Monographs. Dordrecht: D. Reidel, 1984.

Lawton, Roy. *Parkside — Armstrong Siddeley to Rolls Royce, 1939–1994*. Historical Series 39. Derby: Rolls-Royce Heritage Trust 2008.

Layton, Edwin. "Mirror-Image Twins: The Communities of Science and Technology in 19th-Century America." *Technology and Culture* 12, no. 4 (1971): 562–80.

———. "Scientific Technology, 1845–1900: The Hydraulic Turbine and the Origins of American Industrial Research." *Technology and Culture* 20, no. 1 (1979): 64–89.

Leyes, Richard A., and William A. Fleming. *The History of North American Small Gas Turbine Aircraft Engines*. Reston, VA: American Institute of Aeronautics and Astronautics, 1999.

Lipstadt, Deborah E. *The Eichmann Trial*. New York: Schocken, 2011.

Lloyd, Ian. *Rolls-Royce: The Merlin at War*. London: Macmillan, 1978.

Lloyd, Ian, and Peter Pugh. *Hives and the Merlin*. Cambridge: Icon Books, 2004.

MacLeod, Christine. "Concepts of Invention and the Patent Controversy in Victorian Britain." In *Technological Change: Methods and Themes in the History of Technology*, edited by Robert Fox, 137–53. Amsterdam: Harwood Academic, 1996.

———. *Heroes of Invention: Technology, Liberalism, and British Identity, 1750–1914*. Cambridge Studies in Economic History. Cambridge: Cambridge University Press, 2008.

MacLeod, Christine, and Alessandro Nuvolari. "The Pitfalls of Prosopography: Inventors in the *Dictionary of National Biography*." *Technology and Culture* 47 (2006): 757–76.

McGee, David. "Making Up Mind: The Early Sociology of Invention." *Technology and Culture* 36, no. 4 (1995): 773–801.

Merton, Robert K. "Fluctuations in the Rate of Industrial Invention." *Quarterly Journal of Economics* 49, no. 3 (1935): 454–74.

———. "Priorities in Scientific Discovery: A Chapter in the Sociology of Science." *American Sociological Review* 22, no. 6 (1957): 635–59.

———. "The Role of Genius in Scientific Advance." *New Scientist* 259 (1961): 306–8.

———. "Singletons and Multiples in Scientific Discovery: A Chapter in the Sociology of Science." *Proceedings of the American Philosophical Society* 105, no. 5 (1961): 470–86.

Milward, Alan S. *War, Economy, and Society, 1939–1945*. History of the World Economy in the Twentieth Century 5. Berkeley: University of California Press, 1977.

Mokyr, Joel. *The Lever of Riches: Technological Creativity and Economic Progress*. Oxford: Oxford University Press, 1990.

Molella, Arthur P. "The Longue Durée of Abbott Payson Usher." *Technology and Culture* 46, no. 4 (2005): 779–96.

Mönnich, Horst, Anthony Bastow, and William Henson. *The BMW Story: A Company in Its Time*. London: Sidgwick and Jackson, 1991.

Mordell, Donald L. "Better Than a PhD." *Archive* (n.d.): 8–15.

Moult, Eric S. "The Development of the Goblin Engine." *Journal of the Royal Aeronautical Society* 51 (1947): 655–85.

Mowery, David C., and Nathan Rosenberg. *Paths of Innovation: Technological Change in 20th-Century America*. Cambridge: Cambridge University Press, 1999.

Münzberg, H. "Nachruf auf Hermann Oestrich." *Jahrbuch der DGLR* (1973): 274–76.

Nahum, Andrew. *Frank Whittle: Invention of the Jet*. Revolutions in Science. Cambridge: Icon Books, 2004.

———. "Origins of the British Jet Engine." *Science Museum Review* (1987): 19–21.

———. "Two-Stroke or Turbine? The Aeronautical Research Committee and British Aero Engine Development in World War II." *Technology and Culture* 38, no. 2 (1997): 312–54.

———. "World War to Cold War: Formative Episodes in the Development of the British Aircraft Industry, 1943–1965." PhD diss., London School of Economics, 2003.

National Academy of Engineering. "Previous Recipients of the Charles Stark Draper Prize." Accessed October 26, 2010, http://www.nae.edu/Activities /Projects20676/Awards/20681/PastWinners.aspx.

Ndiaye, Pap. *Nylon and Bombs: DuPont and the March of Modern America*. Baltimore: Johns Hopkins University Press, 2007.

Nelson, Richard, Merton Peck, and Edward Kalachek. *Technology, Economic*

Growth, and Public Policy. A Rand Corporation and Brookings Institution Study. Washington, DC: Brookings Instituton, 1967.

Neufeld, Michael J. "The Nazi Aerospace Exodus: Towards a Global, Transnational History." *History and Technology* 28, no. 1 (2012): 49–67.

———. "Rocket Aircraft and the 'Turbojet Revolution.'" In *Innovation and the Development of Flight*, edited by Roger D. Launius, 207–34. College Station: Texas A&M University Press, 1999.

———. *The Rocket and the Reich: Peenemünde and the Coming of the Ballistic Missile Era*. New York: Free Press, 1995.

Neville, Leslie, and Nathaniel Silsbee. *Jet Propulsion Progress: The Development of Aircraft Gas Turbines*. New York: McGraw-Hill, 1948.

Nichelson, Brian J. "Early Jet Engines and the Transition from Centrifugal to Axial Compressors: A Case Study in Technological Change." PhD diss., University of Minnesota, 1988.

Niemann, Harry, Wilfried Feldenkirchen, and Armin Hermann. *Gasturbinen und Flugtriebwerke der Daimler-Benz AG 1952–1960*. Wissenschaftliche Schriftenreihe des Daimlerchrysler Konzernarchivs 6. Vaihingen/Enz: IPa, 2004.

Noble, David F. *Forces of Production: A Social History of Industrial Automation*. New York: Knopf, 1984.

Nowarra, Heinz J. *Heinkel und seine Flugzeuge*. Munich: Lehmann, 1975.

Nye, David E. *The Invented Self: An Anti-Biography, from Documents of Thomas A. Edison*. Odense: Odense University Press, 1983.

Oestrich, Hermann. "Die Entwicklung der Fluggasturbine bei den Bayrischen Motorenwerken Während des Krieges 1939/45." *Flugwelt* 4 (1950): 69–71.

Ogburn, William, and Dorothy Thomas. "Are Inventions Inevitable? A Note on Social Evolution." *Political Science Quarterly* 37, no. 1 (1922): 83–98.

Oldenziel, Ruth. *Making Technology Masculine: Men, Women, and Modern Machines in America, 1870–1945*. Amsterdam: Amsterdam University Press, 1999.

Ordway, F., and Mitchell Sharpe. *The Rocket Team*. London: Heinemann, 1979.

Overy, Richard J. *The Air War: 1939–1945*. Dulles, VA: Brassey's, 2005.

———. *The Dictators: Hitler's Germany and Stalin's Russia*. New York: Penguin Adult, 2005.

———. *Goering: The "Iron Man."* London: Routledge and Kegan Paul, 1984.

Owner, Frank. "9th Barnwell Memorial Lecture." 1962.

"Packard vs RR." Accessed January 17, 2011, http://www.ww2aircraft.net/forum /aviation/merlins-packard-vs-rr-2013-8.html.

Pavelec, Sterling Michael. "The Development of Turbojet Aircraft in Germany, Britain, and the United States: A Multi-National Comparison of Aeronautical Engineering, 1935–1946." PhD diss., Ohio State University, 2004.

———. *The Jet Race and the Second World War*. Westport, CT: Praeger, 2007.

Peden, G. C. *Arms, Economics, and British Strategy: From Dreadnoughts to Hydrogen Bombs*. Cambridge: Cambridge University Press, 2007.

Pestre, Dominique. "Commemorative Practices at CERN: Between Physicist' Memories and Historians' Narratives." *Osiris*, 2nd Series, 14 (1999): 203–16.

Phelps, Stephen. *The Tizard Mission: The Top-Secret Operation That Changed the Course of World War II.* Yardly, PA: Westholme Publishing, 2012.

Postan, M., Denys Hay, and J. Scott. *Design and Development of Weapons: Studies in Government and Industrial Organisation.* History of the Second World War, British Official Histories. London: HMSO, 1964.

Pugh, Peter. *The Magic of a Name: The Rolls-Royce Story; The First 40 Years.* Cambridge: Icon Books, 2000.

———. *The Magic of a Name: The Rolls-Royce Story, Part II: The Power behind the Jets, 1945–1987.* Cambridge: Icon Books, 2001.

Pye, David. *The Nature and Art of Workmanship.* Cambridge: Cambridge University Press, 1968.

Rathjen, W. "'Weniger Gerausch, Keine Vibration!': 50 Jahre Strahltriebwerk." *Kultur und Technik: Zeitschrift des Deutschen Museums München* 13 (1989): 206–15.

Reader, W. J., James Kenneth Weir, and Elizabeth McClure Thomson. *The Weir Group: A Centenary History.* London: Weidenfeld and Nicolson, 1971.

Reich, Leonard S. "Irving Langmuir and the Pursuit of Science and Technology in the Corporate Environment." *Technology and Culture* 24, no. 2 (1983): 199–221.

———. *The Making of American Industrial Research: Science and Business at GE and Bell, 1876–1926.* Studies in Economic History and Policy: The United States in the Twentieth Century. Cambridge: Cambridge University Press, 1985.

Reynolds, John. *Engines and Enterprise: The Life and Work of Sir Harry Ricardo.* Stroud: Sutton, 1999.

Rieger, Bernhard. *The People's Car: A Global History of the Volkswagen Beetle.* Cambridge, MA: Harvard University Press, 2013.

Ritchie, Sebastian. *Industry and Air Power: The Expansion of British Aircraft Production, 1935–41,* Studies in Air Power. London: F. Cass, 1996.

Roland, Alex. *Model Research: The National Advisory Committee for Aeronautics, 1915–1958.* 2 vols. National Aeronautics and Space Administration History Series. Washington, DC: Scientific and Technical Information Branch, National Aeronautics and Space Administration, 1985.

———. "Theories and Models of Technological Change: Semantics and Substance." *Science, Technology, and Human Values* 17, no. 1 (1992): 79–100.

Rolls-Royce. *The Jet Engine.* 4th ed. Derby: Rolls-Royce, 1986.

Rowland, John. *The Jet Man: The Story of Sir Frank Whittle.* Bristol: Western Printing Services, 1967.

Rosenberg, Nathan. *Inside the Black Box: Technology and Economics.* Cambridge: Cambridge University Press, 1982.

Roxbee Cox, Harold. "British Aircraft Gas Turbines: The Ninth Wright Brothers Lecture." *Journal of the Aeronautical Sciences* 13, no. 2 (1946): 53–87.

———. "The Beginnings of Jet Propulsion: The Trueman Wood Lecture." *Royal Society of Arts Journal* 132 (1985): 705–23.

Rubbra, A. A. "Alan Arnold Griffith, 1893–1963." *Biographical Memoirs of Fellows of the Royal Society* 10 (1964): 117–36.

Saunders, Owen. "David Macleish Smith: 9 June 1900–3 August 1986." *Biographical Memoirs of Fellows of the Royal Society* 33 (1987): 605–17.

Saxon, Wolfgang. "Robert O. Schlaifer, 79, Managerial Economist." *New York Times*. July 28, 1994.

Schabel, Ralf. *Die Illusion der Wunderwaffen: Die Rolle der Düsenflugzeuge und Flugabwehrraketen in der Rüstungspolitik des Dritten Reiches*. Beiträge zur Militärgeschichte 35. Munich: Oldenbourg, 1994.

Schatzberg, Eric. *Wings of Wood, Wings of Metal: Culture and Technical Choice in American Airplane Materials, 1914–194*. Princeton, NJ: Princeton University Press, 1999.

Scherenberg, H. "Die Flugmotorenentwicklung bei Daimler-Benz." *Flug-Revue* 5 (1964): 59–60.

Schlaifer, Robert, and S. D. Heron. *Development of Aircraft Engines: Two Studies of Relations between Government and Business*. Boston: Division of Research Graduate School of Business Administration Harvard University, 1950.

Schmookler, Jacob. *Invention and Economic Growth*. Cambridge, MA: Harvard University Press, 1966.

Schumpeter, Joseph Alois. *Capitalism, Socialism, and Democracy*. London: George Allen and Unwin, 1943.

Scranton, Philip. "Technology-Led Innovation: The Non-Linearity of Us Jet Propulsion Development." *History and Technology* 22, no. 4 (2006): 337–67.

———. "Technology, Science, and American Innovation." *Business History* 48, no. 3 (2006): 311–31.

———. "Turbulence and Redesign: Dynamic Innovation and the Dilemmas of US Military Jet Propulsion Development." *European Management Journal* 25, no. 3 (2007): 235–48.

Shacklady, Edward. *The Gloster Meteor*. Macdonald Aircraft Monographs. London: Macdonald, 1962.

Sharp, C. Martin. *D. H.: An Outline of De Havilland History*. London: Faber and Faber, 1960.

Smith, D. M. "The Development of an Axial Flow Gas Turbine for Jet Propulsion." *Proceedings of the Institution of Mechanical Engineers* 157 (1947).

Smith, G. Geoffrey. *Gas Turbines and Jet Propulsion for Aircraft*. 1st ed. London: Flight, 1942.

———. *Gas Turbines and Jet Propulsion for Aircraft*. 4th ed. London: Flight, 1946.

Speer, Albert. *Inside the Third Reich*. London: Orion Publishing Group, 1997.

St. Peter, J. "The History of Aircraft Gas Turbine Development in the US." American Society of Mechanical Engineers International Gas Turbine Institute, 1999.

Stanier, William A. "Henry Lewis Guy: 1887–1956." *Biographical Memoirs of Fellows of the Royal Society* 4 (1958): 99–101.

Staudenmaier, John M. "Recent Trends in the History of Technology." *American Historical Review* 95 (1990): 715–25.

Susskind, Charles, and Martha Zybkow. "The Argument." In *Technological Innovation: A Critical Review of Current Knowledge*, edited by Patrick Kelly and Melvin Kranzberg, ix–xviii. San Francisco: San Francisco Press, 1978.

Taylor, Douglas R. *Boxkite to Jet: The Remarkable Career of Frank B. Halford*. Derby: Rolls-Royce Heritage Trust, 1999.

Taylor, John W. R. *Jane's All the World's Aircraft, 1989–90*. London: Jane's Information Group, 1989.

Thomson, George P., and Arnold A. Hall. "William Scott Farren: 1892–1970." *Biographical Memoirs of Fellows of the Royal Society* 17 (1971): 215–41.

Tooze, Adam. *The Wages of Destruction: The Making and Breaking of the Nazi Economy*. London: Penguin, 2007.

Tucker, Anthony. "Sir Arnold Hall." Accessed January 17, 2011, http://www.guardian.co.uk/news/2000/jan/11/guardianobituaries2.

Usher, Abbott Payson. *A History of Mechanical Inventions*. 1st ed. New York: McGraw-Hill, 1929.

———. *A History of Mechanical Inventions*. Rev. ed. Cambridge, MA: Harvard University Press, 1954.

Uziel, Daniel. "Between Industrial Revolution and Slavery: Mass Production in the German Aviation Industry in World War II." *History and Technology* 22, no. 3 (2006): 277–300.

———. "Der Volksjäger: Rationalisierung und Rationalität von Deutschlands Letztem Jagdflugzeug im Zweiten Weltkrieg." In *Rüstung, Kriegswirtschaft und Zwangsarbeit im "Dritten Reich*," edited by Andreas Heusler, Mark Spoerer, and Helmuth Trischler, 63–82. Munich: MTU Aero Engines, BMW Group; Oldenbourg Wissenschaftsverlag, 2010.

Vajda, Franz-Antal, and Peter Dancey. *German Aircraft Industry and Production, 1933–1945*. Shrewsbury: Airlife, 1998.

Vincenti, Walter G. *What Engineers Know and How They Know It: Analytical Studies from Aeronautical History*. Baltimore: Johns Hopkins University Press, 1990.

———. "The Social Shaping of Technology: Real-World Constraints and Technical Logic in Edison's Electrical Lighting System." *Social Studies of Science* 25, no. 3 (1995): 553–74.

Von Ohain, Hans. "Aufbruch in Den Überschallflug." *Technik & Kultur* 15, no. 2 (1981): 65–72.

Wagner, Wolfgang. *Die Ersten Strahlflugzeuge der Welt*. Die Deutsche Luftfahrt 14. Koblenz: Bernard and Graefe, 1989.

Wagner, Wolfgang, and Don Cox. *The First Jet Aircraft*. History of German Aviation. Atglen, PA: Schiffer, 1998.

Watkins, David. *De Havilland Vampire: The Complete History*. Stroud: Sutton, 1996.

Werner, Constanze. *Kriegswirtschaft und Zwangsarbeit bei BMW*. Perspektiven 1. Oldenbourg: Oldenbourg Wissenschaftsverlag, 2006.

Whitfield, Jakob. "Metropolitan Vickers, the Gas Turbine, and the State: A Socio-Technical History, 1935–1960." PhD diss., University of Manchester, 2012.

White, Lynn. "The Act of Invention: Causes, Contexts, Continuities, and Consequences." *Technology and Culture* 3, no. 4 (1962): 486–500.

Whittle, Frank. "The Early History of the Whittle Jet Propulsion Gas Turbine." *Journal of the Institution of Mechanical Engineers* 152 (1945): 419–35.

———. *Jet: The Story of a Pioneer*. London: Frederick Muller, 1953.

———. *Jet: The Story of a Pioneer*. 2nd ed. London: Pan, 1957.

Whittle, Frank, and Hans von Ohain. *An Encounter between the Jet Engine Inventors: Sir Frank Whittle and Dr. Hans Von Ohain, 3–4 May 1978, Wright-Patterson Air Force Base, Ohio*. History Office, Aeronautical Systems Division, Air Force Systems Command, 1978.

Whyte, Lancelot Law. *Focus and Diversions*. London: Cresset Press, 1963.

———. "Whittle and the Jet Adventure." *Harper's Magazine* 208, no. 1249 (1954).

Wilde, Geoffrey. "Dr. Griffith's CR.1." *Archive* 12, no. 1 (1994): 85–92.

Wilson, Charles, and William Reader. *Men and Machines: A History of D. Napier & Son Engineers Ltd, 1808–1958*. London: Weidenfeld and Nicolson, 1958.

Wilkinson, Paul H. *Aircraft Engines of the World*. New York: Wilkinson, 1945.

Windley, R. O. "Gas Turbine Combustion: A Personal Perspective." *Proceedings of the Institution of Mechanical Engineers: Part G, Journal of Aerospace Engineering* 203, G2 (1989): 79–96.

Wise, George. *Willis R. Whitney, General Electric, and the Origins of U.S. Industrial Research*. New York: Columbia University Press, 1985.

Wright, T. P. "Factors Affecting the Cost of Airplanes." *Journal of Aeronautical Sciences* 3, no. 4 (1936): 122–28.

Young, James O. "Riding on England's Coattails." In *Innovation and the Development of Flight*, edited by Roger D. Launius, 263–98. College Station: Texas A&M University Press, 1999.

Zaidi, Waqar. "The Janus-Face of Techno-Nationalism: Barnes Wallis and the 'Strength of England.'" *Technology and Culture* 49 (2008): 62–88.

Ziegler, Frank. "Germany's Aviation Industry: A Personal Report, Part 1." *Flight* (January 27, 1956): 103–5.

———. "Germany's Aviation Industry: A Personal Report, Part 2." *Flight* (February 3, 1956): 144–45.

Zimmerman, David. *Top Secret Exchange: The Tizard Mission and the Scientific War*. Stroud: Sutton, 1996.

Index